AF388954

Synthesis, Characterization and Application of Hybrid Composites

Synthesis, Characterization and Application of Hybrid Composites

Editor

Ignazio Blanco

MDPI • Basel • Beijing • Wuhan • Barcelona • Belgrade • Manchester • Tokyo • Cluj • Tianjin

Editor
Ignazio Blanco
University of Catania
Italy

Editorial Office
MDPI
St. Alban-Anlage 66
4052 Basel, Switzerland

This is a reprint of articles from the Special Issue published online in the open access journal *Applied Sciences* (ISSN 2076-3417) (available at: https://www.mdpi.com/journal/applsci/special_issues/Hybrid_Composites).

For citation purposes, cite each article independently as indicated on the article page online and as indicated below:

LastName, A.A.; LastName, B.B.; LastName, C.C. Article Title. *Journal Name* **Year**, *Article Number*, Page Range.

ISBN 978-3-03943-062-8 (Hbk)
ISBN 978-3-03943-063-5 (PDF)

Contents

About the Editor

Ignazio Blanco is a full professor of Chemical Foundations of Technologies at the Department of Civil Engineering and Architecture of the University of Catania. His research activities are focused on the following themes: the synthesis and characterization of toughened thermoset blends; the process technology of polymeric fiber-reinforced composites; heat capacities, fusion, and solid-to-solid transition of series of organic molecules; comparative kinetic studies of the thermal degradation of model polymers; comparative kinetic studies of the thermal degradation of nanocomposites; synthesis and characterization of aromatic copolymers; thermal characterization of films used in food packaging applications; synthesis and characterization of nanoparticles; POSS synthesis and characterization; and material characterization for rapid prototyping. To date, he has authored and/or co-authored 110 papers indexed on Scopus, 1 book chapter, 3 book and about 80 presentations in national and international conferences.

Editorial

Synthesis, Characterization and Application of Hybrid Composites

Ignazio Blanco

Department of Civil Engineering and Architecture and UdR-Catania Consorzio INSTM, University of Catania, Viale Andrea Doria 6, 95125 Catania, Italy; iblanco@unict.it

Received: 24 July 2020; Accepted: 5 August 2020; Published: 7 August 2020

In the last century, polymers have dominated the materials market, and in the later part of the 20th century, the possibility to further improve their properties and expand their applications was explored by combining polymers with reinforcing elements. This practice has become common, enabling the definition of the composites as the reference materials of the 21st century and, thus, their occupying of an important portion of the market in the production of modern plastics [1]. The composites are designed by adding a second component to the matrix (generally polymeric) to enhance their properties and/or to achieve specific functions [2]. Among the various composites, the organic–inorganic hybrid materials offer much better performance than their non-hybrid counterparts [3]. The dramatic improvement in physical properties achieved with the incorporation of inorganic particles or nanoparticles into an organic polymeric matrix could bridge the gap between ceramics and polymers.

This Special Issue, which consists of 12 articles, including three review articles, written by research groups of experts in the field, considers recent research on polymeric composite materials.

Since increasing attention is, today, devoted to composite materials with low flammability, Majka and coworkers used the layer-by-layer (LbL) technique to form multilayered protective coatings for polyamide 6/montmorillonite (PA6/MMT) hybrid nanocomposites [4]. Their results confirmed the success in applying the LbL technique to improve the flammability characteristics of the composites under investigation, reducing the maximum point of heat release rate (PHRR).

Samal et al. investigated the interfacial adhesion of a Ni-Ti shape memory alloy with Poly (methyl methacrylate) (PMMA) as the matrix. They studied, by the means of thermomechanical analysis (TMA) and dynamic mechanical analysis (DMA), the surface pattern of the Ni-Ti channeled by a solid-state laser machine. The PMMA matrix, so modified, showed an increase in adhesion of about 80% in comparison to the free plain surface, showing, in the meantime, a reduction in the shape-memory properties of the composite material [5].

Chen and his working group, considering that the traditional fabrication methods of 3D scaffolds and cell-laden hydrogels still face many difficulties and challenges, proposed a new 3D fabrication technique exalting the concept of recycling the unutilized resource [6]. They prepared an innovative fish gelatin (FG) methacrylate polymer hydrogel mixed with a strontium-doped calcium silicate powder (FGSr) 3D scaffold via photo-crosslinking. Their mechanical assessment indicated an increase in the tensile strength of FGSr, thus making it a better candidate for future clinical applications. Furthermore, they observed good biocompatibility with human Wharton jelly-derived mesenchymal stem cells (WJMSC), as well as the enhancement of the osteogenic differentiation of WJMSC.

Weichold et al. have engaged in the challenge of preparing compact engineering composites with a combination of citric acid and glycerol. Polycondensation-type thermoset resins from natural reactants, such as citric and glutaric acid, as well as 1,3-propanediol and glycerol, were prepared and analyzed by thermogravimetric analysis (TGA), not only showing that shredded poultry feathers increased the conversion and the reaction rate of the citric acid/glycerol combination but also that increasing the amount of feathers continuously decreased the number of visible bubbles [7].

Catauro et al., by using the sol-gel method, prepared silica-based hybrid organic/inorganic amorphous composites with polycaprolactone [8]. They spectroscopically characterized the obtained composites and assessed their bioactivity and antimicrobial behavior, aiming at proposing them for bone implants.

Poly(lactic acid) (PLA)-based biocomposites reinforced with flax fibers were designed and prepared by Seggiani and coworkers by melt blending and a co-rotating conical twin-screw extruder. For the obtained materials, the Italian researchers observed good matrix/fiber adhesion and mechanical behavior in terms of break stress and composite stiffness [9].

Lan and colleagues, by mixing ferromagnetic nickel micro-powders with polyurethane (PU) shape memory polymer (SMP)/dimethylformamide solution, prepared a micron-sized protrusive PU SMP composite, curing the mixture under a low magnetic field [10]. Morphology and thermal tests were carried out by the authors to show the hybrid composites' potential applications for micro-electro-mechanical systems (MEMS) and biomedical devices.

Polydimethylsiloxane (PDMS)-based composites were fabricated by Sheng et al. to study their optical transmittance properties. With a simple production process, the Chinese researchers checked the materials' transmittance by varying the microcrystalline graphite powder concentration or the composite film's thickness, thus obtaining a wide range of transmittance properties [11].

Quan and coworkers designed a new hybrid catalyst based on activated carbon (AC)-supported sulfonated cobalt phthalocyanine (AC-CoPcS) [12]. Firstly, they spectroscopically investigated the prepared catalyst to characterize its structure and then its catalytic performance, finding superior behavior in terms of yield and purity with good industrial application prospects.

The first of the three reviews in this Special Issue is by Samal and colleagues, who present the most outstanding advances in the rheological performance of magneto-rheological elastomer (MRE) composites. They focus on filler distribution, arrangement and wettability within elastomer matrices, and their contribution towards the performance of the mechanical response when subjected to a magnetic field are evaluated. Particular attention is devoted to the understanding of their internal micro-structures, filler–filler adhesion, filler–matrix adhesion and the viscoelastic behavior of the MRE composite under static (valve), compressive (squeeze) and dynamic (shear) modes [13].

A short review, on the effect of surfactants on the mechanico-thermal properties of polymer nanocomposites, was then proposed by Shamsuri and Jamil [14], highlighting polymer nanocomposites as a function of the surfactants used in their modification, namely surfactant-modified inorganic nanofiller/polymer nanocomposites and surfactant-modified organic nanofiller/polymer nanocomposites. The effect of surfactants on their mechanical and thermal properties is also shortly reviewed, with an attempt to capture the interest of the polymer composite researchers and encourage the further enhancement of new theories in this research field.

Finally, Qin et al., focus their attention on the progress in hybrid solar cells based on solution-processed organic and semiconductor nanocrystals, analyzing perspectives on the device design [15]. They give a brief introduction to the progress on solution-processed organic/inorganic semiconductor hybrid solar cells, including a summary of the development of hybrid solar cells in recent years, the strategy of creating hybrid solar cells with different structures and the incorporation of new organic hole transport materials, with new insights into device processing for high efficiency.

Conflicts of Interest: The author declares no conflict of interest.

References

1. Blanco, I. The Rediscovery of POSS: A Molecule Rather than a Filler. *Polymers* **2018**, *10*, 904. [CrossRef] [PubMed]
2. Pielichowski, K.; Majka, T.M. *Polymer Composites with Functionalized Nanoparticles: Synthesis, Properties, and Applications*; Elsevier Inc.: Amsterdam, The Netherlands, 2019; pp. 1–504.
3. Lazzara, G.; Cavallaro, G.; Panchal, A.; Fakhrullin, R.; Stavitskaya, A.; Vinokurov, V.; Lvov, Y. An assembly of organic-inorganic composites using halloysite clay nanotubes. *Curr. Opin. Colloid Interface Sci.* **2018**, *35*, 42–50. [CrossRef]

4.   Majka, T.M.; Witek, M.; Radzik, P.; Komisarz, K.; Mitoraj, A.; Pielichowski, K. Layer-by-Layer Deposition of Copper and Phosphorus Compounds to Develop Flame-Retardant Polyamide 6/Montmorillonite Hybrid Composites. *Appl. Sci.* **2020**, *10*, 5007. [CrossRef]

5.   Samal, S.; Tyc, O.; Heller, L.; Šittner, P.; Malik, M.; Poddar, P.; Catauro, M.; Blanco, I. Study of Interfacial Adhesion between Nickel-Titanium Shape Memory Alloy and a Polymer Matrix by Laser Surface Pattern. *Appl. Sci.* **2020**, *10*, 2172. [CrossRef]

6.   Yu, C.-T.; Wang, F.-M.; Liu, Y.-T.; Lee, A.K.-X.; Lin, T.-L.; Chen, Y.-W. Enhanced Proliferation and Differentiation of Human Mesenchymal Stem Cell-laden Recycled Fish Gelatin/Strontium Substitution Calcium Silicate 3D Scaffolds. *Appl. Sci.* **2020**, *10*, 2168. [CrossRef]

7.   Brenner, M.; Popescu, C.; Weichold, O. Anti-Frothing Effect of Poultry Feathers in Bio-Based, Polycondensation-Type Thermoset Composites. *Appl. Sci.* **2020**, *10*, 2150. [CrossRef]

8.   Catauro, M.; Piccolella, S.; Leonelli, C. FT-IR Characterization of Antimicrobial Hybrid Materials through Sol-Gel Synthesis. *Appl. Sci.* **2020**, *10*, 1180. [CrossRef]

9.   Aliotta, L.; Gigante, V.; Coltelli, M.-B.; Cinelli, P.; Lazzeri, A.; Seggiani, M. Thermo-Mechanical Properties of PLA/Short Flax Fiber Biocomposites. *Appl. Sci.* **2019**, *9*, 3797. [CrossRef]

10.   Lan, X.; Huang, W.; Leng, J. Shape Memory Effect in Micro-Sized Shape Memory Polymer Composite Chains. *Appl. Sci.* **2019**, *9*, 2919. [CrossRef]

11.   Wang, Q.; Sheng, B.; Wu, H.; Huang, Y.; Zhang, D.; Zhuang, S. Composite Films of Polydimethylsiloxane and Micro-Graphite with Tunable Optical Transmittance. *Appl. Sci.* **2019**, *9*, 2402. [CrossRef]

12.   Cheng, Z.; Dai, M.; Quan, X.; Li, S.; Zheng, D.; Liu, Y.; Yao, R. Synthesis and Catalytic Activity of Activated Carbon Supported Sulfonated Cobalt Phthalocyanine in the Preparation of Dimethyl Disulfide. *Appl. Sci.* **2019**, *9*, 124. [CrossRef]

13.   Samal, S.; Škodová, M.; Abate, L.; Blanco, I. Magneto-Rheological Elastomer Composites. A Review. *Appl. Sci.* **2020**, *10*, 4899. [CrossRef]

14.   Shamsuri, A.A.; Jamil, S.N.A.M. A Short Review on the Effect of Surfactants on the Mechanico-Thermal Properties of Polymer Nanocomposites. *Appl. Sci.* **2020**, *10*, 4867. [CrossRef]

15.   Xie, S.; Li, X.; Jiang, Y.; Yang, R.; Fu, M.; Li, W.; Pan, Y.; Qin, D.; Xu, W.; Hou, L. Recent Progress in Hybrid Solar Cells Based on Solution-Processed Organic and Semiconductor Nanocrystal: Perspectives on Device Design. *Appl. Sci.* **2020**, *10*, 4285. [CrossRef]

Article

Layer-by-Layer Deposition of Copper and Phosphorus Compounds to Develop Flame-Retardant Polyamide 6/Montmorillonite Hybrid Composites

Tomasz M. Majka *, Monika Witek, Paulina Radzik, Karolina Komisarz, Agnieszka Mitoraj and Krzysztof Pielichowski

Department of Chemistry and Technology of Polymers, Cracow University of Technology, ul. Warszawska 24, 31-155 Kraków, Poland; monika.witek@interia.eu (M.W.); paulinaradzik@chemia.pk.edu.pl (P.R.); karolina.komisarz@doktorant.pk.edu.pl (K.K.); agnieszka.mitoraj@foxfittings.com (A.M.); kpielich@pk.edu.pl (K.P.)
* Correspondence: tomasz.majka@pk.edu.pl

Received: 26 June 2020; Accepted: 17 July 2020; Published: 21 July 2020

Abstract: Nowadays, increasing attention is devoted to the search for polymeric composite materials that are characterized by reduced flammability. In this work, the layer-by-layer (LbL) technique was applied to form multilayered protective coatings for polyamide 6/montmorillonite (PA6/MMT) hybrid nanocomposites. This time, the double layers LbL deposition was used in order to improve the thermal properties or flammability of PA6 materials. Our goal was to check how five, 10, and 15 triple-layer deposition onto the surface of PA6 and PA6/MMT composites influenced these relevant properties. For this reason, disodium H-phosphonate, sodium montmorillonite, and iodo-bis(triphenylphosphino)copper were used for polyelectrolyte solution preparation. It was found that the LbL method could be successfully used to improve the flammability characteristics of polyamide 6-based composites. Nevertheless, the deposition of the copper complex should be combined with other flame retardants—preferentially containing phosphorus—which enable synergistic effects to occur. Moreover, microscopic observations confirmed that the surfaces on which the formation of interwoven fibrous crystal structures was observed had a tendency to protect the entire material against the destructive effects of heat, contributing, among other things, to reduce the maximum point of heat release rate (PHRR).

Keywords: PA6; montmorillonite; flammability; surface modification; thermal properties; flame retardants; hybrid composites; chemical engineering

---

1. Introduction

Due to its interesting properties, such as high mechanical strength, chemical resistance, high abrasion, spinnability, and low thermal expansion, polyamide 6 is one of the most important thermoplastic polymers, widely used as a constructive material [1,2]. The main limitation of polyamide 6 (PA6) applications is its flammability. Among different methods to reduce PA6 flammability, the application of nanoparticles, such as layered silicates, has been reported as an efficient route toward flame-retarded polyamide composites [3]. The use of nanoparticles is also beneficial from the viewpoint of enhanced mechanical, thermal, and barrier properties [3,4].

In order to improve the flame retardancy of polyamides, novel compounds are used in composite materials. He et al. obtained composites of polyamide 6 and aluminum diisobutylphosphinate (APBA), combined with organically modified layered montmorillonite (OMMT) by melt compounding. The results showed synergy between these two compounds in terms of flame retardancy; the highest value of limiting oxygen index (LOI)—36%, was obtained for the sample with a ratio for APBA and OMMT of 5:1 [3,5]. Liu et al. investigated the influence of caged bicyclic pentaerythritol phosphate alcohol (PEPA) on the flame-retardant properties of PA6. The results showed that the LOI values of composites increased with increasing PEPA content [6]. Other compounds which have been recently investigated to reduce the flammability of PA6 were phosphorous–nitrogen flame-retardant and platelet-shaped hexagonal boron nitride. The incorporation of those compounds into polyamide allows the application of the final product in electronic and electrical devices [7]. Moreover, char sulfonic acid and ammonium polyphosphate, metal oxalates, brominated polystyrene/antimony trioxide, or red phosphorus/magnesium hydroxide systems have been applied to reduce the flammability of polyamides [8–13].

The layer-by-layer (LbL) deposition technique offers a direct way to form flame-retardant coatings on the polymers' surface [14,15]. LbL involves applying layers via the immersion of the substrate, alternately in solutions containing positively and negatively charged moieties. The absorption of layers occurs mainly through electrostatic interactions, though other interactions, i.e., the formation of covalent bonds, hydrogen bonding, donor/acceptor interactions, and the formation of stereo-complexes are also possible [16,17]. Considering the traditional approach to incorporating flame retardants in polymer matrices, the application of the LbL technique may prove advantageous. Firstly, it diminishes problems related to the introduction of flame retardants into the matrix. Secondly, as the flame affects the exterior part of the substrate, the presence of layers containing flame retardants may directly impact the combustion process [16]. Recent reports on PA6 flame retardancy improvements using LbL include the coating of polyamide with poly(allylamine) and montmorillonite [15], poly(allylamine) with sodium polyphosphate and titanium(IV) oxide [18,19], anionic sodium phosphate and cationic polysiloxanes [20], chitosan, phytic acid and oxidized sodium alginate [21] with additional cross-linking between the layers by the introduction of sodium tetraborate [22] or chitosan and montmorillonite [23].

In other work [24], membranes with a polyamide thin-film active layer were used in reverse osmosis-based water desalination applications. The incorporation of graphene oxide nanoplatelets in the polyamide layer can alter the surface characteristics, permeability, selectivity, and can enhance the chlorine resistance of these membranes. Polyamide layers with graphene oxide nanoplatelets were synthesized in various sequences. The incorporation of graphene oxide nanoplatelets resulted in an increase in surface hydrophilicity as captured by the change of the water contact angle. Water flux and salt rejection properties of the synthesized membranes have been investigated by using a dead-end cell. The salt rejection ability of membranes increased slightly with the incorporation of GONPs, while the water flux was found to be similar to that observed for the pristine membranes without graphene oxide nanoplatelets.

In the latest works, [25] the layer-by-layer (LBL) deposition of chitosan/melamine/urea and phytic acid onto the acrylic acid-grafted polyamide fabrics, including a pad-dry-cure treatment with chitosan/graphene oxide nanocomposites, was considered to improve the hydrophilicity and durable flame retardancy. The obtained results indicated that these hybrid coatings could significantly improve the flame retardancy as the limiting oxygen index value went up to 25% from 18.5% and could completely stop the melt dripping where the fabric sample with two bilayer deposition and simultaneously treated with chitosan/graphene oxide nanocomposite achieved a V-1 rating. Moreover, a considerable improvement in the thermal stability and char yield was also realized in the thermogravimetric analysis test. Furthermore, the as-prepared hybrid coatings imparted better hydrophilicity as a two bilayer deposition along with a pad-dry-cure treatment by the chitosan/graphene oxide system could boost up the hydrophilicity of the treated fabrics further compared to the only two bilayer deposited fabric sample.

In the work [26], also chitosan and phosphorylated chitosan were deposited onto the polyamide fabric surfaces along with poly-acrylate sodium, via one-pot and layer-by-layer assembly methods in preparing flame-retardant coatings. Subsequently, to stabilize the deposited coatings, some of the fabric samples were treated under UV irradiation, and additionally, a thermal treatment was also carried out for the remaining fabric samples. The LbL assembled fabrics showed a better homogeneity in the coating structure over the one-pot deposited fabrics than as appeared in scanning electron microscopy. Nonetheless, the LbL-treated fabric sample with a higher weight gain% exhibited a greater improvement in the limiting oxygen index, and a reduced peak heat release rate.

Similarly, in Ref. [27], chitosan and phytic acid, along with a sol–gel treatment from (3-aminopropyl) triethoxysilane and boron-doped APTES sol solutions, was followed in constructing a flame-retardant and hydrophilic coating onto polyamide fabric surfaces. Layer-by-layer assembled and simultaneously sol–gel-treated PA fabric samples could be able to stop the melt-dripping in a vertical burning test. The boron-doped APTES sol-treated fabric samples with a 2–5 bilayer deposition showed a reduced peak heat release rate in the cone calorimetry test and revealed an improved thermal stability in the thermogravimetric analysis test. More interestingly, the boron-doped silica sol could significantly increase the hydrophilicity of the treated fabrics compared to only LbL and simultaneously LbL-deposited and silica sol-treated fabric samples.

The obtained results reveal that the LbL technique can be successfully used to deposit coatings providing a better flame retardancy of the polyamide matrices [25–28].

In this work, polyamide 6 nanocomposites with MMT were surface modified via the layer-by-layer deposition of copper and phosphorus compounds. The influence of the deposition technique on the thermal properties and flammability of the surface-modified hybrid materials was examined and discussed.

2. Materials and Methods

2.1. Materials

Tarnamid® T-27 polyamide 6 (PA6) in the form of pellets was obtained from Grupa Azoty S.A. (Tarnów, Poland). It is a medium viscosity type of polymer, designed for the processing of modified pellets by compounding methods.

Cloisite® 20A (BYK Additives&Instruments, Wesel, Germany) montmorillonite (MMT), organophilized with 125 meq/100 g quaternary ammonium salt chloride (composed of ~65% C18, ~30% C16, ~5% C14 mixture of aliphatic chains), was used to prepare the PA6 nanocomposite (REFMMT).

For the ionic solutions' preparation, sodium montmorillonite (MMT-Na$^+$) (Dellite® LVF, Laviosa Chimica Mineraria, Livorno, Italy), H-phosphonate (HP) (Sigma-Aldrich, Steinheim, Germany), iodo-bis(triphenylphosphino)copper (CC) (Sigma-Aldrich, Steinheim, Germany), deionized water (DI) (Hipernet, Kraków, Poland), ethyl acetate (EA) (POCH, Gliwice Poland), hydrochloric acid solution (Chempur, Piekary Śląskie, Poland) and 1M sodium hydroxide solution (POCH, Gliwice, Poland) were used.

2.2. Processing Methods

PA6 pellets, vacuum dried at 80 °C for 6 h, and organophilized layered silicate, were mixed in mass proportion 97:3 by the mechanical shaking of a batch in a plastic container for 15 min. Later, the batch was dosed into a twin-screw counter-rotating laboratory extruder Thermo Scientific Haake PolyLab PTW 16/25 to form a nanocomposite with 150 rpm rotation speed. As a dosing device, Brabender DDSR20 feeder compatible with an extruder, working with 45 rpm, was used. During previous tests in our laboratory, this speed had been found as optimal for volumetric dosing, regardless of any screw rotation speed settings. The temperature from the feed end to the die end ranged from 245 to 260 °C. The obtained streak was cooled in a water bath (18 °C) of 1.5 m length and then pelletized. To erase the thermal history, the same procedure was applied to pristine PA6 resin. After one more vacuum drying, two types of materials (REF and REFMMT) were compression molded in a laboratory press ZAMAK P-200 to form 15.0 × 15.0 × 0.3 cm sheets. In addition, 50–55 g of pellets were filled into a mold and pressed at 250 °C under 20 MPa for 2.5 min. The sheets were cooled for 20 min and cut into 15.0 × 1.3 × 0.3 cm bar specimens, that were then used for fire testing.

2.3. Ionic Solutions Preparation

To obtain a 0.5% (m/v) cationic solution, H-phosphonate (HP) was dissolved in 18.2 MΩ deionized water (DI) and a copper complex—iodo-bis(triphenylphosphino)copper (CC), was dissolved in ethyl acetate. Sodium montmorillonite (MMT-Na$^+$) was dispersed in DI at 0.5% (m/v) to form anionic mixtures. All mixtures were magnetically stirred overnight, and after that, the pH of the HP and CC dispersions were adjusted to 4.0 ± 0.5 by adding 1M hydrochloric acid solution or 1M sodium hydroxide solution. The pH of the MMT-Na$^+$ dispersion was 9.8 ± 0.1 (ζ-potential value—45.1 mV, isoelectric point value 1.05).

2.4. Layer-by-Layer Deposition (LbL)

Before depositing the LbL assemblies on the pellets, both materials were washed with DI. Bar specimens were additionally immersed into a 10% acetone solution prior to the DI washing and then, either the pellets or bars were vacuum dried at 80 °C for 2 h. All of the specimens were alternatively dipped into the positively and negatively charged solutions: to enhance the adhesion, the first dips in MMT-Na$^+$ (1), HP (3), and CC (5) (Figure 1) were set at 5 min, but subsequent dipping lasted for 1 min.

In the case of pellets, to protect from material loss, specially designed polypropylene sieves, closed from the top by aluminum sieves, were used for the dipping procedure. After each adsorption step, the materials were washed with DI for 1 min in order to desorb the weakly adsorbed surface modification and dried for 30 min at 80 °C. Some part of an excessive amount of water was taken away using tissue paper.

The scheme of the described procedure is shown in Figure 1. Each cycle was repeated until 5, 10 and 15 triple layers (L) were built on each specimen type (for instance, 5L denotes PA6 with 5 triple layers, deposited on the surface of the sample and 10LMMT denotes the PA6/MMT nanocomposite with 10 triple layers deposited by using the LbL technique).

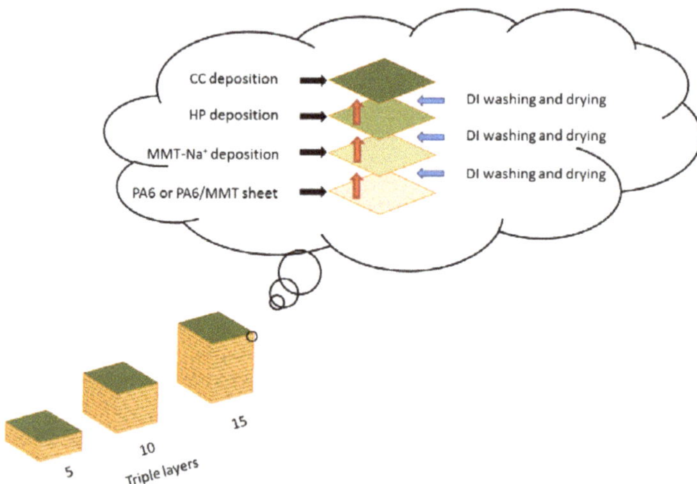

Figure 1. The procedure of the triple-layer cycle of the layer-by-layer deposition on the polyamide 6 (PA6) or PA6/ montmorillonite (MMT) surface, including: deionized water (DI) washing and drying at 80 °C; dipping in MMT-Na$^+$ dispersion; DI washing and drying at 80 °C; dipping in H-phosphonate (HP) solution; DI washing and drying at 80 °C; and dipping in iodo-bis(triphenylphosphino)copper (CC) solution.

2.5. Thermogravimetric Analysis

The thermogravimetric analysis was performed under the air atmosphere using a Netzsch TG 209 F1 Libra thermogravimetric analyzer. The measurements were conducted in the temperature range from 25 to 600 °C at a heating rate of 10 °C/min, in the synthetic air atmosphere with an airflow rate of 15 cm^3/min. Samples of ~5 mg mass were placed in open corundum pans.

In addition, the measurements were carried out in an inert atmosphere in the temperature range from 25 to 600 °C at a heating rate of 1 °C/min, in order to compare the results of the thermal analysis with the measurements made by microscale combustion calorimetry (MCC).

2.6. Flammability Testing

Vertical (VFT) and Horizontal (HFT) flammability tests were performed according to ISO 1210:1992(E) (UL-94) standard using bar specimens. Total burning time and rate of burning (unless the material showed a self-extinguishing effect under specified test conditions) were measured and the digital photographs were taken to depict the characteristics of burning together with residues. Limiting oxygen index (LOI) measurements were performed according to DIN EN ISO 4589-2 using the Concept Fire Testing Oxygen Index Module apparatus that let adjust the concentration of oxygen in a gas stream with 0.1% by volume accuracy. The nitrogen/oxygen mixture rate flow was set at 10 l/min.

The combustion properties were also tested by the means of microscale combustion calorimetry (MCC) according to the ASTM D7309 method A at a synthetic air mixture volume ratio 80/20 (nitrogen/oxygen). The mass of the samples was ca. 5 mg, whereas the heating rate was 1 °C/min.

2.7. Scanning Electron Microscopy (SEM)

For the morphological analysis of the 15L and 15LMMT samples, a Jeol JSM-6010LA scanning electron microscope was used. Pictures of the gold-sputtered samples were obtained at an accelerating voltage of 8–20 kV.

2.8. Surface Roughness and 3D Scanning

The surface roughness profile was tested using a HOMMELWERKE 3D optical profilometer. The basic parameters of roughness were defined in accordance with the PN-87/M-04256/02, PN-87/M-04250, and PN-87/M-04256/01 standards. The measurement for each sample was carried out three times at different places on their surface.

Three-dimensional surface imaging was performed using a Keyence VHX-900F microscope using 100× magnification.

3. Results and Discussion

3.1. Thermogravimetric Analysis

For compositions containing tri-layers deposited with the layer-by-layer method, with the increased number of layers, the temperature of 5% mass loss decreased, independently of the employed substrate (Figure 2). The highest values were observed for the 5L and 5LMMT samples. The thermogravimetry (TG) curves demonstrating the first and the second step of thermal degradation for all samples did not exhibit significant differences.

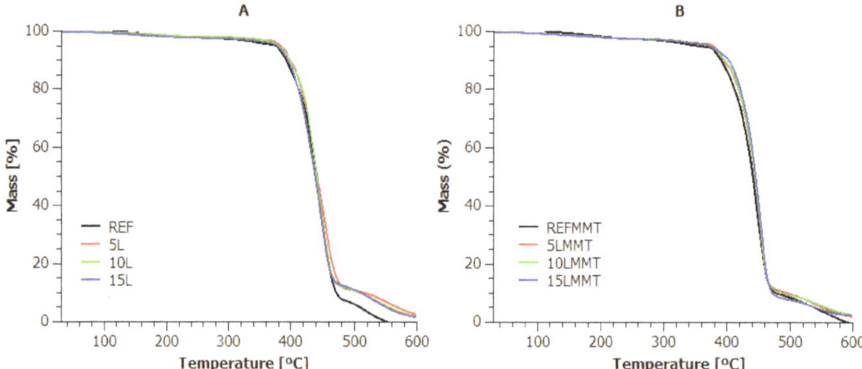

Figure 2. Thermograms of the materials with 5, 10, and 15 triple layers, deposited on the surface of (**A**) the PA6 and (**B**) the PA6/MMT nanocomposites.

The analysis of the parameters of thermal degradation showed that in the low-temperature range, the thermal stability decreases with the increase in the number of deposited layers. It was also concluded that the application of any number of layers by the LBL method on the surface of PA6/MMT composites did not affect their thermal stability above 440 °C. The adsorption of five layers on the pure matrix as well as 10 layers on the nanocomposite allowed for obtaining the largest amount of charred residues in comparison to the rest of the analyzed materials (Table 1).

Table 1. Temperatures at which 5%, 10%, 20%, and 50% mass loss occurs for the PA6 and the PA6/MMT nanocomposites stabilized with triple layers.

Sample	$T_{5\%}$ (°C)	$T_{10\%}$ (°C)	$T_{20\%}$ (°C)	$T_{50\%}$ (°C)	T_{max} (°C)	Residue at 600 °C (%)
REF	374	391	412	440	445	0.9
5L	383	399	417	442	454	2.4
10L	381	399	418	442	449	1.8
15L	377	395	412	439	450	1.5
REFMMT	336	384	409	438	445	1.8
5LMMT	379	395	418	444	454	2.0
10LMMT	373	397	418	444	460	2.6
15LMMT	372	403	421	445	453	2.5

In those materials, a relationship between the decreased mass of the obtained residue and an increase in the number of layers deposited on the surface occurred. Analogously to the results obtained by other research teams [29,30], such results suggest that the application of the triple layers might have a positive impact on the flame retardancy of the pure and filled polyamide matrix, respectively. The barrier effect was particularly pronounced with the application of the least number of deposited layers (5L). The adsorption of the tri-layers on the surface of polyamide materials could reduce the air permeability into the interior of the composite, as a result of which the greater mass of residue in comparison to the reference material was noted. It was concluded that the deposition of layers may have affected the mechanism of thermal degradation of the analyzed compositions, but did not alter the number of steps.

In order to present the influence of the number of deposited tri-layers on the overall thermal stability of the obtained materials, the column graph detailing the values of the additional thermogravimetric index, OSE (*overall stabilization effect*) (Figure 3), calculated on the basis of the formula (1) given below [31], was created:

$$\text{OSE} = \sum\nolimits_{30}^{600}(\text{mass\% PA6/MMT at T temperature} - \text{mass\% pure PA6 at T temperature}) \tag{1}$$

The purpose of the OSE index is to assess the materials' thermal stability in the whole measurement range since its value is determined on the basis of the surface area below the mass loss curve as a function of temperature.

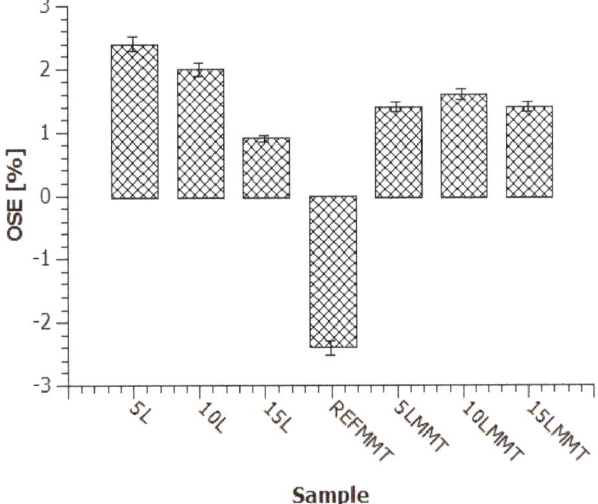

Figure 3. Overall thermal stabilization effect of the triple layers on polyamide 6.

The positive values of the OSE index indicate that all the samples with deposited triple layers are more thermally stable than polyamide 6. The negative value observed for the REFMMT material confirms the existence of the destabilization effect caused by the implementation of the nanofiller in the form of montmorillonite. The best effects were observed after the deposition of five tri-layers via the layer-by-layer method on the surface of the polyamide matrix without reinforcement. Further deposition contributed to a decrease in the value of the OSE index. In the case of the reinforced polymer matrix, the adsorption of tri-layers contributed to the unification of the average effect of thermal stabilization to the value of 1.4%. Thermogravimetric analysis results revealed that the application of the LbL method for the deposition of five triple layers effectively enhances the thermal stability of polyamide materials up to 400 °C.

3.2. Limiting Oxygen Index (LOI)

LOI is defined as the minimum oxygen concentration in the oxygen/air mixture that causes the material to burn. The measurement of the oxygen limit value was started from the reference sample for which 21% was chosen as the reference point [32]. In Table 2, the LOI values obtained for the materials, on which five to 15 triple layers were applied by the LbL method, are presented.

Analyzing the results of the coated polyamide matrices, a decrease in the oxygen index value was observed along with the increase in the number of triple layers applied, regardless of the type of deposition method used. Most likely, this effect was influenced by the way the layers were applied—predominantly on the surface. The application of the LbL method for the adsorption of materials constituting the polyelectrolyte solution and suspensions caused a uniform movement of the flame front towards the holder. The combustion rate for the 5L, 10L, and 15L composites was much smaller than for the reference material.

Table 2. Limiting oxygen index (LOI) value summary for the coated polymeric materials using the layer-by-layer (LbL) method.

Sample	LOI (% Oxygen)	ΔLOI (% Oxygen)
REF	19.8	-
5L	19.6	−0.2
10L	19.4	−0.4
15L	19.3	−0.5
REFMMT	19.7	-
5LMMT	19.9	0.2
10LMMT	19.9	0.2
15LMMT	20.1	0.4

LOI values for PA6 coated by using the LbL method show no flame-retardant action of the deposited layers. In general, the deposition of triple layers adversely affects the flammability of the pristine polymer matrix.

The incorporation of MMT in an amount of 3% by weight resulted in a decrease in the LOI value by 0.1% compared to pure polyamide 6, which is within the measurement error. The presence of triple layers on the surface of the nanocomposite increased the value of the LOI; the highest LOI improvement was achieved for 15LMMT.

A comparison of the obtained results shows that the use of the LbL method for embedding the HP layers, MMT, and CC on the surface of the PA6 nanocomposite promotes the ignition of the polymeric material. During the combustion of the samples reinforced with MMT, the flame evenly covered the surface, and after its self-extinguishing, no visible smoke appeared, which was observed for the pristine polyamide. This insignificant effect could be related to both the introduction of aluminosilicate into the polyamide matrix, affecting the "closure" of the surface of the incinerated sample by creating a dense protective coating, as well as the depositing of phosphorus and copper compounds on the surface of the PA6/MMT nanocomposite.

The MMT suspension and polyelectrolyte solutions/dispersions introduced into the matrix were most likely thermally decomposed first, resulting in the formation of carbonaceous char preventing the transfer of heat, energy, and oxygen between the flame and the material, thereby reducing the smoke emission and toxic decomposition products [33].

3.3. Vertical (VFT) and Horizontal (HFT) Flammability Tests (UL-94)

Characteristic parameters of the combustion process, carried out using the UL-94 method in horizontal (HB or HFT) and vertical (VB or VFT) mode, are summarized in Tables 3 and 4. They enable us to qualify the polymer materials obtained into individual flammability classes in accordance with ISO 1210. In the case of a horizontal system, as the initial value of ignition time, the time when the flame reached the beginning of the measurement line, determined by the start line, was considered.

Both reference materials (REF, REFMMT) showed different behavior in the flame during the measurements, however, each of them was completely burned. The incorporation of MMT into the polymer matrix led to a four-fold reduction in the number of detached fragments of the molten material. The reason for this phenomenon can be the reduced mobility of polymer chains in the presence of aluminosilicate packages. During the combustion of the nanocomposite, the formation of a protective barrier was observed in the form of charcoal, resulting from the combustion of dispersed MMT layers in a polyamide matrix. Moreover, in this case, the flame reached a much higher height. The molten parts of the REF and REFMMT materials broke off together with the flame under the influence of the acting force of gravity, nevertheless, the time intervals of the falling fragments of the polyamide matrix were noticeably smaller.

Analyzing the combustion process of the unreinforced polymer matrix coated with triple layers using the LbL method, it was found that an increase in the degree of deposition of solution materials and polyelectrolyte dispersions resulted in an increase in the average combustion rate (Table 3). The consequence of the deposition of the triple layers in the number of ten was the crossing of the starting line by the flame, nevertheless, this material did not completely burn. The main cause of this phenomenon was found in the rapid flame breakdown together with the alloy, interrupting the process of maintaining the heat. In addition, a few seconds before the extinguishing of the material itself, the appearance of air bubbles on the surface of the burning material was observed. In contrast to the REF sample, the detachment of the molten fragments of the composition took place at regular intervals.

Table 3. Results of the UL-94 combustion carried out in a horizontal position (HB) for the composites covered by using the LbL method.

Sample	Burning Time (s)	Length of the Burnt Sample (mm)	Burning Rate (mm/min)	Number of Drops	Type of Standard Class
REF	441 ± 2	100	14 ± 1	280	I
5L	-	Did not reach the starting line *	-	-	II
10L	264 ± 2	60	14 ± 1	151	III
15L	271 ± 2	100	22 ± 1	179	I
REFMMT	430 ± 2	100	13 ± 1	66	IV
5LMMT	255 ± 2	100	24 ± 1	19	V
10LMMT	266 ± 2	100	23 ± 1	11	V
15LMMT	275 ± 2	100	22 ± 1	26	V

* the flame did not exceed the 25 mm line (measuring start line) at which the measuring section began

The adsorption of triple layers on the surface of the nanocomposite contributed up to about a 38% reduction in the burning time of the 10LMMT composition in relation to the REFMMT material. There was a tendency to increase the average burning time of the materials with an increase in the coverage of the PA6/MMT system. The effect of this dependence was also reducing the average burning rate with the increase in the deposited triple layers. Nevertheless, applying the next five triple layers reduced the burning rate by only one unit.

In order to qualify the obtained composites for individual fire classes, the ISO 1210 standard was applied, according to which the material belongs to the horizontal position (HB) class, when after removing the applied flame, one of the following scenarios occurs:

a. The material does not burn visibly, and the flame front does not reach the measuring line beginning (25 mm); and/or

b. The flame front reaches the entire range of the measuring section (100 mm) at the same time not exceeding the measurement end line; and/or

c. The flame front exceeds the measuring range, and the combustion speed for the samples with a thickness of 3–13 mm is not greater than 40 mm/min [34].

Observations made in the course of the study proved that all the materials met at least one of the above criteria and can be given the HB flame class.

The research carried out in the horizontal mode showed the total combustion of MMT-reinforced materials, although in this case, the flame also detached with the molten fragments of the nanocomposite. The list of the parameters obtained by the reference samples with all the compositions, on which 5–15 triple layers were deposited, presented a significant increase in the average burning rate and a reduced number of detached fragments of modified materials. Most likely, the compounds included in the polyelectrolyte solution and dispersions during their decomposition were combined with the free radicals formed during the decomposition process. The consequence of this reaction could be the dehydrogenation of the polymer, as a consequence of which a protective barrier of coke was formed on the surface of the sample [21].

The parameters obtained by the combustion of materials in a vertical position are presented in Table 4. The main criteria that allow giving the tested materials a specific combustibility class were the combustible properties of the sample, determined as the burning time of the material, as well as the nature of the fragments of the burning material. In this case, the classes V0, V1, V2 were distinguished. The first and second of them refer to the materials whose falling parts of the polymer melt do not contribute to the ignition of the cotton sheet placed under the analyzed composite. These classes differed from each other by burning time.

Due to the presented criteria, materials 10L and 15L were qualified for the best fire classes, because the molten fragments of these samples did not initiate the ignition set up under the cotton sheet under study. The shorter burning time of the 10L composite contributed to giving it the FV-0 class. Compositions marked REFMMT, 5LMMT, 10LMMT, 15LMMT could not be characterized in the vertical test according to the standard ISO 1210: 1992, due to the fact that the flame reached the mounting bracket. The class of the reference materials REF qualified for the FV-2 group, because the first detached piece of material resulted in cotton ignition.

Table 4. Results of the UL-94 combustion carried out in a vertical position (VB) for the composites covered by using the LbL method.

Sample	Burning Time (s)	Length of the Burnt Sample (mm)	Burning Rate (mm/min)	Number of Drops	Type of Standard Class
REF	40	27	41	22	FV-2
5L	242	125	31	121	-
10L	47	31	40	32	FV-0
15L	58	44	46	45	FV-1
REFMMT	148	125	50	13	-
5LMMT	86	125	87	12	-
10LMMT	80	125	94	4	-
15LMMT	66	125	114	4	-

The analysis of the combustion process showed a significant increase in the average burning rate of composites compared to the horizontal mode. In the case of the layer-by-layer method, the average burning time and the length of the burnt sample were reduced. Fittings, into which the matrix aluminosilicate was introduced in the form of MMT, with triple layers embedded in both methods, were completely burned. During the test, the flame applied to the surface of these samples during 10 seconds, evenly lit both sides of the nanocomposite, contributed to the mutual drip of the alloy towards the cotton sheet placed under the analyzed sample. The described behavior of the material led to a noticeable reduction in the number of detached parts of the material and an increase in the average burning rate in relation to the samples not filled with MMT.

Tests carried out in both vertical and horizontal arrangements showed the effect of the degree of coverage of materials on their behavior during combustion. For all fittings, an increase in the burning rate was observed in relation to the reference compositions. Applying triple layers on the surface of the nanocomposite resulted in the characteristic detachment of the larger parts of the sample. On the basis of the conducted study, it was found that the LbL method of deposition of layers affected the improvement of the fire class of samples. In the case of a PA6 matrix, the molten composite fragments did not contribute to the inflammation of cotton.

3.4. Microscale Combustion Calorimetry (MCC)

On the basis of the microcalorimetry test, it was possible to obtain information on the heat release rate (HRR) as a function of temperature and time.

The effect of introducing MMT in an amount of 3% by weight to the polyamide matrix was shown by the point of heat release rate (PHRR) value of 670.2 W/g. Of all the obtained results, this value was the highest. In addition, it was achieved at the same time and temperature as in the REF material. It may suggest that the introduction of nanofiller adversely affected the thermal effect of PA6 decomposition. However, the effect of achieving a higher peak on the HRR curve in relation to virgin polymer is often caused by the release of volatiles evolving through the decomposition of low-molecular-weight organic compounds used for the MMT modification, such as ammonium salts [35].

The HRR curves of coated materials using the LbL method are shown in Figure 4. The adsorption of triple layers on the surface of a pure polyamide matrix did not contribute to a visible change in the PHRR value, because even for 15L material it is about 0.65%. However, a significant change in the time and temperature values of reaching the maximum heat release rate was observed. With the increase in the number of deposited layers, the peaks shifted towards lower temperatures. The difference between the material temperature values REF and 15L was 16 °C (Table 5). In addition, the time to reach the maximum point of the HRR curve for this material was decreased by 18 s.

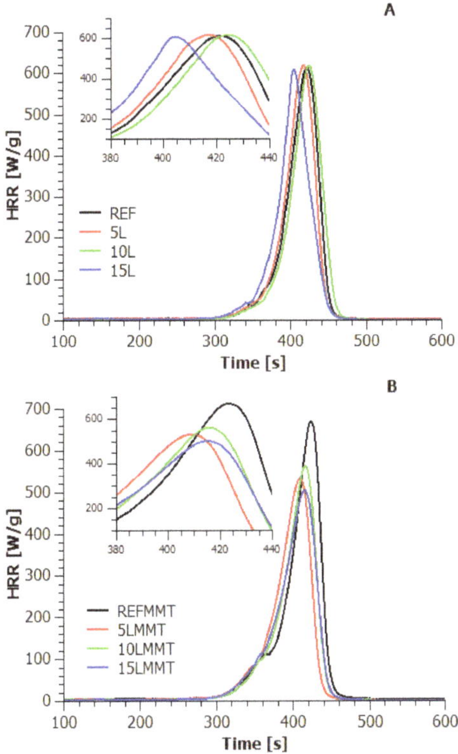

Figure 4. Heat release rate (HRR) as a function of time of the polymeric materials coated using the LbL method: (**A**) the samples without nanofiller; and (**B**) the samples with nanofiller.

Table 5. Point of heat release rate (PHRR) parameters for the materials coated utilizing the LbL method.

Sample	PHRR (W/g)	Temperature to Reach PHRR (°C)	Time to Reach PHRR (s)
REF	613.6	473	422
5L	619.4	468	417
10L	618.8	472	424
15L	609.6	457	404
REFMMT	670.2	471	423
5LMMT	533.3	469	409
10LMMT	564.2	470	416
15LMMT	504.4	468	415

The application of the triple layers to the reinforced aluminosilicate matrix of polyamide 6 resulted in a visible reduction in the rate of heat release. The introduction of 15 triple layers onto the surface of the nanocomposite reduced the PHRR value by 18% compared to a non-modified matrix with no significant effect at the temperature at which the PHHR was achieved. The effect of reducing the PHRR of this material was even greater when it was compared with the REFMMT sample (the observed flame-retardant effect was as high as 25%).

The high temperature of the conducted study affected the total destruction of the protective barrier created. Whilst analyzing the course of all the pyrolysis and combustion microcalorimetry decomposition profiles, no changes in the shape of the peaks were observed. The application of 5–15 triple layers by LbL on a pristine PA6 matrix did not have a visible effect on reducing the maximum heat release rate. In the case of a reinforced polyamide matrix, this effect was much larger. The synergistic effect of MMT and the compounds constituting the triple-layer deposited on the polyamide 6 surface probably had an impact on reducing the maximum amount of heat released. During the combustion of the material, the embedded layers created a protective barrier preventing the thermal decomposition of the surfactant used for the organophilization of MMT. It can be assumed that during the thermal decomposition of the compound based on iodo-bis(triphenylphosphino)copper, phosphoric acid anhydride or its derivatives are produced, which in turn are the source of free radicals that can recombine with the radicals produced as a result of the degradation of PA6 [36]. In addition, iodine anions could inhibit the reactions occurring in the flame, and copper atoms prevent the formation of harmful phosphine [37]. Most probably, the applied disodium H-phosphonate layer and sodium MMT became a catalyst for carbonization, and the phosphorus compounds reacted with carbon, resulting in a reduction in PHHR. In addition, the deposition of sodium MMT together with iodo-bis(triphenylphosphino)copper could contribute to the dispersion of copper compounds and the formation of Cu–P–Si bonds during the thermal decomposition, as a result of which the produced coating was characterized by greater durability [15,38,39].

In order to get additional knowledge on the degradation course of the composite materials studied, the deconvolution of HRR = f (T) curves of REF, REFMMT, 15L, and 15LMMT was performed (Figure 5).

On the basis of the performed separation of signals, a four-stage mechanism of decomposition of samples without organofiller and the three-stage mechanism of decomposition of samples with organofiller could be postulated. For all of the presented materials, the rate of heat release during the first stage of degradation was visibly lower in relation to the second and third stages. For the REF and REFMMT materials, there were significantly smaller differences between the second and third signals compared to the 15L and 15LMMT samples. The application of triple layers to the surface of reinforced as well as unreinforced polyamide matrix could not have an influence on the change of the degradation mechanism. Differences presented in Figure 5C,D were probably caused by a lack of organophilized MMT in the PA6 matrix, which played the role of stabilizer in the range of 450–500 °C.

Based on the curves obtained as a result of the HRR test, the total heat release (THR) was calculated. The HRR results referred to the sample based solely on polyamide 6 (assuming its THR value was 100%) are shown in Figure 6.

The introduction of MMT as a filler resulted in a significant increase in the total amount of heat released. In comparison to the reference material, its energy emission rate increased by as much as 12.2%. The effect of depositing the triple layers on the surface of the nanocomposite was lowering the THR value. There was a 4.0% decrease in THR of the composition in a sequence of 15LMMT in combination with an unmodified polyamide matrix.

The analysis of the obtained results of the maximum speed and the total amount of heat released proved the effectiveness of the layer-by-layer method to obtain the effect of reduced flammability. The best results in relation to the reference sample were noted for the 15LMMT material. In the case of other materials, the PHRR and THRR values obtained through the use of pyrolysis and combustion microcalorimeter have shown a similar trend.

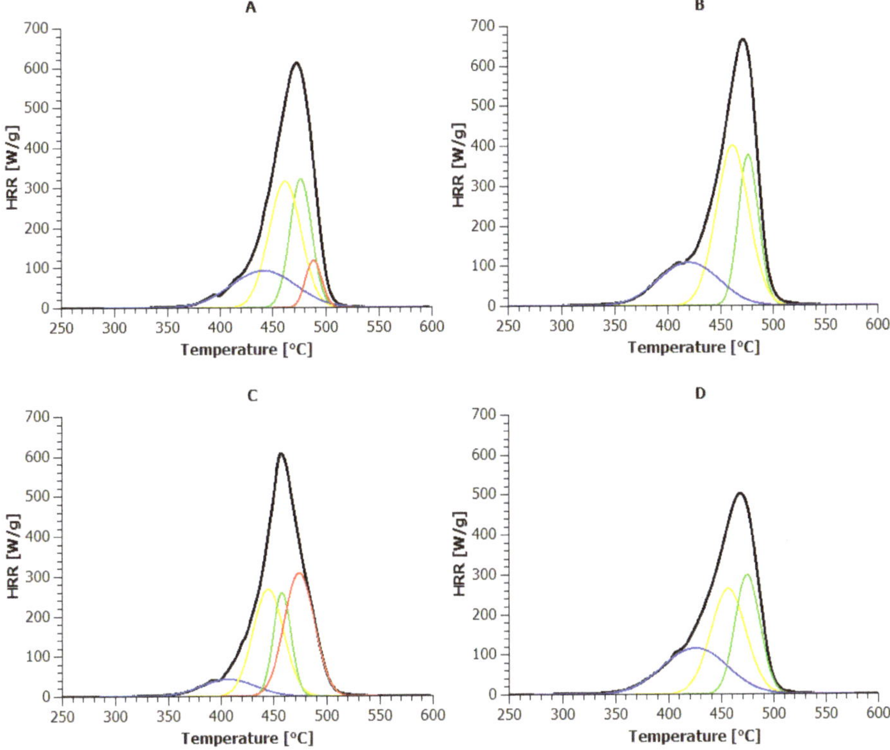

Figure 5. HRR = f (T) profiles and the deconvoluted curves from the primary data obtained using microcalorimetry for: (**A**) REF; (**B**) REFMMT; (**C**) 15L; and (**D**) 15LMMT.

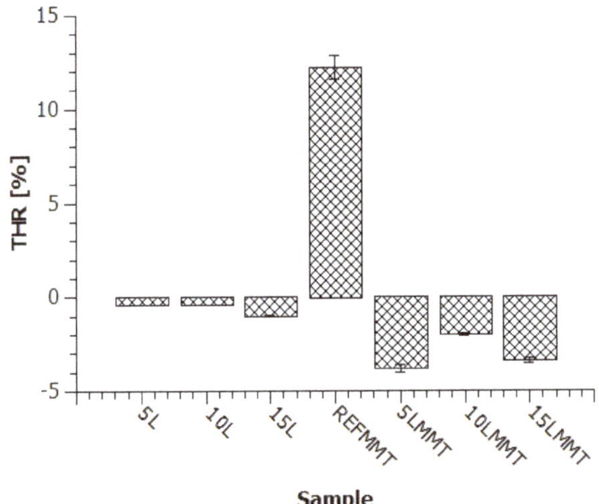

Figure 6. The total amount of heat released (THR) emitted by PA6 coated using the LbL method, compared with the REF material.

3.5. TG vs. MCC

We performed TG measurements at the same heating rates as in the MCC experiments, and the results of both experimental techniques are compared in Figures 7 and 8. Based on the procedure proposed in work [40], for the TG and MCC data, the reaction rates were calculated from the Formulas (2)–(4), respectively:

$$\dot{r}_m = -(dm/dt)/(m_0 - m_\infty) \tag{2}$$

$$\dot{r}_q = \dot{q}/\overline{\Delta q\prime} \tag{3}$$

$$\Delta q\prime = \dot{q}/\dot{r}_m \tag{4}$$

where:

$\dot{r}_m$ the reaction rate of weight loss, s^{-1};
dm/dt the weight derivative over time;
m_0 the initial mass of the sample, kg;
m_∞ the final mass of the sample, kg;
$\dot{r}_q$ the reaction rate of heat release, s^{-1};
$\dot{q}$ the specific heat release rate, W/g;
$\Delta q'$ the heat of combustion per unit mass of the sample, J/g;
$\overline{\Delta q\prime}$ the total heat of combustion per unit mass of the sample, J/g.

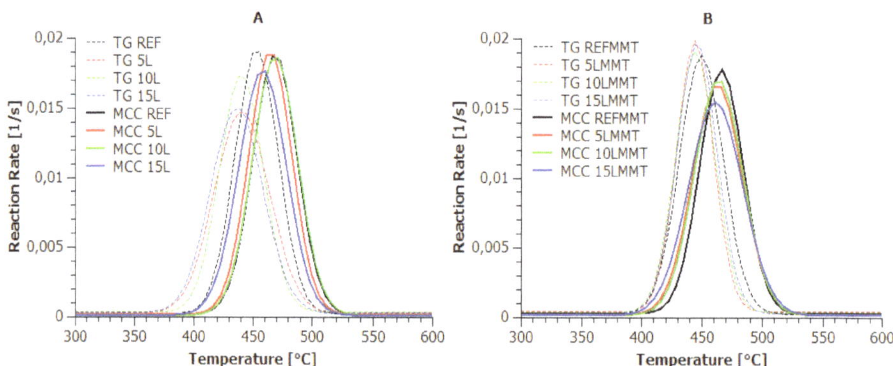

Figure 7. Comparison of the thermogravimetry (TG) and microscale combustion calorimetry (MCC) measurements by determining the reaction rates detected by TG and MCC as a function of temperature during the thermal decomposition of the PA6 samples (heating rate 1 °C/min): (**A**) the samples without nanofiller; and (**B**) the samples with nanofiller.

Such a comparison leads to the conclusion that most of the sample mass is lost during the main step, and also that most of the heat release occurs in the oxidation of volatiles produced in the same step. This indicates that the heats of combustion of the volatiles produced in the main step are very different for the samples with and without nanofiller. The heat of combustion evaluated by dividing the measured heat release rate by the measured mass loss rate is plotted in Figure 8 as a function of sample temperature.

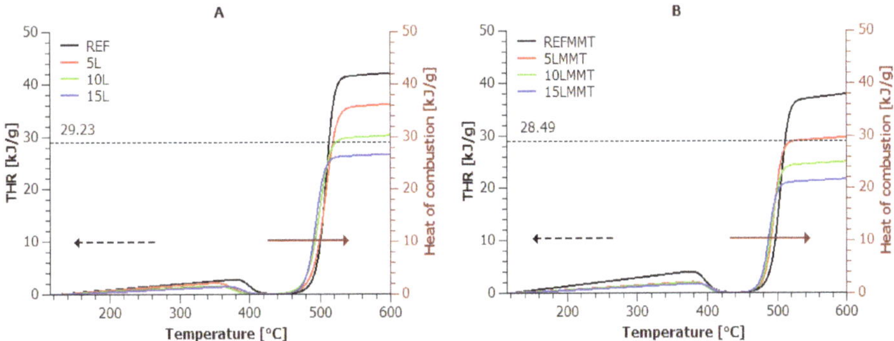

Figure 8. Comparison of the TG and MCC measurements by determining the heat of combustion as a function of temperature during the thermal decomposition of the PA6 samples (heating rate 1 °C/min): (**A**) the samples without nanofiller; and (**B**) the samples with nanofiller.

It can be seen that the volatiles released up to 400 °C have a very low calorific value (~2.5 kJ/g for the samples without MMT, and ~3.7 kJ/g for the samples with MMT, approximately), while most combustible volatiles (with the heat of combustion in the range of 20–40 kJ/g) are produced at higher sample temperatures around 500–600 °C. It is worthy to note that the integral heat of combustion differs significantly from these values (~29.23 kJ/g for the samples without MMT and ~28.49 kJ/g for the samples with MMT, approximately). We noticed that, with increasing applied layers, the heat of combustion in the final stage of combustion is reduced to almost half, with 15 layer deposition in both types of samples, with or without MMT. However, the treated samples have a slightly better char forming ability, which was also observed in both the TG and MCC investigations.

3.6. Morphological Analysis

The SEM investigations of the 10L and 10LMMT samples (Figure 9) confirm that covering the surface of the PA6 and the PA6/MMT composites with ten triple layers did not contribute to the formation of a regular structure. The deposition of ten triple layers on the surface of the composites resulted in a highly porous structure. The obtained microscopic images show numerous fibers with uneven distribution on the surface of the tested material penetrating each other. It should be noted that the LbL method may cause the partial reorganization of crystalline and amorphous phases on the surface.

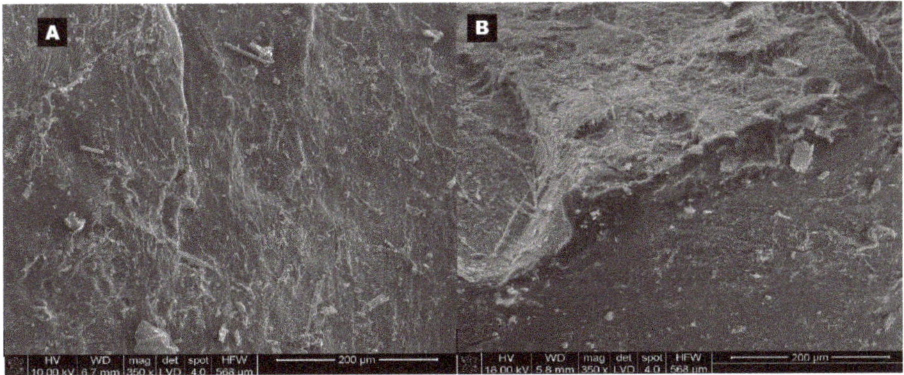

Figure 9. SEM investigations of (**A**) the 15L and (**B**) the 15LMMT samples with the magnification of 350×.

It may lead to the formation of local polycrystalline anisotropic structures, visible in Figure 9A,B, built of interpenetrating micro- and nanofibers. However, greater surface coverage was observed for the composites containing MMT. Such observations could confirm the impact of the presence and distribution of the mineral filler on the deposition of appropriate layers. The reason for this could be the local and temporary generation of weak, secondary binding forces between the nanofiller particles dispersed in the matrix and the particles of the compound present in the applied solution or in polyelectrolyte dispersion. Microscopic observations confirm the supposition that the surfaces on which the formation of (even local) interwoven fibrous crystal structures was observed showed a tendency to protect the entire material against the destructive effects of heat, contributing, among other things, to the reduction of the maximum point of heat release rate (PHRR).

The adsorption of the compounds constituting the solution and polyelectrolyte dispersions resulted in the unification of the R_t values (Table 6) of both the pure and montmorillonite-reinforced polyamide matrix. The value of this parameter for compositions 5L, 10L and 15L oscillated around 4.70 µm, while 5LMMT, 10LMMT, 15LMMT were about 3.60 µm. The deposition of the triple layers the by the LbL technique on pure polyamide 6 significantly reduced the R_a parameter, which suggested an improvement in the quality of the obtained surface. The material became smoother, indicating that the triple layers filled the pores on the surface of PA6.

Table 6. Roughness parameters for the materials coated utilizing the LbL method, where: R_a—the arithmetical mean roughness; R_{max}—the maximum peak height; R_t—the linear distance between the top of peak and base of the valley; and R_z—the height of roughness presented on a scale of ten points.

Sample	R_t (µm)	R_{max} (µm)	R_z (µm)	R_a (µm)
REF	5.04 ± 1.59	4.54 ± 1.08	3.21 ± 0.54	0.58 ± 0.23
5L	4.54 ± 1.90	4.15 ± 1.95	2.67 ± 0.86	0.36 ± 0.18
10L	4.69 ± 1.50	4.60 ± 1.46	2.34 ± 0.50	0.28 ± 0.09
15L	4.79 ± 1.59	4.38 ± 1.02	2.20 ± 0.83	0.31 ± 0.37
REFMMT	3.35 ± 0.66	3.15 ± 0.83	2.05 ± 0.06	0.23 ± 0.02
5LMMT	3.95 ± 0.55	3.93 ± 2.26	2.25 ± 0.55	0.23 ± 0.06
10LMMT	3.64 ± 0.58	3.56 ± 0.47	2.43 ± 0.04	0.25 ± 0.01
15LMMT	3.56 ± 0.59	3.56 ± 0.59	2.99 ± 0.45	0.24 ± 0.18

In the case of the nanocomposite samples, the value of the parameter mentioned above did not significantly change concerning the reference material reinforced with montmorillonite. In addition, for these samples, the distances between the peak and the valley were the same as those determined for the reference materials.

The analysis of surface roughness proved that the use of the LBL method resulted in the even adsorption of the polyelectrolyte solution and dispersions (roughness profile curves—Figures 10 and 11), contributing to a significant reduction in the plane unevenness in the case of the composite. The average value of the R_a parameter obtained by the REFMMT, 5LMMT, 10LMMT, 15LMMT materials was about 0.23 µm. According to the literature, it is the value that characterizes the surfaces after grinding and then polishing [41].

Thanks to the 3D mapping of the 15L (Figure 10) and 15LMMT (Figure 11) sample surfaces, it was possible to determine the height of the triple layers applied. The blank film REF (PA6) had a thickness of 28 µm. The application of the five layers 5L increased the material thickness to about 63 µm. The application of another five layers to obtain 10L increased the thickness of the material to about 105 µm. The treatment of the sample with subsequent layers to obtain 15L increased the thickness of the sample profile to over 145 µm (max 147.17 µm). This means that, on average, one triple layer had a thickness close to 40 µm.

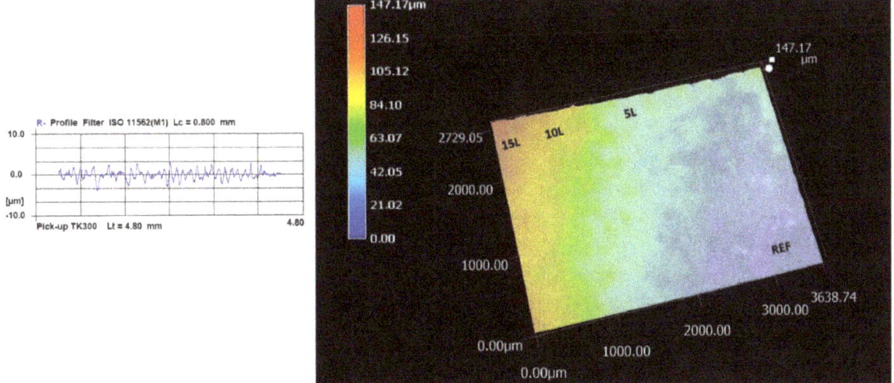

Figure 10. Two-dimensional roughness profile (**left**) and the 3D surface structure (**right**) for the sample 15L.

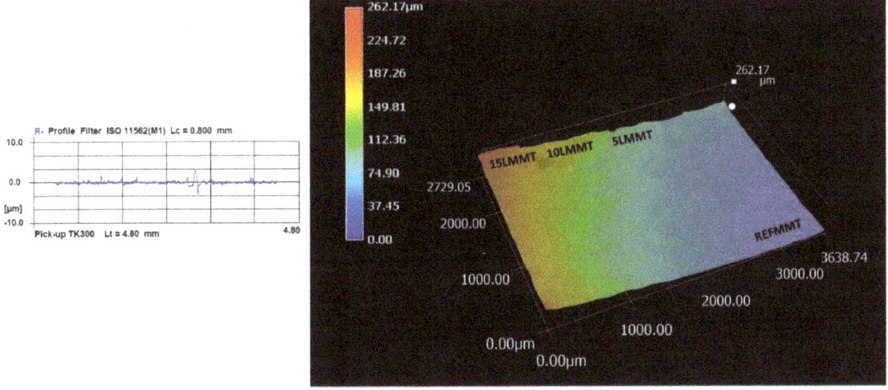

Figure 11. Two-dimensional roughness profile (**left**) and the 3D surface structure (**right**) for the sample 15LMMT.

However, the REFMMT film (PA6/MMT) had a thickness of 55 μm. The application of the 5LMMT triple layers increased the thickness to about 112 μm. The deposition of another five layers to obtain 10LMMT increased the thickness of the material to about 187 μm. The last treatment cycle to obtain 15LMMT increased the thickness of the composite sample to over 260 μm (max 262.17 μm). The calculated average value of one triple layer's thickness was close to 69 μm.

The most important thing is that in both cases, the smallest thickness value was obtained by the first triple layer, and the next two had a similar value (35/42/42 for REF and 57/75/75 for REFMMT). This means that the composite surface containing a quaternary anion salt capable of producing short-range interactions between the applied compound and the surface of the material, has a much higher affinity for layer deposition. This would explain the behavior of the sample during the combustion, primarily at high temperatures of 400–550 °C. Since the new impacts were created, although they do not significantly affect the rate of reaction of sample decomposition, they certainly change its visible mechanism, e.g., by reducing the PHRR value or the heat of combustion just in the final phase of decomposition.

4. Conclusions

The application of multilayered protective coatings by using the LbL technique on PA6 is a promising way to reduce the flammability of this important engineering polymer. In the course of our systematic study, it has been found that the LOI test did not show a significant effect of reducing the flammability of the 15LMMT material. The UL-94 test, both in the vertical (VB) and horizontal (HB) test, made it possible to isolate the polyamide compositions coated with triple layers in the amount of 5–15, as the materials of the best combustibility class compared to the others. In horizontal mode, these composites were not completely burned, and the triple layers deposited on their surface affected the lack of ignition of cotton.

Further information on the flame-retarding effect was obtained as a result of the microcalorimetric analysis. For the polyamide matrix and PA6/MMT nanocomposites, a tendency to reduce the heat release rate (HRR) with an increasing degree of coverage of the polymer matrix was found. The deposition of 15 triple layers on the surface of the nanocomposite using the LbL method, resulted in a 20% reduction in the maximum point of HRR. This improvement in fire resistance was most likely due to the synergistic effect caused by the introduction of aluminosilicate into the polyamide matrix and the simultaneous application of H-phosphonate, montmorillonite and a copper complex, whose action was based on the formation of a protective barrier during the combustion of the composite. Interestingly, the effect of reducing the PHRR of this material was even greater when it was compared with the REFMMT sample—the observed flame-retardant effect was as high as 75%.

Based on the discussion of the results, it was found that the layer-by-layer method could be successfully used to considerably improve the flammability characteristics of the polyamide 6-based hybrid composites, under specific conditions.

Author Contributions: Conceptualization, T.M.M. and M.W.; methodology, T.M.M. and K.P.; software, T.M.M. and K.K.; validation, T.M.M., K.K. and K.P.; formal analysis, T.M.M. and A.M.; investigation, T.M.M., A.M. and M.W.; resources, K.P.; data curation, P.R.; writing—original draft preparation, T.M.M. and P.R.; writing—review and editing, T.M.M., P.R. and M.W.; visualization, K.K.; supervision, K.P.; project administration, T.M.M.; All authors have read and agreed to the published version of the manuscript.

Funding: This research received no external funding.

Conflicts of Interest: The authors declare no conflict of interest.

References

1. Liu, K.; Li, Y.; Tao, L.; Xiao, R. Preparation and characterization of polyamide 6 fibre based on a phosphorus-containing flame retardant. *RSC Adv.* **2018**, *8*, 9261–9271. [CrossRef]

2. Majka, T.M.; Leszczyńska, A.; Kandola, B.K.; Pornwannachai, W.; Pielichowski, K. Modification of organo-montmorillonite with disodium H-phosphonate to develop flame retarded polyamide 6 nanocomposites. *Appl. Clay Sci.* **2017**, *139*, 28–39. [CrossRef]

3. He, W.-T.; Liao, S.-T.; Xiang, Y.-S.; Long, L.-J.; Qin, S.-H.; Yu, J. Structure and Properties Study of PA6 Nanocomposites Flame Retarded by Aluminium Salt of Diisobutylphosphinic Acid and Different Organic Montmorillonites. *Polymers* **2018**, *10*, 312. [CrossRef]

4. Horrocks, R.; Sitpalan, A.; Zhou, C.; Kandola, B.K. Flame retardant polyamide fibres: The challenge of minimising flame retardant additive contents with added nanoclays. *Polymers* **2016**, *8*, 288. [CrossRef] [PubMed]

5. He, W.; Zhu, H.; Xiang, Y.; Long, L.; Qin, S.; Yu, J. Enhancement of flame retardancy and mechanical properties of polyamide 6 by incorporating an aluminum salt of diisobutylphosphinic combined with organoclay. *Polym. Degrad. Stab.* **2017**, *144*, 442–453. [CrossRef]

6. Liu, T.; Wang, R.; Dong, Z.-F.; Zhu, Z.-G.; Zhang, X.-Q.; Liu, J.-G. Role of caged bicyclic pentaerythritol phosphate alcohol in flame retardancy of PA6 and mechanism study. *J. Appl. Polym. Sci.* **2018**, *135*, 46236. [CrossRef]

7. Zhong, Y.; Zhang, L.; Fischer, A.; Wang, L.; Drummer, D.; Wu, W. The effect of hBN on the flame retardancy and thermal stability of P-N flame retardant PA6. *J. Macromol. Sci. Part A* **2018**, *55*, 17–23. [CrossRef]

8. Yuan, L.; Feng, S.; Hu, Y.; Fan, Y. Effect of char sulfonic acid and ammonium polyphosphate on flame retardancy and thermal properties of epoxy resin and polyamide composites. *J. Fire Sci.* **2017**, *35*, 521–534. [CrossRef]

9. Wendels, S.; Chavez, T.; Bonnet, M.; Salmeia, K.A.; Gaan, S. Recent Developments in Organophosphorus Flame Retardants Containing P-C Bond and Their Applications. *Materials* **2017**, *10*, 784. [CrossRef]

10. Holdsworth, A.F.; Horrocks, A.R.; Kandola, B.K.; Price, D. The potential of metal oxalates as novel flame retardants and synergists for engineering polymers. *Polym. Degrad. Stab.* **2014**, *110*, 290–297. [CrossRef]

11. Li, M.; Cui, H.; Li, Q.; Zhang, Q. Thermally conductive and flame—Retardant polyamide 6 composites. *J. Reinf. Plast. Compos.* **2016**, *35*, 435–444. [CrossRef]

12. Zope, I.S.; Dasari, A.; Guan, F.; Yu, Z.Z. Influence of metal ions on thermo-oxidative stability and combustion response of polyamide 6/clay nanocomposites. *Polymer* **2016**, *92*, 102–113. [CrossRef]

13. Zhan, Z.; Xu, M.; Li, B. Synergistic effects of sepiolite on the flame retardant properties and thermal degradation behaviors of polyamide 66/aluminum diethylphosphinate composites. *Polym. Degrad. Stab.* **2015**, *117*, 66–74. [CrossRef]

14. Iler, R.K. Multilayers of colloidal particles. *J. Colloid Interface Sci.* **1966**, *21*, 569–594. [CrossRef]

15. Kadir, A.; Abdelghani, L.; Vincent, B.; Maude, J.; Serge, B.; Valérie, T.; David, R. Polyallylamine-montmorillonite as super flame retardant coating assemblies by layer-by layer deposition on polyamide. *Polym. Degrad. Stab.* **2013**, *98*, 627–634.

16. Qiu, X.; Li, Z.; Li, X.; Zhang, Z. Flame retardant coatings prepared using layer by layer assembly: A review. *Chem. Eng. J.* **2018**, *334*, 108–122. [CrossRef]

17. Malucelli, G. Surface-Engineered Fire Protective Coatings for Fabrics through Sol-Gel and Layer-by-Layer Methods: An Overview. *Coatings* **2016**, *6*, 33. [CrossRef]

18. Apaydin, K.; Laachachi, A.; Ball, V.; Jimenez, M.; Bourbigot, S.; Toniazzo, V.; Ruch, D. Intumescent coating of (polyallylamine-polyphosphates) deposited on polyamide fabrics via layer-by-layer technique. *Polym. Degrad. Stab.* **2014**, *106*, 158–164. [CrossRef]

19. Apaydin, K.; Laachachi, A.; Ball, V.; Jimenez, M.; Bourbigot, S.; Ruch, D. Layer-by-layer deposition of a TiO_2-filled intumescent coating and its effect on the flame retardancy of polyamide and polyester fabrics. *Colloids Surf. A Pchysicochem. Eng. Asp.* **2015**, *469*, 1–10. [CrossRef]

20. Mosurkal, R.; Muller, W.S. Layer-By-Layer Assembly of Halogen-Free Polymeric Materials on Nylon/Cotton. *Fire Mater.* **2014**, *40*, 206–2018.

21. Kundu, C.; Wang, W.; Zhou, S.; Wang, X.; Sheng, H.; Pan, Y.; Song, L.; Hu, Y. A green approach to constructing multilayered nanocoating for flame retardant treatment of polyamide 66 fabric from chitosan and sodium alginate. *Carbohydr. Polym.* **2017**, *166*, 131–138. [CrossRef] [PubMed]

22. Kundu, C.K.; Wang, X.; Song, L.; Hu, Y. Borate cross-linked layer-by-layer assembly of green polyelectrolytes on polyamide 66 fabrics for flame-retardant treatment. *Prog. Org. Coat.* **2018**, *121*, 173–181. [CrossRef]

23. Majka, T.M.; Cokot, M.; Pielichowski, K. Studies on the thermal properties and flammability of polyamide 6 nanocomposites surface-modified via layer-by-layer deposition of chitosan and montmorillonite. *J. Anal. Calorim.* **2018**, *131*, 405–416. [CrossRef]

24. Abbaszadeh, M.; Krizak, D.; Kundu, S. Layer-by-layer assembly of graphene oxide nanoplatelets embedded desalination membranes with improved chlorine resistance. *Desalination* **2019**, *470*, 114116. [CrossRef]

25. Rahman, M.Z.; Kundu, C.K.; Nabipour, H.; Wang, X.; Song, L.; Hu, Y. Hybrid coatings for durable flame retardant and hydrophilic treatment of Polyamide 6.6 fabrics. *Prog. Org. Coat.* **2020**, *144*, 105640. [CrossRef]

26. Kundu, C.K.; Wang, X.; Song, L.; Hu, Y. Chitosan-based flame retardant coatings for polyamide 66 textiles: One-pot deposition versus layer-by-layer assembly. *Int. J. Biol. Macromol.* **2020**, *143*, 1–10. [CrossRef]

27. Kundu, C.K.; Wang, X.; Liu, L.X.; Song, L.; Hu, Y. Few layer deposition and sol-gel finishing of organic-inorganic compounds for improved flame retardant and hydrophilic properties of polyamide 66 textiles: A hybrid approach. *Prog. Org. Coat.* **2019**, *129*, 318–326. [CrossRef]

28. Apaydin, K.; Laachachi, A.; Fouquet, T.; Jimenez, M.; Bourbigot, S.; Ruch, D. Mechanistic investigation of a flame retardant coating made by layer-by-layer assembly. *RSC Adv.* **2014**, *4*, 43326–43334. [CrossRef]

29. Xiao, L.; Xu, L.; Yang, Y.; Zhang, S.; Huang, Y.; Bielawski, C.W.; Geng, J. Core-Shell Structured Polyamide 66 Nanofibers with Enhanced Flame Retardancy. *ACS Omega* **2017**, *2*, 2665–2671. [CrossRef]

30. Wu, H.; Krifa, M.; Koo, J.H. Flame retardant polyamide 6/nanoclay/intumescent nanocomposite fibers through electrospinning. *Text. Res. J.* **2014**, *84*, 1106–1118. [CrossRef]

31. Katsoulis, C.; Kandare, E.; Kandola, B.K. The effect of nanoparticles on structural morphology, thermal and flammability properties of two epoxy resins with different functionalities. *Polym. Degrad. Stab.* **2011**, *96*, 529–540. [CrossRef]

32. Castrovinci, A.; Camino, G. Fire-Retardant Mechanisms in Polymer Nano-Composite Materials. In *Multifunctional Barriers for Flexible Structure: Textile, Leather and Paper*, 1st ed.; Camino, G., Duquesne, S., Magniez, C., Eds.; Springer: Berlin/Heidelberg, Germany, 2007; Volume 97, pp. 87–108.

33. Zhang, W.; He, X.; Song, R.; Jiao, Q.; Yang, R. The influence of the phosphorus-based flame retardant on the flame retardancy of the epoxy resins. *Polym. Degrad. Stab.* **2014**, *109*, 209–217. [CrossRef]

34. Technical Committee: ISO/TC 61/SC 4 Burning Behavior, ISO 1210: 1992 Plastics—Determination of the Burning Behaviour of Horizontal and Vertical Specimens in Contact with a Small-Flame Ignition Source, ICS: 13.220.40 Lgnitability and Burning Behaviour of Materials and Products 83.080.01 Plastics in General. Available online: www.iso.org (accessed on 11 February 2018).

35. Singla, P.; Mehta, R.; Upadhyay, S.N. Clay Modification by the Use of Organic Cations. *Green Sustain. Chem.* **2012**, *2*, 21–25. [CrossRef]

36. Salmeia, K.A.; Gaan, S.; Malucelli, G. Recent Advances for Flame Retardancy of Textiles Based on Phosphorus Chemistry. *Polymers* **2016**, *8*, 319. [CrossRef] [PubMed]

37. Roth, M.; König, A.; Deglmann, P.; Uske, K.; Minges, C. BASF SE, Flame-Retardant Polyamides with Pale Color. U.S. Patent 20140080949A1, 20 March 2014.

38. Faraji, S.; Rahim, A.A.; Mohamed, N.; Sipaut, C.S. A study of electroless copper-phosphorus coatings with the addition of silicon carbide (SiC) and graphite (Cg) particles. *Surf. Coat. Technol.* **2011**, *206*, 1259–1268. [CrossRef]

39. Yi, D.; Yang, H.; Zhao, M.; Huang, L.; Camino, G.; Frache, A.; Yanga, R. A novel, low surface charge density, anionically modified montmorillonite for polymer nanocomposites. *RSC. Adv.* **2017**, *7*, 5980–5988. [CrossRef]

40. Berkowicz, D.; Majka, T.M.; Żukowski, W. The pyrolysis and combustion of polyoxymethylene in a fluidised bed with the possibility of incorporating CO_2. *Energy Convers. Manag.* **2020**, *214*, 112888. [CrossRef]

41. Whitehouse, D. Surface and their measurements. In *Kogan Page Science Paper Edition*; Butterworth-Heinemann: London, UK, 2006.

Article

Study of Interfacial Adhesion between Nickel-Titanium Shape Memory Alloy and a Polymer Matrix by Laser Surface Pattern

Sneha Samal [1],*, Ondřej Tyc [1], Luděk Heller [1], Petr Šittner [1], Monika Malik [2], Pankaj Poddar [2], Michelina Catauro [3] and Ignazio Blanco [4],*

1 Institute of Physics of Czech Academy of Sciences, Na Slovance 1999/2, 182 21 Prague, Czech; tyc@fzu.cz (O.T.); heller@fzu.cz (L.H.); sittner@fzu.cz (P.Š.)
2 National Chemical laboratory, Pune 411008, India; m.malik@ncl.res.in (M.M.); p.poddar@ncl.res.in (P.P.)
3 Department of Engineering, University of Campania "Luigi Vanvitelli", Via Roma 29, I-81031 Aversa, Italy; michelina.catauro@unicampania.it
4 Department of Civil Engineering and Architecture and UdR-Catania Consorzio INSTM, University of Catania, Viale Andrea Doria 6, 95125 Catania, Italy
* Correspondence: samal@fzu.cz (S.S.); iblanco@unict.it (I.B.)

Received: 22 February 2020; Accepted: 20 March 2020; Published: 23 March 2020

Abstract: The aim of this article is to investigate the interfacial adhesion of Ni-Ti shape memory alloy with a polymer matrix of Poly (methyl methacrylate) (PMMA). The surface pattern on Ni-Ti plates was channeled by a solid state laser machine. The laser machine allows for creating channels on the Ni-Ti surface for infiltration of the PMMA matrix, which could be attached as an intra-surface locking pattern to the Ni-Ti surface. The influence of the PMMA matrix on the surface of the NiTi plate was evaluated by thermomechanical analysis (TMA) and dynamic mechanical analysis (DMA). The surface characterization was carried out by an optical microscope on the PMMA/NiTi composite after mechanical testing. During mechanical testing, the polymer displays the multiple cracks in the longitudinal direction that result in slipping and fracture. TMA and DMA analyses were performed on the Ni-Ti- and PMMA-coated Ni-Ti ribbon to observe elasticity and the storage modulus for both samples. Better adhesion than 80 % was observed in the Ni-Ti surface, in the laser surface pattern, in comparison to the free plain surface. However, the polymer acts as mechanical backing that caused a reduction in the shape-memory properties of the composite material.

Keywords: adhesion; NiTi plate; polymer; PMMA/NiTi composites; mechanical properties; surface features

1. Introduction

The application of Ni-Ti shape memory alloys has gained considerable interest in the medical field due to the application of the pseudo elastic nature of the material in the medical devices and implants [1]. Ni-Ti shape memory alloys (SMA) are used in medical technology for applications such as stent-grafts and guide wires [2]. However, as this material of Ni-Ti-SMAs has received increased attention in the medical field due to its functional properties, the release of Ni content towards allergy issues raises serious concern for bio medical applications [3]. To avoid Ni release from Ni-Ti- SMAs, the polymer coating is gaining interest in the areas of Ni-Ti alloys. This type of application could benefit from hybrid systems like polymer-coated shape memory composites. Hybrid composites, consisting of binary or more elemental systems involving polymer and Ni-Ti, have been developed in the last 5 years [4]. Some researchers have studied polymer composite actuators with SMA alloys and taking the advantages of the glass transition temperature of the polymer and the phase transition temperature of Ni-Ti alloys [5]. The effect of temperature on the phase transition in the hybrid composite opens up wide potential in shape setting behavior [6–9] in the field of actuators and in the mechanical response of the composite. However, the mechanical behavior of hybrid composites reveals that the interfacial adhesion between the Ni-Ti ribbon and polymer interface plays a crucial role in adhesion property, thus controlling the thermal and mechanical response of the hybrid composite [10,11]. Good adhesion between the polymer and the Ni-Ti ribbon on the surface quality holds a crucial step prior to the potential application of hybrid composites. This is because a hybrid composite of the shape memory alloy with polymer provides actuator motion with deflection, and, relying on their mechanical response, these composites of thin films are promising candidates for actuators in MEMs [12].

Ni-Ti shape memory alloys are most successful because of their good structural and functional properties and they can be combined with polymers, forming functional engineering composites (FECs). The growing demand in the sensors and actuators field could be fulfilled by FECs using shape-memory properties, such as bistable behaviour [13]. Ni-Ti shape memory alloys show thermodynamic behavior of three kinds of shape memory effect such as 1-way, 2-way, and pseudo elasticity. All shape memory effects depend on the phase transformation during high and low temperature processes. Polymeric materials are not as strong as alloys but have a number of attractive features, such as low density, good flexibility, good chemical stability, and low electrical conductivity.

Poly (methyl methacrylate) (PMMA) was chosen because of its high tensile strength and higher elastic (Young's modulus) and shear modulus of 3.2 and 1.7 GPa, respectively.

A number of parameters need to be considered prior to the composites design, including the alloy's surface conditions, the alloy's microstructure, the morphology of the polymer, the build of the metal–polymer interface, the influence of thin layers acting as a coupling agent between Ni-Ti and the polymer, the influence of different processing condition, and the influence of the thermo-mechanical load history. An alloy–polymer composite shows shape memory effect during cooling and heating. On cooling, the polymer acts as a bias spring and leads to shape change. On heating, a shape change occurs against the elastic deformation of the polymer. As a result, the composite as a whole could perform a two-way effect by reinforcing the one-way effect of Ni-Ti ribbon into the polymer matrix. The polymer matrix, in return, could act as a bias force into the material. A strong alloy–polymer interface is required to transfer the stress inside the polymer onto the shape memory alloy on cooling. The deformation of the interface during cyclic shape change in service can lead to delamination, resulting in a loss of memory. The physical nature of the polymer–alloy interface is thus a key point and a good bonding between the polymer and the shape memory alloy is required. The interaction of shape memory alloys and polymers can occur in various ranges starting from chemical to mechanical adhesion.

Research has been carried out to improve the adhesion property of Ni-Ti alloys' surface quality with polymer using silane as a coupling agent [14]. There are various methodologies to improve the interfacial adhesion, either chemically, or mechanically or using a coupling agent [15–17]. In this study, we investigate the interfacial adhesion of Ni-Ti with poly (methyl methacrylate) (PMMA) by the mechanical grooving method obtained by the laser pattern method. A mechanical test is one of the way to determine the adhesion of the polymer on the surface of the Ni-Ti ribbon. The effect of adhesion and its role towards mechanical, thermal, and physical properties have been investigated.

2. Experimental

2.1. Materials

Commercial psuedoelastic Ni-Ti ribbons, thickness 0.35 mm, with an Ni content of 50.8 % were used in this study (SAES, New Hartford, USA). Poly (methyl methacrylate) was considered here for coating material on the surface of the Ni-Ti ribbon. PMMA was chosen because of its high tensile strength and its higher elastic (Young's modulus) and shear modulus of 3.2 and 1.7 GPa, respectively. The elastic modulus of Ni-Ti ribbon in the austenite phase is 80 GPa, whilst in the martensitic phase it is 40 GPa. Commercial powder of PMMA (Particle size: 48 μm, M.W: 550 kg/mol) was purchased from Sigma-Aldrich. A solid state laser RD 20 was used for mechanical surface grooving on the Ni-Ti surface.

2.2. Methods

The thermo-mechanical response of the composite was investigated by thermo-mechanical analysis (TMA) by using a LINSEIS L75 Cryo (Linseis, Sel, Germany) to observe the deflection in terms of displacement during the cooling and heating profiles. The displacement of the Ni-Ti ribbon and its corresponding composite, on one and both sides, were considered at a constant load of 200 mN with a strain rate of 0.01 m/s. The temperature was chosen in the range of −150 °C to +150 °C.

Dynamical mechanical analysis (DMA) was used to investigate the effect of temperature, stress, and frequency on the mechanical response of the Ni-Ti ribbon and its corresponding thermos-elastic transformation. An 8000 Perkin Elemer DMA machine was employed to carry out the test with a frequency of 0.1 Hz and a heating and cooling rate of 2 K/min in the temperature range of −150 °C to +150 °C.

Differential scanning calorimetry (DSC) analysis was carried out by a Pyris Diamond DSC (Perkin Elmer) for the determination of the phase transition and glass transition temperature, respectively. The sample was heated at 10 °C/min from room temperature up to 180 °C, and then maintained for one minute at this temperature before starting with cooling at the same scanning rate. At the phase transformation temperature, the Austenitic (A_s) and Martensitic (M_s) peaks were identified as −7.1 °C and −59.5 °C, respectively (Figure 1).

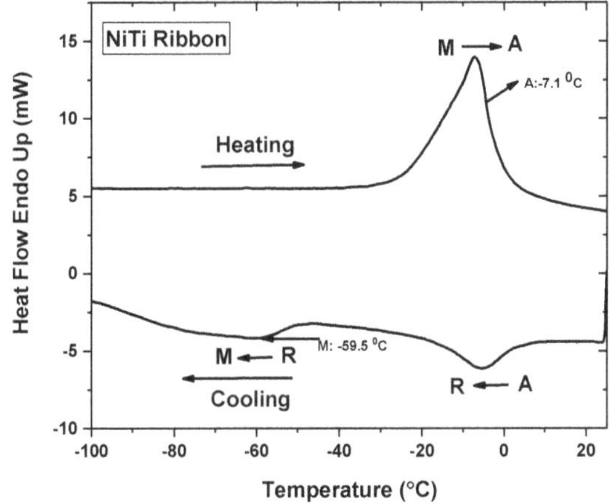

Figure 1. Phase transformation temperature of Ni-Ti ribbon with austenitic (As) and martensitic (Ms) phases.

The glass transition temperature (T_g) for the PMMA was 66.7 °C, with a specific heat capacity of C_p: 0.32 J/g °C (Figure 2).

The tensile behavior of the composite and its components was evaluated by means of an E10K Instron machine (Instron, Norwood, MA, USA), with strain controlled mode. The strain rate was chosen at 0.01% S with the steps of the loading and unloading stage. The stress vs. strain curves of the Ni-Ti ribbon and its hybrid composite were obtained at room temperature. The surface features of the composite, and the corresponding adhesion of the polymer on the Ni-Ti surface, was examined by a Zeiss Imager Z1 Optical Microscope (Zeiss, Oberkochen, Germany).

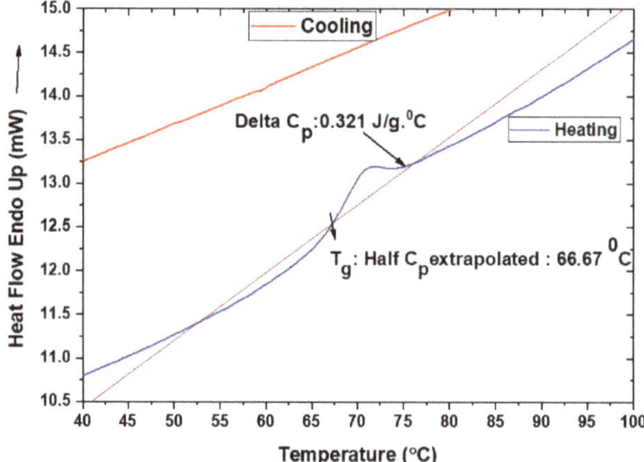

Figure 2. Differential scanning calorimetry curves: the blue color for heating poly (methyl methacrylate), and the red color one for cooling poly (methyl methacrylate).

The pseudoelastic Ni-Ti ribbons (cold rolled, straight annealed), with cross sectional dimensions of 0.35×30 mm, were prepared from the plate of dimension 300×300 mm by laser cutting. The surface of the Ni-Ti ribbon was patterned by laser engraving, with variable depth, using various power sources, such as laser heat and scanning speed. The effect of laser fluency energy was 20% of the maximum power 4 W, with a frequency of 20 kHz and a laser of 100 ns pulse length along the length of the Ni-Ti ribbon. The aim of the laser pattern was to provide a channel for polymer infiltration, thus generating a better interface between the coating layer with the interlock mechanism.

Spin coating has been considered as an efficient technique for the deposition of the polymer, such as natural rubber, polystyrene, and poly (methyl methacrylate) in very thin films on a flat surface [16]. The intensity of the laser beam with respect to the depth of the channels is reported in Figure 3, showing the laser pattern surface of Ni-Ti and the Gaussian profile distribution of the laser beam considered for engraving the channels on the surface.

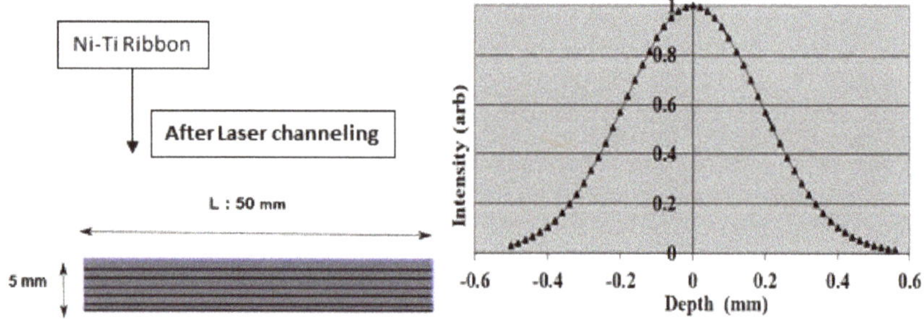

Figure 3. Channel on the Ni-Ti ribbon and its laser intensity versus depth of the Gaussian profile for the surface pattern.

The spin coating of PMMA on the surface of Ni-Ti ribbon was carried out using a standard spin coater (Spin Coater, Germany, Model No. SUSS MicroTec Lithography GmbH, Type DELTA 10 IT, Germany). PMMA powder was diluted with toluene (Sigma Aldrich) in the ratio of 1:10 for a concentrated solution. The suspension of the PMMA powder and solvent underwent stirring using a magnetic stirrer for 12 h at room temperature. The prepared solution was used for the coating of Ni-Ti ribbon in the spin coater for the homogenized coating of the composite. Figure 4 displays the scheme of the spin coater, the inner view shows the holder for substrate in position using a vacuum chuck, and the motor allows the rotation of the sample after the polymers are in the position of the Ni-Ti ribbon.

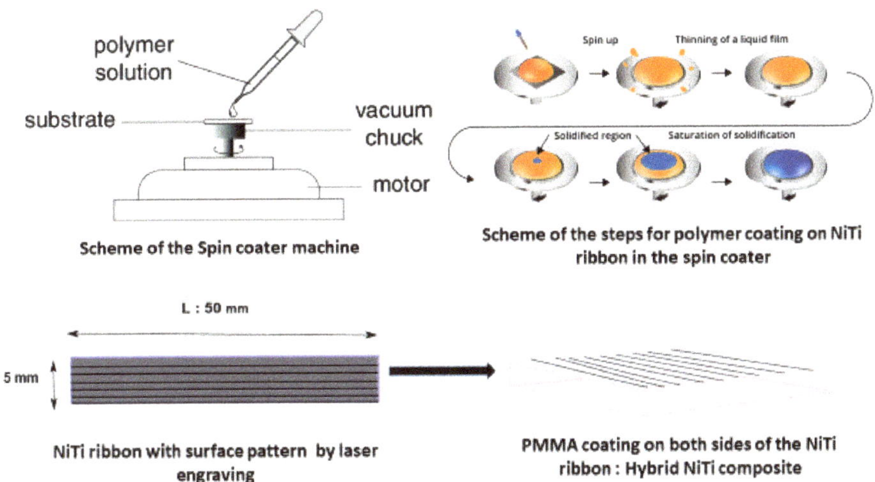

Figure 4. Scheme of the Ni-Ti ribbon with channels and poly (methyl methacrylate) (PMMA) polymer by spin coating.

The rotation's speed was controlled and chosen for a certain interval of time for a uniform distribution of the polymer on the surface of the Ni-Ti ribbon. After the deposition, the polymer solidifies on the substrate and the hybrid composite undergoes curing at 60 °C inside the vacuum furnace overnight. A batch of composite samples was prepared to choose various RPM, on the one side and both sides surface, for interface analysis and mechanical response. Furthermore, a batch of the composite coated on the virgin Ni-Ti samples, without laser lines, was prepared by spin-coating. The residence time was chosen 5 and 2 times to maintain the thicker and thinner coating layers with the alternative one side and both sides of the ribbon. After spin coating, the samples underwent curing at 60 °C inside the vacuum furnace overnight. Table 1 shows the specification of the composite prepared by the spin coater.

Table 1. Specification of the various samples obtained by spin coating: the speed of the substrate (v); the coating surface (CS); the residence time (t_r) of the coating; the number of the coating layer (CL); the coating thickness d (μm); the quality.

Samples	v/(RPM)	CS	t_r/min	CL	d/(μm)	Quality
1	Ni-Ti ribbon with the channel of the engraved line without any coating					
2	100		7	4 times (one surface)	0.18	Good
3	200	One side	5	3 times (one surface)	0.17	Medium
4	300		2	2 times (one surface)	0.15	×
5	100		7	2 × 4 times	0.36	Good
6	200	Both Sides	5	2 × 3 times	0.34	Medium
7	300		2	2 × 2 times	0.30	×

3. Results and Discussion

3.1. Micro Structural Characterization of the Composite of the Ni-Ti ribbon and Polymer

First, PMMA was spin coated on the Ni-Ti ribbon surface without the laser channel, showing complete detachment after curing. With a thick polymer's coating layer, there is delamination of the polymer from the surface that leads to a complete separation of the polymer layer from the surface of the Ni-Ti ribbon. Otherwise, a thin coating layer results in the peeling off of the polymer in the localized area from the Ni-Ti ribbon after curing (Figure SM1). The optical images of the Ni-Ti ribbon surface were taken before and after the spin coating was examined for the adhesion. The surface of the Ni-Ti ribbon was smooth without any distortion. After the laser channel, the polymer was introduced into the surface of the Ni-Ti ribbon by the spin-coating method. The surface of the composite was examined, which showed the adhesion of the polymer into the channel that holds the polymer effectively. Figure 5 shows the Optical Micrograph (OM) images of the Ni-Ti ribbon surface and the composite after PMMA infiltrates into the channel.

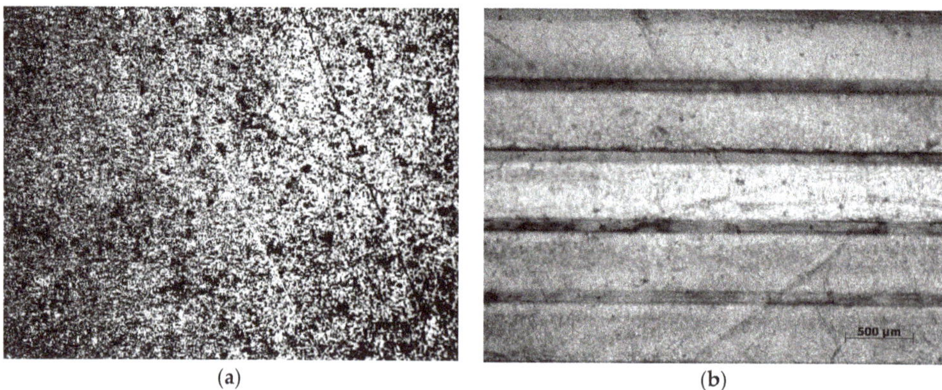

(a)　　　　　　　　(b)

Figure 5. Optical Micrograph (OM) images of the (**a**) Ni-Ti ribbon and (**b**) poly (methyl methacrylate)-Ni-Ti ribbon composite.

3.2. Thermo-Mechanical Characterization of Composite

The thermo-mechanical response of the prepared materials was investigated by TMA to observe the deflection in terms of displacement during the cooling and heating profile. In the investigated range (-150–$150\,^{\circ}$C), Ni-Ti ribbon displayed the deflection (L) in the maximum range of -1200 mm; however, on the coating of the polymer to one side, the surface leads to a deflection decrease towards -350 mm. This decrease in the behavior of the deflection contributes towards additional load in the three-point bending test.

The decrease in deflection behavior was more significant after the coating on both sides, probably because on cooling and heating ranges polymer has significantly adhered to the substrate without any delamination or breakage.

Figure 6 reports the deflection behavior in ΔL as a function of the temperature and time for the Ni-Ti ribbon and PMMA composites, based on the definition of the coefficient of thermal expansion (CTE), it is clear that a large value corresponds to a phase change in the material. The change in the displacement of the Ni-Ti ribbon compared to composite shows a large difference due to the presence of the polymer coating on the surface of the alloy. The polymer coating acts as a back force, showing the good adhesion of the polymer during the heating and cooling cycles.

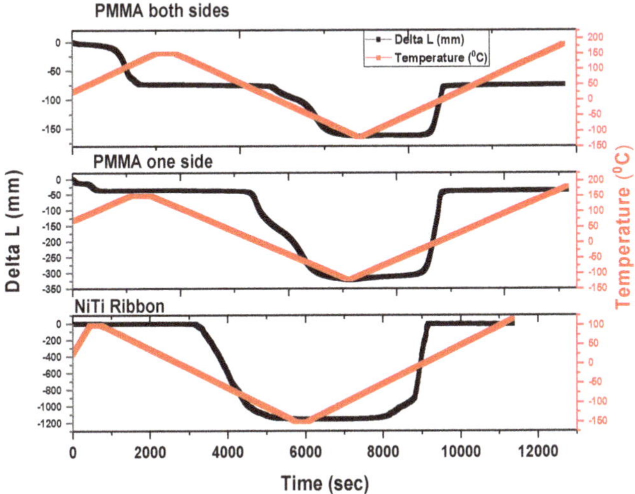

Figure 6. Displacement as a function of time and temperature for the Ni-Ti ribbon and PMMA composite.

The behavior of the hysteresis curve for the composite showed no significant difference in respect of the coating thickness on the one side and both side samples. The coefficient of thermal expansion remains unchanged for both of the composites. Figures 7 and 8 display the response of the Ni-Ti ribbon and composite as a function of the temperature and their corresponding expansion, respectively. The values are significantly reduced from the Ni-Ti ribbon. The polymer still sustains, during cooling and heating cycles, without any delamination from the Ni-Ti surface. The heating profile shows the combination of the austenite and martensitic peaks; as a result, a bi-furcated peak has been observed. On the cooling segment, the austenite and martensitic peaks, with the conversion from B_2 to R, and the $B1_9'$ combined peaks are observed. For the Ni-Ti ribbon–polymer composite, the values of the displacement (L) reduce in comparison to the Ni-Ti ribbon as a virgin sample response (Figure 7). As mentioned before, the polymer acts as a load to the Ni-Ti surface, which could backlog force.

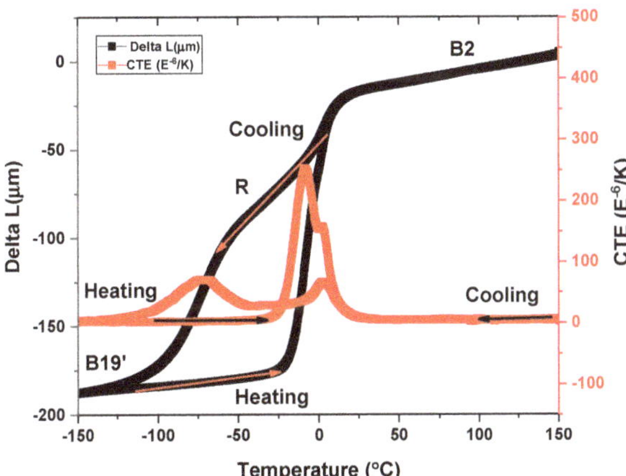

Figure 7. Thermomechanical analysis (TMA) curves of the Ni-Ti ribbon showing displacement (L) and the coefficient of thermal expansion (CTE) versus the temperature profile.

The thickness of the composite increases with the load applied on the surface, and, as a result, the displacement response reduces. The rule of mixture applies here as the combined effect from the matrix polymer as well as the ribbon (Figure 8).

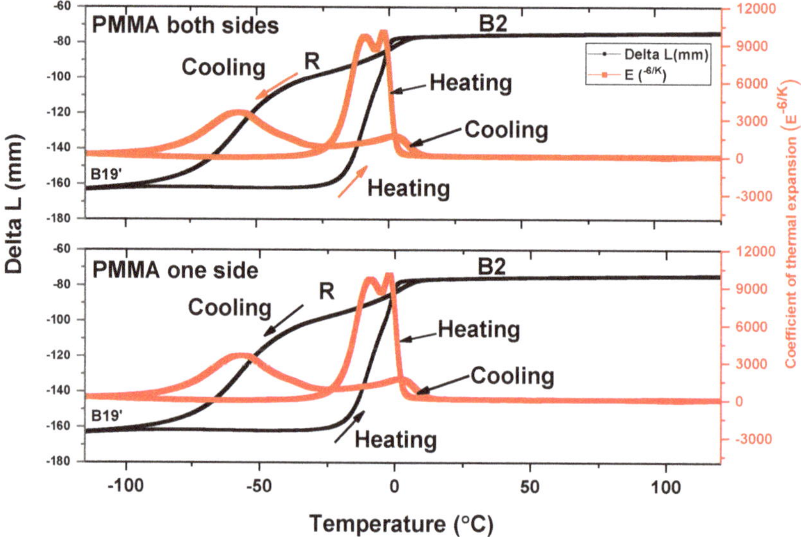

Figure 8. Hysteresis behavior of the PMMA-Ni-Ti composite (one side and both sides) surface as a function of the temperature.

The coefficient of thermal expansion (CTE) of the PMMA composite with one side showed the lower effect of PMMA on the functional behavior of the shape memory alloy (Figure 9).

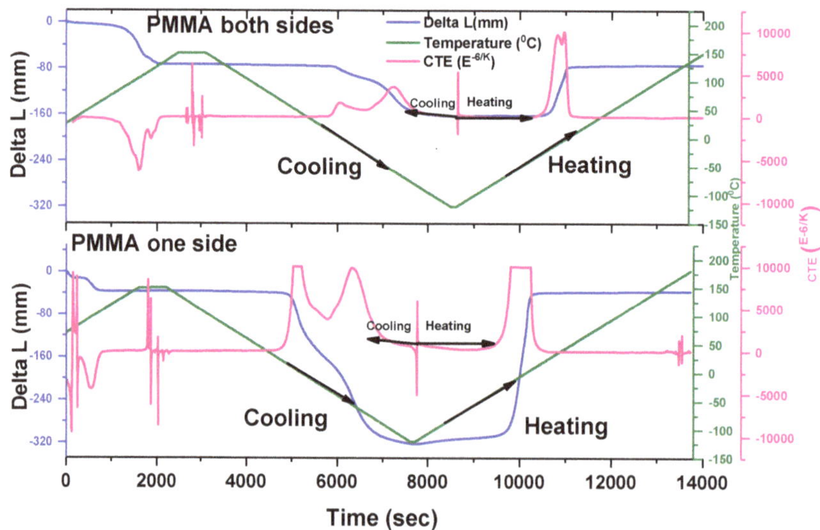

Figure 9. The coefficient of thermal expansion (CTE) response of the composite as a function of time.

The peaks are more prominent during the cooling and heating segment. However, the coating of PMMA on both sides of the Ni-Ti ribbon induces some mechanical effects, and, as a result, the peaks are less prominent during the heating and cooling segment. The mechanical behavior of the composite during cooling and heating shows no delamination of the coating from the substrate.

3.3. Dynamical Mechanical Characterization of the Composite

The effect of temperature, stress, and frequency on the mechanical response of the Ni-Ti ribbon and its corresponding thermo-elastic transformation was evaluated by DMA in the temperature range of −150 °C to +150 °C.

Figure 10 shows the DMA response of the Ni-Ti ribbon and the corresponding hybrid composite. The storage modulus and Tan δ of the prepared materials were determined as a function of the temperature. The Tan δ value corresponds to the internal friction of the sample, which is calculated from the dissipated energy of the sample (E') to the ratio of the storage modulus (E'') of the material. The transformation temperatures from the martensitic phase ($B_{19'}$) to the austenitic one (B_2), during the heating of the Ni-Ti ribbon, and from the austenitic to the martensitic phase, during the cooling of the Ni-Ti ribbon, are depicted from the Figure 10.

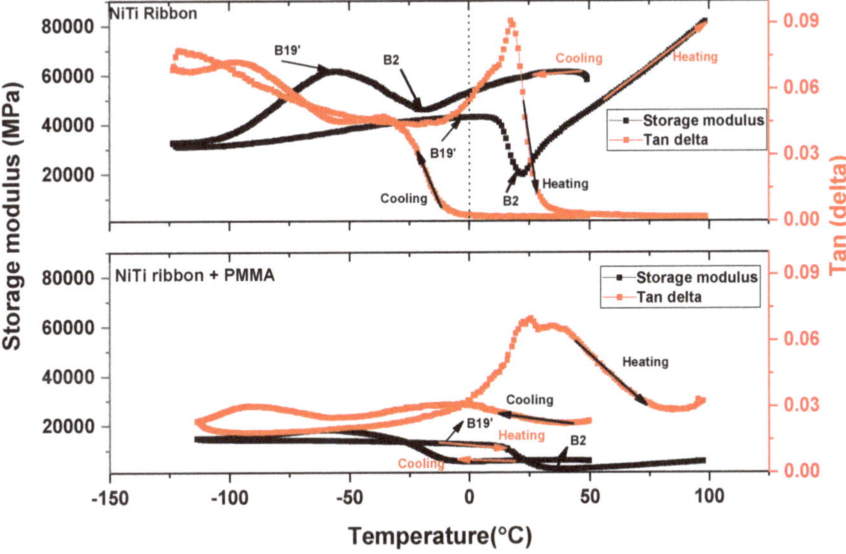

Figure 10. Dynamic mechanical analysis (DMA) response of the Ni-Ti ribbon and PMMA Ni-Ti ribbon composite.

The polymeric matrix showed the high damping and low modulus in comparison to the higher value of the modulus of the Ni-Ti ribbon. The DMA of the composite showed the combination of the effect of the Ni-Ti ribbon and polymer. As the Ni-Ti ribbon and polymer showed the reverse effect in their response, the composite showed the predominate behavior of the polymer with reduced storage modulus and Tan δ values. The presence of the polymer was confirmed from the reduced hysteresis response of the Ni-Ti ribbon–polymer composite. The heating–cooling profile of the Ni-Ti ribbon–PMMA for the storage modulus showed the transformation of $B_{19'}$ to the B_2 phase with a reduced hysteresis area. This draws attention to the fact that the laser line promotes an increase in surface area by creating a channel for better adhesion into the valley of the Ni-Ti surface. This improves adhesion during the interlocking of the polymer into the depth of the channel without any chemical interaction between the Ni-Ti ribbon or polymer.

3.4. Tensile Study of the Ni-Ti Ribbon and Composite

The tensile study of the Ni-Ti ribbon and its corresponding composite were carried out in an Instron machine using the strain-controlled mode at room temperature (Figure 11). The stress of 400 MPa was reached at the strain of 8% with the transformation of the austenitic to martensitic super elastic plateau upon loading and the martensitic to austenitic phase upon unloading with an un-recovered strain of 0.1%. However, the composite with one surface coating showed a reduction in the stress of 280 MPa with a strain of 4.8% upon loading, with a similar value for the un-recovered strain of 0.1%. The composite with the both-side PMMA coating displayed the stress of 220 MPa with a strain of 3.5%. The composites and Ni-Ti ribbon recovered back to their original shape after unloading with an un-recovered strain of 0.1%. The interface of the polymer with the Ni-Ti ribbon was then examined after the tensile test.

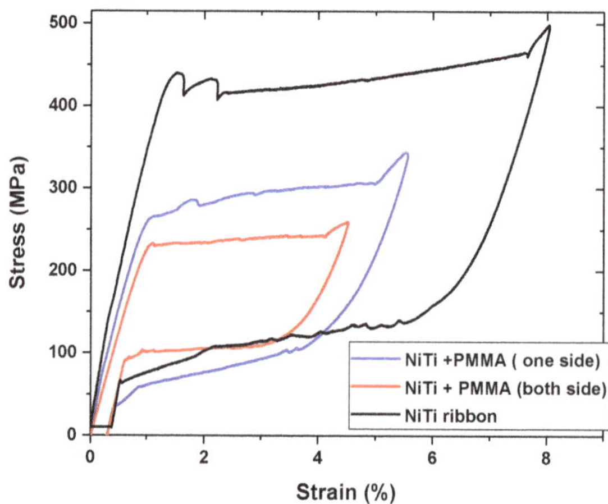

Figure 11. Stress–strain curves for the Ni-Ti ribbon and its corresponding hybrid composite.

3.5. Micro Structural Characterization of the Polymer-Ribbon Interface after the Tensile Test

The adhesion of the polymer on the Ni-Ti ribbon surface after the tensile test was examined by optical microscopy. The surface image of the composite after the tensile test shows that the adherence of the polymer still holds good on the surface of the Ni-Ti ribbon. In Figure 12a–d, it is possible to observe the various regions of the composite's surface after the tensile test. Figure 12a shows the transverse direction of the polymer tearing up during the tensile test. Notwithstanding this, the adhesion remains constant across the surface of Ni-Ti ribbon. Figure 12b shows the tearing of the polymer from both edges in the transverse direction towards the surface of the composite. Figure 12c shows the clamping area and its surroundings. The clamping zone shows the rupture of the polymer from the Ni-Ti alloy surface; however, the close area shows a transverse crack line along the surface of the composite. Figure 12d displays the image of transverse and longitudinal cracks along the surface of the composite. During the tensile test, although some cracks and the tearing of the polymer were exhibited, the adhesion is still very compatible. The polymer was completely un-detached from the surface, thus confirming that the adhesion of the polymer on the surface of the Ni-Ti ribbon still displays good results.

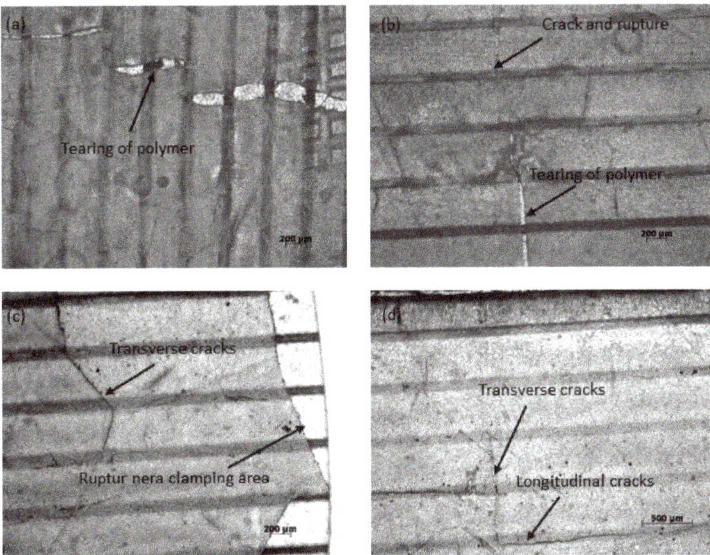

Figure 12. Optical microscopy surface images of the PMMA composite after the tensile test.

The mechanical roughness created by the laser beam on the surface of the Ni-Ti ribbon stands well in the adhesion of the polymer on the Ni-Ti ribbon surface. The channels provide a good path for the infiltration of the polymer onto the surface, providing, in the meantime, a valley of polymer intact towards the surface with the interlocking mechanism of the two material.

4. Conclusions

A hybrid composite material exhibits better adhesion of the polymer on the Ni-Ti ribbon surface by applying mechanical roughness to the surface. In this work, mechanical roughness was created by laser engraving the surface of the Ni-Ti ribbon. The channel allows for the infiltration of the polymer matrix into the surface of the Ni-Ti ribbon. This interlocking mechanism of the polymer into the alloy surface provides stronger interfacial adhesion of the polymer without any breakage. The composite showed improved adhesion behavior in comparison to the starting material. However, the polymer induces backlog on the performance of the mechanical response of the material. A better adhesion of the polymer on the Ni-Ti surface may improve the composite properties in the areas of the sensors and actuators on controlling the movement during thermal cycles. The cooling and heating cycles induce the hardening and softening mechanism of the polymer that adheres to the Ni-Ti surface. The composite behavior reflects on the combination of the mechanism of the polymer and the Ni-Ti ribbon under thermal cycles. Mechanical tests were also performed to observe the adhesion of the polymer into the channel of the Ni-Ti surface. During good adhesion, the composite could act, as in the field of actuators, using shape memory effect rather than mechanical strength. The bending and shape recovery of the composite as the function of heating and cooling temperature could stand as a potential technique in the transducer and actuator field of application.

Supplementary Materials: The following are available online at http://www.mdpi.com/2076-3417/10/6/2172/s1, Figure SM1: Images of the polymer PMMA on Ni-Ti surface without laser lines showing a detachment from the surface.

Author Contributions: S.S., P.S. and I.B. conceived and designed the study; M.M. and P.P. performed the experiments; Discussion on article write up and necessary modification was carried out by all the authors in a team effort. All authors have read and agreed to the published version of the manuscript.

Funding: This work was carried out within the Institute of Physics and Institute of Plasma Physics under the Solid-21 project (SOLID21: CZ.02.1.01/0.0/0.0/16_019/0000760, SOLID21-Fyzika pevných látek pro 21. Století, Fyzikální ústav AV ČR, v. v. i. (2018–2023). Ignazio Blanco is grateful to the University of Catania within the "Bando-CHANCE" n° 59722022250, for supporting the project HYPERJOIN-HYBRID HIGH PERFORMANCE INNOVATIVE JOINTS.

Conflicts of Interest: The authors declare no conflict of interest.

References

1. Saburi, T. Ti-Ni shape memory alloys. In *Shape Memory Materials*; Otsuka, K., Waymann, C.M., Eds.; Cambridge University Press: Cambridge, UK, 1999; pp. 49–96. ISBN 9780521663847.

2. Kapoor, D. Nitinol for medical applications: A brief introduction to the properties and processing of nickel titanium shape memory alloys and their use in stents. *Johns. Matthey Technol. Rev.* **2017**, *61*, 66–76. [CrossRef]

3. Ryhänen, J.; Niemi, E.; Serlo, W.; Niemelä, E.; Sandvik, P.; Pernu, H.; Salo, T. Biocompatibility of nickel-titanium shape memory metal and its corrosion behavior in human cell cultures. *J. Biomed. Mater. Res.* **1997**, *35*, 451–457. [CrossRef]

4. Lomas-González, O.; López-Cuellar, E.; López-Walle, B.; José de Araujo, C.; Reyes-Melo, E.; Gonzalez, C.H. Thermomechanical behavior of a composite based on a NiTi ribbon with a magnetic hybrid polymer. *Mater. Today Proc.* **2015**, *2*, S785–S788. [CrossRef]

5. López-Walle, B.; López-Cuellar, E.; Reyes-Melo, E.; Lomas-González, O.; De Castro, W.B. A Smart Polymer Composite Based on a NiTi Ribbon and a Magnetic Hybrid Material for Actuators with Multiphysic Transduction. *Actuators* **2015**, *4*, 301–313. [CrossRef]

6. Samal, S.; Heller, L.; Brajer, J.; Tyc, O.; Kadrevek, L.; Sittner, L. Laser Annealing on the Surface Treatment of Thin Super Elastic NiTi Wire. *IOP Conf. Ser. Mater. Sci. Eng.* **2018**, *362*, 012007. [CrossRef]

7. Samal, S.; de Prado, E.; Tyc, O.; Sittner, P. Shape Setting in super-elastic NiTi ribbon. *IOP Conf. Ser. Mater. Sci. Eng.* **2019**, *461*, 012075. [CrossRef]

8. Samal, S.; de Prado, E.; Manak, J.; Tyc, O.; Heller, L.; Sittner, P. Internal stresses and plastic strains inroduced into surface layers of bent NiTi ribbon by low temperature shape setting. In Proceedings of the ASM International-International Conference on Shape Memory and Superelastic Technologies, Konstanz, Germany, 13–17 May 2019; pp. 100–101. ISBN 978-151089272-9.

9. Samal, S.; Heller, L.; Brajer, J.; Manak, J.; Sittner, P. Gradient of hardness and Young's modulus over crosssection of laser annealed NiTi wires. *Esomat Proc.* **2018**, *362*, 012007.

10. Samal, S.; Stuchlík, M.; Petrikova, I. Thermal behavior of flax and jute reinforced in matrix acrylic composite. *J. Therm. Anal. Calorim.* **2018**, *131*, 1035–1040. [CrossRef]

11. Samal, S.; Škodová, M.; Blanco, I. Effects of Filler Distribution on Magnetorheological Silicon-Based Composites. *Materials* **2019**, *12*, 3017. [CrossRef] [PubMed]

12. Tobushi, H.; Hayashi, S.; Pieczyska, E.; Date, K.; Nishimura, Y. Three-way actuation of shape memory composite. *Arch. Mech.* **2011**, *63*, 443–457.

13. Chillara, V.S.C.; Dapino, M.J. Shape memory alloy-actuated bistable composites for morphing structures. *SPIE Proc.* **2018**, *10596*, 1–8. [CrossRef]

14. Smith, N.A.; Antoun, G.G.; Ellis, A.B.; Crone, W.C. Improved adhesion between a nickel-titanium shape memory alloy and a polymer matrix via silane coupling agents. *Compos. Part A Appl. Sci. Manuf.* **2004**, *35*, 1307–1312. [CrossRef]

15. Ehrenstein, G.W. *Polymeric Materials. Structure, Properties, Applications. Hanser, Munich*; Carl Hanser Verlag: Munich, Germany, 2004; ISBN 978-3-446-21461-3.

16. Wingfield, J.R.J. Treatment of composite surfaces for adhesive bonding. *Int. J. Adhes. Adhes.* **1993**, *13*, 151–156. [CrossRef]

17. Extrand, C.W. Spin coating of very thin polymer films. *Polym. Eng. Sci.* **1994**, *34*, 390–394. [CrossRef]

Article

Enhanced Proliferation and Differentiation of Human Mesenchymal Stem Cell-laden Recycled Fish Gelatin/Strontium Substitution Calcium Silicate 3D Scaffolds

Chun-Ta Yu [1], Fu-Ming Wang [1], Yen-Ting Liu [2,3], Alvin Kai-Xing Lee [2,3], Tsung-Li Lin [4,5,†] and Yi-Wen Chen [6,7,*,†]

[1] Graduate Institute of Applied Science and Technology, National Taiwan University of Science and Technology, Taipei 10607, Taiwan; D10422605@ntust.edu.tw (C.-T.Y.); mccabe@mail.ntust.edu.tw (F.-M.W.)

[2] School of Medicine, China Medical University, Taichung 40447, Taiwan; u102001416@cmu.edu.tw (Y.-T.L.); Leekaixingalvin@gmail.com (A.K.-X.L.)

[3] 3D Printing Medical Research Center, China Medical University Hospital, Taichung 40447, Taiwan

[4] Department of Orthopedics, China Medical University Hospital, Taichung 40447, Taiwan; d18144@mail.cmuh.org.tw

[5] Department of Sports Medicine, China Medical University, Taichung 40447, Taiwan

[6] Graduate Institute of Biomedical Sciences, China Medical University, Taichung 40447, Taiwan

[7] 3D Printing Medical Research Institute, Asia University, Taichung 41354, Taiwan

* Correspondence: evinchen@gmail.com; Tel.: +886-4-22052121; Fax: +886-4-24759065

† These authors contributed equally to this work.

Received: 31 January 2020; Accepted: 14 March 2020; Published: 22 March 2020

Abstract: Cell-encapsulated bioscaffold is a promising and novel method to allow fabrication of live functional organs for tissue engineering and regenerative medicine. However, traditional fabrication methods of 3D scaffolds and cell-laden hydrogels still face many difficulties and challenges. This study uses a newer 3D fabrication technique and the concept of recycling of an unutilized resource to fabricate a novel scaffold for bone tissue engineering. In this study, fish-extracted gelatin was incorporated with bioactive ceramic for bone tissue engineering, and with this we successfully fabricated a novel fish gelatin methacrylate (FG) polymer hydrogel mixed with strontium-doped calcium silicate powder (FGSr) 3D scaffold via photo-crosslinking. Our results indicated that the tensile strength of FGSr was almost 2.5-fold higher as compared to FG thus making it a better candidate for future clinical applications. The in-vitro assays illustrated that the FGSr scaffolds showed good biocompatibility with human Wharton jelly-derived mesenchymal stem cells (WJMSC), as well as enhancing the osteogenesis differentiation of WJMSC. The WJMSC-laden FGSr 3D scaffolds expressed a higher degree of alkaline phosphatase activity than those on cell-laden FG 3D scaffolds and this result was further proven with the subsequent calcium deposition results. Therefore, these results showed that 3D-printed cell-laden FGSr scaffolds had enhanced mechanical property and osteogenic-related behavior that made for a more suitable candidate for future clinical applications.

Keywords: fish gelatin methacrylate; strontium-doped calcium silicate; bone regeneration; cell-laden scaffold; bioprinting; recycling material

---

1. Introduction

Extensive bone tissue defects caused by trauma, surgical resection and bone diseases require complicated bone tissue repair and regeneration, but complete regeneration continues to be a great challenge in clinical medicine as bone tissues are known to have limited self-regenerating capabilities [1].

Furthermore, bone regeneration is a complex physiological process regulated by osteoclast resorption with osteoblast bone formation and the entire process includes inflammatory reactions, endochondral bone formation and bone remodeling. Currently, for clinical cases, the golden method to repair bone defects is to use autologous bone grafts as this method avoids disease spreading and immunological rejection. However, autografts have some major drawbacks such as limited availability and invasive harvesting that requires additional surgery to harvest healthy bones from another site. Furthermore, additional surgeries mean increased surgical risks and complications [2]. Allograft is an alternative to bone repair; however, it is highly limited due to rejection and lack of availability. To solve this problem, researchers had attempted to develop novel biomaterials including ceramics, polymers and composites as an alternative choice for bone regeneration [3–5]. In order to promote bone healing, the material must be biocompatible, biodegradable and able to promote bone regeneration which mimics the distinctive property of natural bone [6]. With the emergence of newer 3D printing techniques, we are now able to fabricate scaffolds with customized shape, pore sizes and architectures that are similar to native bone [7,8].

For many years, bioceramics has emerged as promising biomaterial for bone regeneration [9,10]. Calcium phosphate is one of the most popular bioceramics which had been repeatedly shown to provide good osteoconduction activities [9,10]. However, it is lacking in terms of biodegradability and biocompatibility, thus limiting its potential in clinical applications. To solve this problem, researchers attempted to combine calcium phosphate with various types of osteogenesis-related growth factors in order to enhance bone regeneration [11]. Recently, a calcium silicate-based biomaterial was investigated as a bioactive material for bone substitutes and it was found that calcium silicate (CS) had far better biodegradation and osteoconduction characteristic than calcium phosphate-based materials, thus resulting in enhanced bone regeneration [6,12]. There were several studies indicating that CS-based materials were able to release Si ions gradually immersion and Si ions were found to not only increase regeneration of surrounding cells and tissues, but also induce bone mineralization by activating osteoblasts [13,14]. In addition, several studies had attempted to incorporate different ions (such as strontium and copper) into CS-based materials in order to fabricate CS scaffolds with varied and diverse functions [15,16], of which strontium (Sr) is a promising ion constantly shown to have beneficial effects for bone regeneration [17]. Furthermore, Sr-contained CS cements fabricated via the sol-gel process was found to not only enhance mechanical properties but also promote cell behaviors [18]. Moreover, the molecular mechanisms of the Sr-contained ceramic scaffold influenced osteogenic differentiation was affected by the influx of Sr ions by the Ca channel [19].

With the emergence of 3D printing technologies, it is now possible to design personalized bone substitutes specifically designed to suit a patient's needs [20]. Numerous studies have since been conducted to attempt to find suitable 3D printable biomaterials for the fabrication of potential bone substitutes via the extrusion method [21–23]. The extrusion method usually involves high temperature heating, and thus it would be impossible to blend cells or growth factors into materials as they would be denatured or damaged [24]. Therefore, several studies were developed with the simple process for growth factor or molecular coating on 3D printed scaffolds [25,26]. By using photo-curable polymers, we are able to allow cell encapsulation and bioprinting simultaneously [27]. Among the most commonly used material is gelatin-methacryl (GelMa), which is a photopolymerizable natural polymer material widely used in biomedical and tissue engineering applications [27–29]. The excellent flexibility and biocompatibility of GelMa showed that GelMa could be used for cell encapsulation and that GelMa could be manipulated to regulate cell behaviors, such as migration, proliferation, and differentiation [30]. In this study, we will use gelatin sourced from fish collagen that provided in the recycling of an unutilized resource and successive the high value-added products [31].

The main aim of this study was to recycle waste materials and use them to fabricate a useful material. Gelatin could be obtained from fish wastes, and in this study we mixed it with SrCS ceramic powder (FGSr) and synthesized it into a light-curing hydrogel. Initial properties and characteristics of FGSr such as material composition, mechanical properties, degradation rate and ion release were evaluated, after which human Wharton jelly derived mesenchymal stem cells (WJMSC) were loaded into both FG and FGSr bio-scaffolds, whereby biocompatibility, alkaline phosphatase activity and calcium deposition were evaluated to determine FGSr's potential for bone regeneration applications.

2. Results and Discussion

2.1. Scaffold Fabrication

Printed FG scaffolds appeared to be clear and transparent in color whilst FGSr was whitish in color (Figure 1). The photo image of FG and FGSr scaffold showed that the surfaces were rather smooth and that the struts were rather continuous with no excessive tears or breakage at the middle or the ends. Therefore, it could be concluded that the addition of 2% SrCS didn't affect printing resolution and scaffold formation. Instead, the addition of SrCS seemed to enhance stability of printed FG scaffolds as the lower layers of pure FG scaffolds were unable to withstand the weight of the upper FG stacks. As seen in Figure 1, the left-hand side of the FG scaffold was slightly toppled compared to FGSr. Similarly, the addition of Sr into FG did not affect the geometry of the pores, and all pores were found to have even edges with a diameter of 500 μm and interconnected with one another. The geometry of scaffolds and pores were found to be significant in increasing cellular responses, and reports had been made stating that the pore sizes were able to regulate cellular behaviors [32]. In addition, both of the scaffolds were immersed in simulated body fluid (SBF) for 7 days and post-immersion images of the scaffolds were taken and shown in Figure 1. As seen, most of the bottom layers of the FG scaffold were degraded, leaving behind only the top two layers. On the other hand, the overall structure and shape of the FGSr scaffold was still intact with several localized degradations. Therefore, the addition of SrCS enhanced printing stability thus allowing us to print and stack layered scaffolds according to our desired designs that were hypothesized to be advantageous in enhancing cellular activities without any worries of structural compromise. Furthermore, initial results showed that addition of SrCS slowed down degradation of scaffold, which is a desirable property to have as it allowed for proper tissue growth and regeneration before degradation.

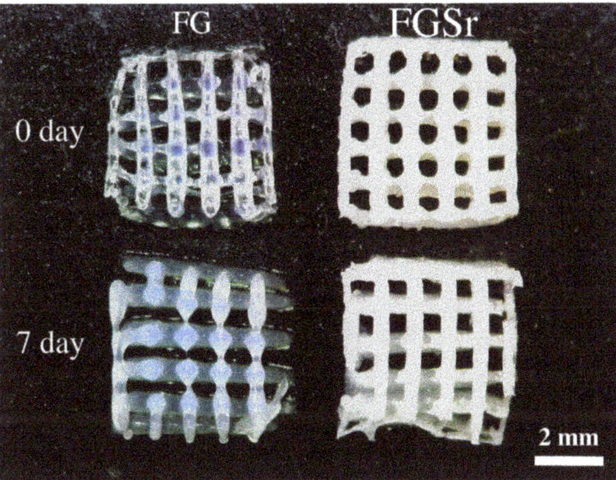

Figure 1. The top-view photograph of 3D-printed scaffolds before and after immersed in SBF.

Figure 2 showed the thermogravimetric analysis (TGA), differential scanning calorimetry (DSC), Fourier transform infrared (FTIR) and X-ray diffraction (XRD) analysis results for both FG and FGSr. TGA of 10 °C/min up to 700 °C was done to confirm for the content of both scaffolds and the results were shown in Figure 2A. The three major episodes of derivative weight loss were observed in all the TGA runs, including elimination of physiosorbed water at a low temperature from 20–200 °C, the decomposition of organic components from 200–400 °C and combustion of the residual organic matrix from 400–800 °C. Both the scaffolds stabilized after 400 °C with FGSr and FG having lost approximately 64% and 76% of weight respectively. SrCS had a higher melting point, as well as a higher temperature insulation property, and the ceramic primarily accounts for the remaining weight of the scaffolds when heated to 800 °C. In addition, it was hypothesized that the difference in weight between both the scaffolds was attributed to SrCS. The above results clearly indicated that SrCS was successfully incorporated into FG without disruption and loss of the original structural characteristics of FG and thus further results were required to observe for characteristics of FGSr scaffolds. A differential scanning calorimetry (DSC) analysis was performed on the FG and FGSr hydrogels in the range 30–400 °C in order to consider the effect of the presence of the Sr-CS ceramic on the glass-transition temperature and on the degree of crystallinity of the neat FG. According to the DSC thermograms shown in the Figure 2B, the endothermic peak at 182.7 °C corresponded to the denaturation temperature of gelatin [33]. The FGSr composite showed the higher denaturation temperature at 108.8 °C. The improved thermal stability is due to the interaction of ceramic and FG during photocuring process of the scaffold. The FTIR spectrum of FG and FGSr specimens containing characteristic peaks resembled both spectrum of FG and calcium silicate (Figure 2C). FG show the peaks at approximately 1635 cm^{-1} (amide-I, C = O stretching), 1535 cm^{-1} (amide-II, N-H bending) and 1242 cm^{-1} (amide-III, C-N and N-H stretching) [34]. The band at 1130 to 1260 cm^{-1} is attributed to the asymmetric stretching vibration of SiO4, while siloxane (Si–O–Si) stretching at 919 to 1024 cm^{-1} [35]. As seen in Figure 2D, FGSr had multiple peaks at 29.7° (C3S), 33.3° (Sr2SiO4), 35° (Ca(OH)2) and 32.8°, 35.2° and 40° (SrSiO3) which strongly indicates that SrCS was present in the hydrogel scaffold. Furthermore, this result also implied that the addition of SrCS did not adversely affect the original structural properties of FG, thus suggesting that several beneficial properties of FG that were desired were retained in the FGSr scaffolds [15].

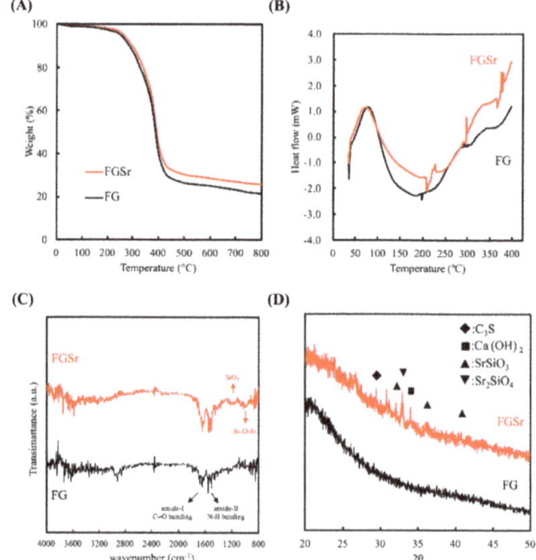

Figure 2. (**A**) TGA, (**B**) DSC, (**C**) FTIR and (**D**) XRD for specimens prepared with a series of FG and FGSr scaffolds.

The Dynamic mechanical analysis (DMA) is conducted to consider the reinforcing effect brought by the addition of Sr ceramic in FG gel. The temperature dependent of the storage modulus, loss modulus, and tan δ for different scaffolds were show in Figure 3. The storage modulus of FGSr indicated the significant improvement compared to that of FG that demonstrates the elastic behavior of FGSr scaffold towards deformation is superior to that of FG scaffold. In addition, for the curves of tan δ for versus temperature, the glass transition temperature (Tg) of FGSr scaffold was obviously shifted to the higher temperature because of the Sr ceramic powder increased the crosslink density that restricted the motion of FG molecular chains [36]. Representative stress-strain curves of both FG and FGSr scaffolds were shown in Figure 4. As seen, it was clearly indicated that FGSr had significantly higher mechanical properties as compared to neat FG scaffolds with approximately 9.2 KPa for FGSr and 3.8 KPa for FG. There was almost a 2.5-fold increase for the FGSr scaffold which makes it a better candidate for further clinical applications. A major disadvantage of FG is that it has lower mechanical strength as compared to its porcine or bovine counterparts due to FG having different amino acid composition and molecular weight distribution. In addition to the photo-crosslinking curing mechanism of FGSr, we speculate that SrCS will release ions to allow FGSr hydrogel to crosslink and provide more chances for the reaction at the beginning, which can also improve the stability of the hydrogel scaffold after photocuring. Similar studies were demonstrated that high-mechanical properties polymer-based hydrogels were manufactured using a sequential physical combined chemical crosslinking process [37,38]. Therefore, the addition of SrCS powder can improve the mechanical properties of the scaffold thus its clinical availability will be relatively improved in the future. SEM was used to observe for the morphology and microstructures of the scaffolds before and after immersion and the images were shown in Figure 5. As seen from the W0 images, FG had fine sharp folds and creases on its surface whilst FGSr had rough and irregular contours on its surface. Initial hypothesis suggested that FGSr would be able to enhance regeneration as it was reported by others that the degree of surface roughness is directly correlated to cellular activities and behaviors [39]. After 4 weeks of immersion, it could be seen from the SEM images that FG had a smooth surface whilst FGSr had coarse apatite aggregations covered on its entire surface. This supported the initial hypothesis that FGSr could support cellular activities due to its coarse and rough surface. In addition, reports made by others suggested that the amount of apatite precipitation and aggregation could be used as an indirect estimation of further bone tissues regeneration. The EDS analysis of the different scaffolds at W0 showed that FGSr had 5.25% of Si, 0.46% of Sr and 0.41% of Ca thus further indicating that SrCS components were present in the 3D-printed hydrogel scaffolds which further confirmed the results above. Similarly, studies published by others indicated that traces of Si and Sr ions in the surrounding fluids were able to enhance cellular proliferation and differentiation. Therefore, we believe that FGSr scaffold was able to achieve higher levels of proliferation, differentiation and thus enhanced regeneration [3].

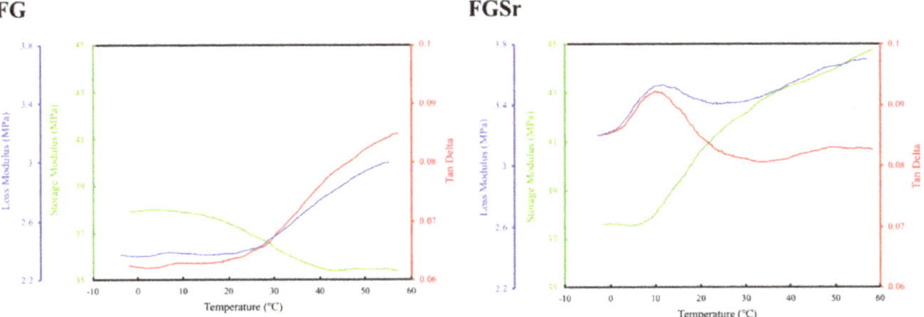

Figure 3. Dynamic mechanical analysis curves of 3D-printed FG and FGSr scaffolds at loss modulus, storage modulus, and tan δ.

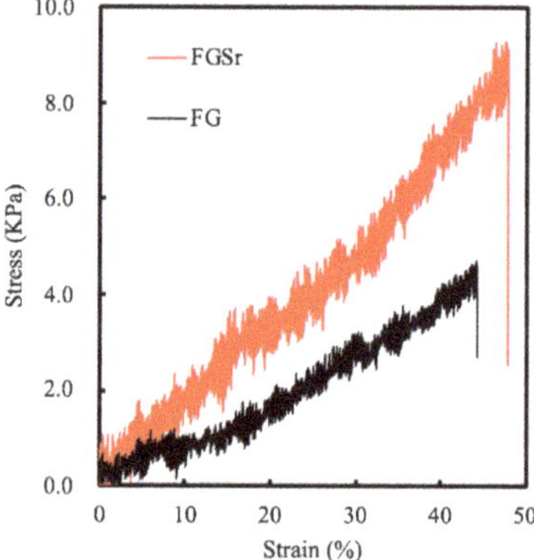

Figure 4. Tensile stress–strain curves of 3D-printed FG and FGSr scaffolds.

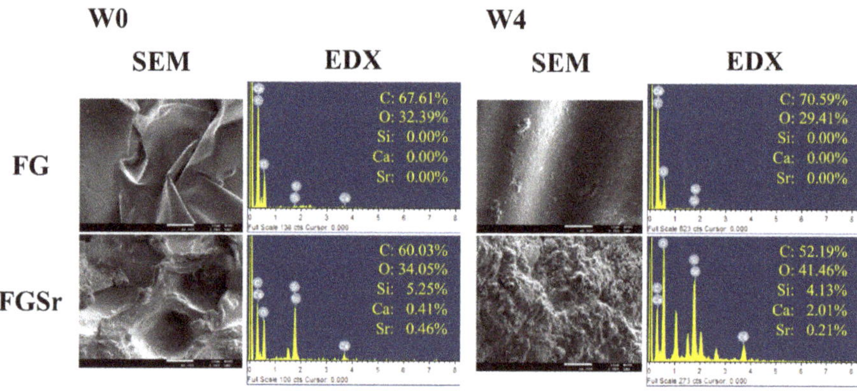

Figure 5. Microstructure images of the FG and FGSr scaffolds and evolution of different ions by energy dispersive spectroscopy before and after immersion in SBF. The scale bar is 10 μm.

2.2. Immersion Behavior

The physicochemical stability of the hydrogel scaffold is crucial during the bone tissue regeneration process, as the degradation rate might match bone ingrowth and promote subsequent remodeling and functional revive [40]. Figure 5 indicated the weight loss profile of the FG and FGSr scaffolds when immersed in SBF at 37 °C for a period of 4 weeks. As seen, FGSr was noted to degrade approximately 30% at a faster rate as compared to FG at week 1. As mentioned above, FG was noted to have fewer amino acids as compared to mammalian gelatin thus resulting in lower strength and higher swelling properties which leads to higher degradation [41]. In the FGSr scaffold, we speculate that the reason for the faster degradation rate may be in the light-cured FG matrix because the addition of SrCS powder affects the curing process and changes the ionic bonding between the FG matrix that accelerated the degradation rate. Moreover, it was probable that the dissolution of SrCS and FGelMa in SBF resulted in weight loss [42]. Two critical factors regarding degradation rates need to be considered when deciding on feasibility of scaffold. Firstly, there is a need to consider the rate of the extracellular matrix secretion and production of encapsulated cells versus degradation rate. Secondly, there is a need to consider residual weight and bone formation weight. According to Figure 6, FGSr had a residual weight of approximately 40% left after 4 weeks of immersion. Reports had been made stating that complete bone regeneration requires six to twelve weeks. Therefore, to fully consider this result, there is a need to consider for the regeneration rate and remodeling rate of encapsulated cells and surrounding matrix.

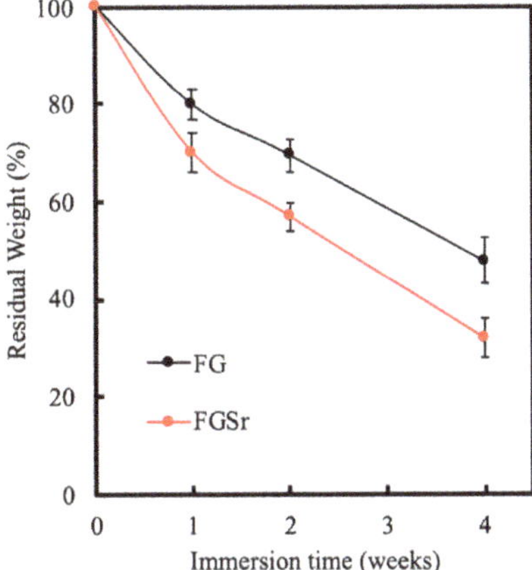

Figure 6. The weight loss of FG and FGSr scaffolds immersed in SBF for different time-points.

Ion release from scaffolds were shown to influence cellular behaviors and activities and the release profiles of several ions were shown in Figure 6. It was reported that appropriate amounts of Si ions (<2 mM) in surrounding fluid was able to stimulate collagen production and enhance proliferation. In addition, Sr was also reported to enhance cellular proliferation and osteogenic differentiation via expression of early and late osteoblastic biomarkers such as ALP and OC. From Figure 7, it was clearly indicated that there was a steady release rate of Si and Sr ions into the surrounding fluid from the FGSr scaffolds whilst FG scaffolds had no Si and Sr release at all. For FGSr, the concentrations of all Ca, Si, and Sr ions in DMEM were significantly higher ($p < 0.05$) than FG scaffolds at all time points. The ion concentrations of FGSr in DMEM were approximately at 1.67 mM, 0.93 mM, and 0.50 mM for Ca, Si, and Sr ions, respectively. Interestingly, there was also a gradual release of Ca from both FG and FGSr scaffolds which gradually declined on the immersion period. This release profile was similar to results shown by others, and it was hypothesized that Ca ions were utilized to form apatite formation thus showing a decline in concentrations over time. In total, our study's findings showed that FGSr had various release profiles of Ca, Si and Sr ions and thus might play a role in enhancing cellular activities which make it a better candidate than FG alone. Several studies indicated the ideal bioscaffold must be able to release various ions in a sustained manner in order to accomplish and maintain the therapeutic effect, and yet, at the same time, this release profile should be controllable to avert harmful effects [43]. In the previous study, the extracellular ionic variation was regulated the cellular behaviors such as migration, proliferation, and differentiation [44]. The detailed amount of ions quantity nor release rates in various solutions such as SBF, PBS and culture medium at which it will stimulate in-vivo osteogenesis is not yet definitively determined, though such values derived from several studies for Si (0.02 to 50 ppm) and Sr (8.7 to 87.6 ppm) had been previously studied and discussed [45].

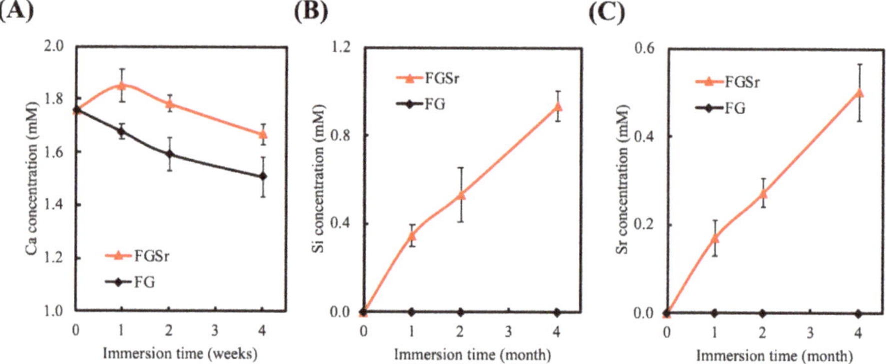

Figure 7. (**A**) Ca, (**B**) Si, and (**C**) Sr ion concentrations in SBF after immersion for different durations. Data presented as mean ± SEM, $n = 6$ for each group.

2.3. Cell Proliferation

Biocompatibility of cell-laden FGSr scaffolds were considered using MTT (Figure 8) and live/dead (Figure 9) assays. Quantification results were shown as percentage of viable cells normalized to the control FG. It is a common knowledge that FG is non-toxic to cells as it is a natural component derived from fish. The cytotoxicity assay demonstrated that FGSr was non-toxic to cells and there were significant differences between cell viability at all time points of 1, 3 and 7 days. After 7 days of culture, it can be noted that there was almost a 1.7-fold increment in cellular viability between FGSr and FG. Therefore, this result concluded that not only was FGSr non-toxic, it was also able to increase cellular proliferation and viability. As seen, there was considerable live cells (green stained) noted in all images at all time points and there was no dead cell (red stained) noted. This result proved that both FG and FGSr was non-toxic to cells. At the same time, it can be noted that there were noticeably more green stains on FGSr at all time points as compared to FG. In recent years, several studies demonstrated the CS-contained scaffolds were able to improve cellular behaviors and activities via raising Si ion influx through L-type Ca channels thus leading to enhance phosphorylation of MAPK pathways in primary cells [46]. Thus, we deduce that the FGSr hydrogel scaffold is favourable for WJMSC growth even when the cells are kept homogenously afloat in the 3D microenvironment [30].

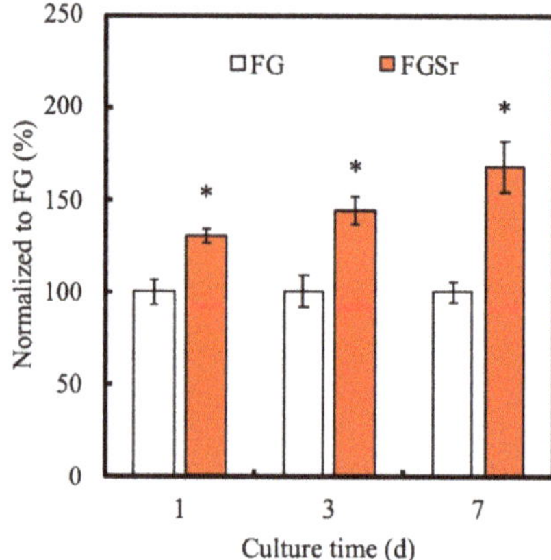

Figure 8. Proliferation of WJMSC on FG and FGSr scaffolds for 1, 3, and 7 days. Data presented as mean ± SEM, $n = 6$ for each group. "*" indicates a significant difference ($p < 0.05$) when compared to FG.

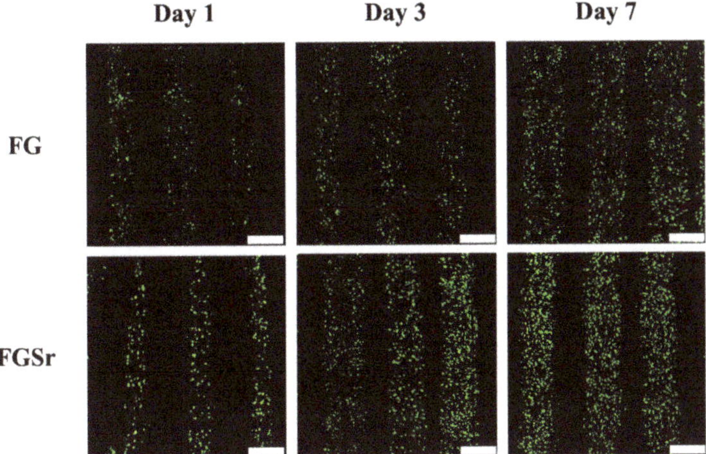

Figure 9. The image of live/dead assay results of WJMSC-laden FG and FGSr scaffolds for different time-points. Scale bar: 400 μm.

2.4. Osteogenesis

ALP was used as an early marker of maintenance of osteoblastic differentiation and was reported to be critically involved in the early processes of bone matrix mineralization and thus can be used as the marker for prediction of subsequent bone formation. As seen from Figure 10, FGSr was able to induce significantly higher amount of ALP at all time points. Our results were similar to that Wu [47], who demonstrated that Si up-regulated the osteogenic-related gene expression of COL, ALP, and RUNX2 in the stem cells. Moreover, Si and Sr ions can activate Ca sensing receptor and improve osteogenic-related proteins production [48]. Also, Shaltooki et al. reasoned that the scaffolds reinforced with Sr-bioglass exhibited a significant increase in the ALP activity in comparison with the scaffolds containing pure bioglass [49]. In addition, it has been demonstrated that the interaction of Sr with other metal ions can induce and accelerate ALP activity [50]. Therefore, it was hypothesized that FGSr scaffolds were able to induce higher amounts cellular differentiation and bone tissue secretion as compared to FG.

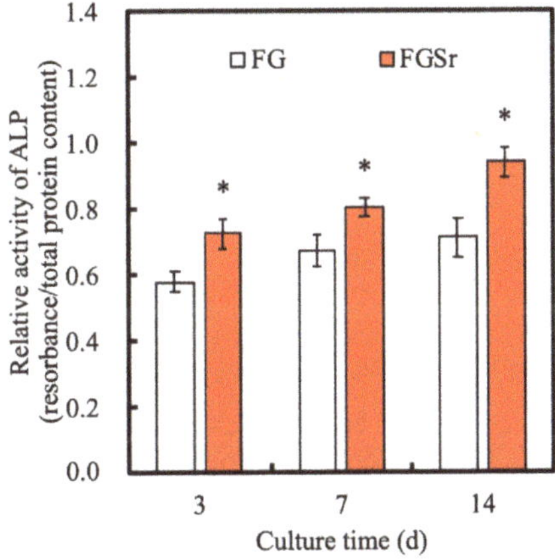

Figure 10. ALP profile of the WJMSCs-laden FG and FGSr scaffolds in osteogenic differentiation medium. "*" indicates a significant difference ($p < 0.05$) when compared to FG.

The extracellular matrix mineralization was the ultimate osteogenesis differentiation step of stem cells and always evaluated as a definitive hallmark of progressive differentiation of stem cells into mature osteoblasts [51]. Both qualitative and quantitative evaluation of calcium deposition of MSCs on both FG and FGSr scaffolds were done and the results were shown in Figure 11. As seen, the stains were taken on after 7 and 14 days of culture to observe for any time-dependent Ca deposition (Figure 11A). The staining intensity of FGSr after 7 and 14 days of culture was dramatically enhanced as compared to FG alone. Furthermore, Ca depositions were only seen mainly along the edges of the struts for FG whilst FGSr was entirely covered with Ca depositions as seen from the stains. However, there seemed to be no increment in intensity between day 7 and 14. This suggest that majority of the Ca secretion was during the initial 7-day and appeared not to increase with culture time. It was previously reported that stem cells only start depositing Ca during their late proliferative stages thus the extent of Ca deposition is dependent strongly on the duration of cell culture. A quantitative determination of calcium deposition was conducted to cross confirm the qualitative results. As seen from Figure 11B, FGSr had significantly higher Ca content as compared to FG at all time points, which was in good agreement with the qualitative results of Ca deposition and was also in line with the ALP activity above. Since the capacity of Ca deposition could be used as a marker for presence of mature osteoblasts, it could be concluded that FGSr allowed for a faster differentiation of cells into osteoblast phenotypes thus allowing for accelerated Ca deposition. Xie et al. illustrated Sr salts could direct effect osteoblast behaviors in vitro, with Sr acting as an inhibitor of osteoclast function and with the particularly marked outcome on bone mineralization [52]. Our result was in good agreement with the results above in showing that FGSr had higher potential for bone regenerative purposes.

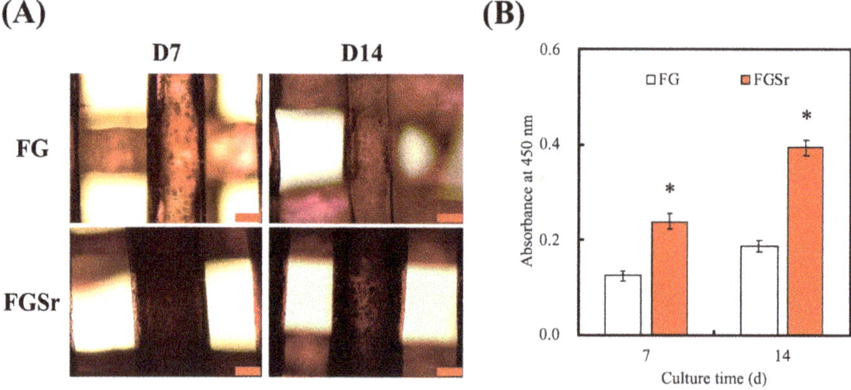

Figure 11. (**A**) Alizarin Red S staining and (**B**) quantification of calcium mineral by WJMSC-laden scaffolds. Scale bar: 200 μm. "*" indicates a significant difference ($p < 0.05$) when compared to FG scaffold. Data presented as mean ± SEM, $n = 3$ for each group.

3. Materials and Methods

3.1. FGSr Scaffold Fabrication

Sr-contained CS powder was synthesized using methods established by the previous report [3]. Analytically graded 65% calcium oxide (CaO, Sigma-Aldrich, St. Louis, MO, USA), 20% silicon dioxide (SiO2, Sigma-Aldrich), 10% strontium oxide (SrO, Sigma-Aldrich), and 5% alumina oxide (Al2O3, Sigma-Aldrich) was mixed evenly and sintered at 1400 °C for 2 h. After cooling to room temperature, the Sr-doped CS powder was wet grinding with 99.5% ethanol in a planetary ball mill (Retsch PM-100, Retsch GmbH, Germany) for 8 h. After this, the mixture was dried in an oven for 12 h.

10% (*w/v*) gelatin from cold water fish skin (Sigma-Aldrich) was fully dissolved in phosphate buffer saline (PBS, Invitrogen, Grand Island, NY, USA) at 50 °C. Then, we added methacrylate (Ma,

Sigma-Aldrich) into the gelatin solution (MA/gelatin ratio = 0.6) under stirring condition and kept reacting for 3 h. After, we added twice the volume of water which was pre-heated to 50 °C into the FGMa solution. After reaction, the mixture was centrifugated at 40 °C and under 8000 rpm to remove the debris, and the supernatant was dialyzed against deionized water using 10 kDa MWCO dialysis tubing (Thermo Fisher Scientific, Waltham, MA, USA) to remove any unreacted Ma. Purified FG was then lyophilized for at least 72 h to obtain sponge foam, after which the products were stored at −20 °C.

The FG powder was subsequently dissolved in the deionized water (10%) and stirred at 50 °C for 10 min. Then 0.25% Irgacure 2959 photoinitiator (I2959, Sigma-Aldrich) was added and used as a bio-ink. The study will use "FG" as the code for this group. Then add SrCS powder slowly dropped into FG solution (SrCS concentration: 5%), stir for 1 h and use it, and this group code is FGSr. A self-designed Pluronic® F-127 (F127, Sigma-Aldrich) mold was first printed using the extrusion-based 3D printer (BioX, Cellink, Gothenburg, Sweden) before depositing GelMA hydrogels into the mold. The uncured FG and FGSr were then exposed to 365 nm UV light (Spot UV irradiation units Spot Cure Series, SP11, Ushio, Japan) for 90 s.

3.2. Chemicophysical Properties Analysis

The ceramic content of the composite was determined by thermogravimetric analysis (TGA, Netzsch STA 449C, Bavaria, Germany). The samples were analyzed in aluminum pans under a nitrogen purge and heated from 100 °C to 800 °C with a heating rate of 10 °C/min. The thermal properties of FG and FGSr specimens were characterized through a TA-Q20 DSC (Thermal Analysis Co., USA). The measurements were carried out at a heating rate of 10 °C/min from 30 °C to 400 °C. To confirm the stability of our product and examine the crystallization phase and crystallinity of FG and FGSr composites, we used X-ray diffractometry (XRD) to perform diffraction analysis. Fourier transform infrared spectroscopy (FTIR, Bomem DA8.3, Hartman & Braun, Frankfurt, Germany) was used for the study of a functional group that analyzed from a 4000–800 cm^{-1} range. The measurement of tensile strength of the wet specimens was conducted by the EZ-Test machine (Shimadzu, Kyoto, Japan). The specimen was fabricated into a dumb-bell shape and subsequently stretched from both ends at a rate of 1 mm/sec until the sample was destroyed. Six assays were conducted for each sample, and recorded the data of distance (mm) and load (N). The morphology of the scaffold was investigated under a scanning electron microscope (SEM) whereas the element analysis was conducted by energy dispersive spectrometer (EDS).

3.3. In Vitro Immersion Test

In this study, we evaluated the weight loss and ions released profile of FG and FGSr immersed in simulated body fluid (SBF). The composition of the SBF is similar to that of human body plasma that consisted of 7.9949 g NaCl, 0.2235 g KCl, 0.147 g K2HPO4, 0.3528 g NaHCO3, 0.071 g Na2SO4, 0.2775 g CaCl2, and 0.305 g MgCl2·6H2O. All of them were dissolved in 1000 mL distilled water, then we adjusted its pH value to 7.4 by adding tris(hydroxymethyl)aminomethane (Tris) and hydrochloric acid (HCl). The FG and FGSr scaffolds were soaked in the SBF at 37 °C for a preset time period, and the degradation rates were calculated using the formula below:

$$\text{Degradation rate (\%)} = (W0 - Wt)/W0 \times 100\%.$$

In addition, we measure the release profiles of Ca, Si, and Sr ions released from scaffolds immersed in SBF by an inductively coupled plasma-atomic emission spectrometer (ICP-AES, Perkin-Elmer OPT 1MA 3000DV, Shelton, CT, USA).

3.4. Cell Proliferation

To investigate the cytotoxicity of the FG and FGSr scaffolds, we used the human Wharton's jelly mesenchymal stem cells (WJMSC) that were obtained from the Bioresource Collection and Research

Center (BCRC, Hsin-Chu, Taiwan) that cultured with the commercially available mesenchymal stem cell medium (#7501, Sciencell, Carlsbad, CA, USA) to passage 4–8. The WJMSC suspensions containing 1×106 cells were mixed into 1 mL of FG or FGSr bioink. After mixing to homogeneity, the cell-laden FG or FGSr were loaded into a printing tube and followed the previous printing process. The cell-laden scaffolds were further cultured in medium which was changed every three days. After different time-periods, we applied 3-(4,5-Dimethylthiazol-2-yl)-2,5-diphenyltetrazolium bromide (MTT) to evaluate the cell survival rate in hydrogel. We washed the scaffolds three times by PBS and added MTT (0.5 mg/mL) thereafter. After 4 h, we withdrew the MTT and then added DMSO to react for 10 min. In the end, we extracted 100 µL solution from different culture medium and distributed it to ELISA 96 well plate. The absorbance at 570 nm was measured by the spectrophotometer. We repeated the procedure three times in six isolated specimens in each group, and the mean value represented the result of each group. In addition, we used LIVE/DEAD assay respectively to quantify cell survival rate after co-printing of cell and hydrogel. We examined the cell survival after the cell being cultured for 1, 3, and 7 days. First, the medium was discarded and the specimens were washed with phosphate buffer solution (PBS) for three times. Then, we added Annexin V-FITC/propidium iodide assay kit to conduct fluorescent staining and observed the cells with Confocal Spectral Microscope (Leica TCS SP8, Wetzlar, Germany). The fluorescent staining method marked the live cell membrane green and the dead cell membrane red, respectively. We calculated the percentage of green-staining cells to total cell counts to get the cell survival rate by applying Image J image software. We repeated the procedure three times in isolated specimens in each group, and the mean value represented the result of each group.

3.5. ALP Activity

To investigate the ability of osteogenic differentiation, the WJMSC-laden scaffolds were cultured in the osteogenic medium (StemPro™ osteogenesis differentiation kit, Invitrogen) for 3, 7 and 14 days. First, the cell-laden scaffolds were lysed by NP40 Cell Lysis Buffer (Sigma) and centrifuged for 15 min at 6000 rpm. The activity of ALP was measured by p-nitrophenylphosphate (pNPP) (Sigma-Aldrich). We mixed each sample with pNPP and 1 M diethanolamine buffer for 30 min, and stopped the reaction by adding 5 N NaOH. The absorbance at 405 nm of the reactant was measured by the spectrophotometer. All data was normalized to data obtained from a bicinchoninic acid (BCA) protein assay kit (Thermo Fisher Scientific, Waltham, MA, USA) which uses BSA as the standard.

3.6. Mineralization

The mineralized nodule formation and calcium deposition behaviors of the cell-laden scaffolds were considered after cultured for 7 and 14 days. To examine the result, the scaffold was fixed by 4% paraformaldehyde (Sigma-Aldrich) for 15 min and stained by 0.5% Alizarin Red S (Sigma-Aldrich) at pH 4.2 for 30 min. Subsequently, they were observed under 200x stereomicroscope (BH2-UMA, Olympus, Tokyo, Japan). Furthermore, we washed the specimen with PBS and the dye on the surface of the specimen was dissolved in an aqueous solution containing 20% methanol and 10% acetic acid. After 15 min, the liquid was transferred to a 96-well plate, and we quantified the Alizarin red S by measuring the absorbance at 450 nm.

3.7. Statistical Analysis

One-way variance statistical analysis was used to assess significant differences in each group, and Scheffe's multiple comparison test was used for each specimen. $p < 0.05$ was considered statistically significant.

4. Conclusions

In this study, FG-based scaffolds were successfully fabricated using 3D printing techniques. Addition of Sr into FG improved printing quality of scaffolds as seen from the results. FGSr was seen to degrade at a higher rate in the initial stages compared to FG scaffolds; however, it was hypothesized that the rate was sufficient to allow bone regeneration. The tensile strength of FGSr scaffolds were shown to be 2.5 times higher than FG scaffolds, thus making it a better candidate for clinical applications. In addition, in-vitro studies demonstrated that FGSr was able to enhance cellular proliferation, differentiation, and mineralization as compared to FG. Therefore, we hypothesized that these 3D-printed cell-laden FGSr scaffolds may be a promising and potential candidate for bone defect repair in the future.

Author Contributions: Data Curation, F.-M.W.; Formal Analysis, Y.-T.L.; Funding Acquisition, Y.-W.C. and T.-L.L.; Investigation, A.K.-X.L.; Methodology, Y.-T.L. and A.K.-X.L.; Writing—Original Draft, T.-L.L. and C.-T.Y.; Writing—Review & Editing, Y.-W.C. All authors have read and agreed to the published version of the manuscript.

Funding: The authors acknowledge receipt grants from the Ministry of Science and Technology (MOST 108-2218-E-039-001 and 108-2321-B-039-001) of Taiwan and the China Medical University Hospital (DMR-109-076) of Taiwan.

Acknowledgments: Experiments and data analysis were performed in part through the use of the Medical Research Core Facilities Center, Office of Research & Development at China Medical University.

Conflicts of Interest: The authors declare no conflict of interest.

References

1. Wu, Y.H.; Chiu, Y.C.; Lin, Y.H.; Ho, C.C.; Shie, M.Y.; Chen, Y.W. 3D-printed bioactive calcium silicate/poly-ε-caprolactone bioscaffolds modified with biomimetic extracellular matrices for bone regeneration. *Int. J. Mol. Sci.* **2019**, *20*, 942. [CrossRef] [PubMed]

2. Li, J.J.; Dunstan, C.R.; Entezari, A.; Li, Q.; Steck, R.; Saifzadeh, S.; Sadeghpour, A.; Field, J.R.; Akey, A.; Vielreicher, M.; et al. A novel bone substitute with high bioactivity, strength, and porosity for repairing large and load-bearing bone defects. *Adv. Healthc. Mater.* **2019**, *8*, 1801298. [CrossRef] [PubMed]

3. Chiu, Y.C.; Shie, M.Y.; Lin, Y.H.; Lee, K.X.; Chen, Y.W. Effect of strontium substitution on the physicochemical properties and bone regeneration potential of 3D printed calcium silicate scaffolds. *Int. J. Mol. Sci.* **2019**, *20*, 2729. [CrossRef] [PubMed]

4. Cheng, C.H.; Chen, Y.W.; Lee, K.X.; Yao, C.H.; Shie, M.Y. Development of mussel-inspired 3D-printed poly (lactic acid) scaffold grafted with bone morphogenetic protein-2 for stimulating osteogenesis. *J. Mater. Sci. Mater. Med.* **2019**, *30*, 78. [CrossRef]

5. Tsai, C.H.; Hung, C.H.; Kuo, C.N.; Chen, C.Y.; Peng, Y.N.; Shie, M.Y. Improved bioactivity of 3D printed porous titanium alloy scaffold with chitosan/magnesium-calcium silicate composite for orthopaedic applications. *Materials* **2019**, *12*, 203. [CrossRef]

6. Lin, Y.H.; Chiu, Y.C.; Shen, Y.F.; Wu, Y.H.; Shie, M.Y. Bioactive calcium silicate/poly-ε-caprolactone composite scaffolds 3D printed under mild conditions for bone tissue engineering. *J. Mater. Sci. Mater. Med.* **2018**, *29*, 11. [CrossRef]

7. Li, T.; Zhai, D.; Ma, B.; Xue, J.; Zhao, P.; Chang, J.; Gelinsky, M.; Wu, C. 3D printing of hot dog-like biomaterials with hierarchical architecture and distinct bioactivity. *Adv. Sci.* **2019**, *4*, 1901146. [CrossRef]

8. Lau, N.C.; Tsai, M.H.; Chen, D.W.; Chen, C.H.; Cheng, K.W. Preparation and characterization for antibacterial activities of 3D printing polyetheretherketone disks coated with various ratios of ampicillin and vancomycin salts. *Appl. Sci.* **2020**, *10*, 97. [CrossRef]

9. Kruppke, B.; Wagner, A.-S.; Rohnke, M.; Heinemann, C.; Kreschel, C.; Gebert, A.; Wiesmann, H.-P.; Mazurek, S.; Wenisch, S.; Hanke, T. Biomaterial based treatment of osteoclastic/osteoblastic cell imbalance—Gelatin-modified calcium/strontium phosphates. *Mater. Sci. Eng. C Mater. Biol. Appl.* **2019**, *104*, 109933. [CrossRef]

10. Grima, L.; Díaz-Pérez, M.; Gil, J.; Sola, D.; Peña, J.I. Generation of a porous scaffold with a starting composition in the CaO–SiO2–MgO–P2O5 system in a simulated physiological environment. *Appl. Sci.* **2020**, *10*, 312. [CrossRef]

11. Xia, Y.; Chen, H.; Zhao, Y.; Zhang, F.; Li, X.; Wang, L.; Weir, M.D.; Ma, J.; Reynolds, M.A.; Gu, N.; et al. Novel magnetic calcium phosphate-stem cell construct with magnetic field enhances osteogenic differentiation and bone tissue engineering. *Mater. Sci. Eng. C Mater. Biol. Appl.* **2019**, *98*, 30–41. [CrossRef] [PubMed]

12. Li, K.; Lu, X.; Razanau, I.; Wu, X.; Hu, T.; Liu, S.; Xie, Y.; Huang, L.; Zheng, X. The enhanced angiogenic responses to ionic dissolution products from a boron-incorporated calcium silicate coating. *Mater. Sci. Eng. C Mater. Biol. Appl.* **2019**, *101*, 513–520. [CrossRef] [PubMed]

13. Huang, K.H.; Lin, Y.H.; Shie, M.Y.; Lin, C.P. Effects of bone morphogenic protein-2 loaded on the 3D-printed MesoCS scaffolds. *J. Formos. Med. Assoc.* **2018**, *117*, 879–887. [CrossRef]

14. Huang, K.H.; Chen, Y.W.; Wang, C.Y.; Lin, Y.H.; Wu, Y.H.; Shie, M.Y.; Lin, C.P. Enhanced capability of BMP-2-loaded mesoporous calcium silicate scaffolds to induce odontogenic differentiation of human dental pulp cells. *J. Endod.* **2018**, *44*, 1677–1685. [CrossRef] [PubMed]

15. Huang, T.H.; Kao, C.T.; Shen, Y.F.; Lin, Y.T.; Liu, Y.T.; Yen, S.Y.; Ho, C.C. Substitutions of strontium in bioactive calcium silicate bone cements stimulate osteogenic differentiation in human mesenchymal stem cells. *J. Mater. Sci. Mater. Med.* **2019**, *30*, 68. [CrossRef] [PubMed]

16. Bejarano, J.; Caviedes, P.; Palza, H. Sol-gel synthesis and in vitro bioactivity of copper and zinc-doped silicate bioactive glasses and glass-ceramics. *Biomed. Mater.* **2015**, *10*, 025001. [CrossRef]

17. Gao, C.; Zhao, K.; Wu, Y.; Gao, Q.; Zhu, P. Fabrication of strontium/calcium containing poly(γ-glutamic acid)–organosiloxane fibrous hybrid materials for osteoporotic bone regeneration. *RSC Adv.* **2018**, *8*, 25745–25753. [CrossRef]

18. Liu, W.C.; Hu, C.C.; Tseng, Y.Y.; Sakthivel, R.; Fan, K.-S.; Wang, A.N.; Wang, Y.M.; Chung, R.J. Study on strontium doped tricalcium silicate synthesized through sol-gel process. *Mater. Sci. Eng. C Mater. Biol. Appl.* **2020**, *108*, 110431. [CrossRef]

19. Su, T.R.; Huang, T.H.; Kao, C.T.; Ng, H.Y.; Chiu, Y.C.; Hsu, T.T. The calcium channel affect osteogenic differentiation of mesenchymal stem cells on strontium-substituted Calcium silicate/poly-ε-caprolactone scaffold. *Processes* **2020**, *8*, 198. [CrossRef]

20. Hollister, S.J. Scaffold design and manufacturing: From concept to clinic. *Adv Mater.* **2009**, *21*, 3330–3342. [CrossRef]

21. Park, S.; Lee, H.J.; Kim, K.S.; Lee, S.; Lee, J.T.; Kim, S.Y.; Chang, N.H.; Park, S.Y. In vivo evaluation of 3D-printed polycaprolactone scaffold implantation combined with β-TCP powder for alveolar bone augmentation in a beagle defect model. *Materials* **2018**, *11*, 238. [CrossRef] [PubMed]

22. Lee, H.; Yang, G.H.; Kim, M.; Lee, J.; Huh, J.; Kim, G. Fabrication of micro/nanoporous collagen/dECM/silk-fibroin biocomposite scaffolds using a low temperature 3D printing process for bone tissue regeneration. *Mater. Sci. Eng. C Mater. Biol. Appl.* **2018**, *84*, 140–147. [CrossRef] [PubMed]

23. Shao, H.; Ke, X.; Liu, A.; Sun, M.; He, Y.; Yang, X.; Fu, J.; Liu, Y.; Zhang, L.; Yang, G.; et al. Bone regeneration in 3D printing bioactive ceramic scaffolds with improved tissue/material interface pore architecture in thin-wall bone defect. *Biofabrication* **2017**, *9*, 025003. [CrossRef] [PubMed]

24. Naghieh, S.; Sarker, M.D.; Sharma, N.K.; Barhoumi, Z.; Chen, X. Printability of 3D printed hydrogel scaffolds: Influence of hydrogel composition and printing parameters. *Appl. Sci* **2020**, *10*, 292. [CrossRef]

25. Kao, C.T.; Lin, C.C.; Chen, Y.W.; Yeh, C.H.; Fang, H.Y.; Shie, M.Y. Poly(dopamine) coating of 3D printed poly(lactic acid) scaffolds for bone tissue engineering. *Mater. Sci. Eng. C Mater. Biol. Appl.* **2015**, *56*, 165–173. [CrossRef]

26. Chen, C.C.; Yu, J.; Ng, H.Y.; Lee, K.X.; Chen, C.C.; Chen, Y.S.; Shie, M.Y. The physicochemical properties of decellularized extracellular matrix-coated 3D printed poly(ε-caprolactone) nerve conduits for promoting Schwann cells proliferation and differentiation. *Materials* **2018**, *11*, 1665. [CrossRef]

27. Xiao, W.; Li, J.; Qu, X.; Wang, L.; Tan, Y.; Li, K.; Li, H.; Yue, X.; Li, B.; Liao, X. Cell-laden interpenetrating network hydrogels formed from methacrylated gelatin and silk fibroin via a combination of sonication and photocrosslinking approaches. *Mater. Sci. Eng. C Mater. Biol. Appl.* **2019**, *99*, 57–67. [CrossRef]

28. Anada, T.; Pan, C.C.; Stahl, A.M.; Mori, S.; Fukuda, J.; Suzuki, O.; Yang, Y. Vascularized bone-mimetic hydrogel constructs by 3D bioprinting to promote osteogenesis and angiogenesis. *Int. J. Mol. Sci.* **2019**, *20*, 1096. [CrossRef]

29. Zhong, M.; Liu, X.; Wei, D.; Sun, J.; Guo, L.; Zhu, H.; Wan, Y.; Fan, H. A facile approach for engineering tissue constructs with vessel-like channels by cell-laden hydrogel fibers. *Mater. Sci. Eng. C Mater. Biol. Appl.* **2019**, *101*, 370–379. [CrossRef]

30. Chen, Y.W.; Shen, Y.F.; Ho, C.C.; Yu, J.; Wu, Y.H.; Wang, K.; Shih, C.T.; Shie, M.Y. Osteogenic and angiogenic potentials of the cell-laden hydrogel/mussel-inspired calcium silicate complex hierarchical porous scaffold fabricated by 3D bioprinting. *Mater. Sci. Eng. C Mater. Biol. Appl.* **2018**, *91*, 679–687. [CrossRef]

31. Dong, Y.; Chen, H.; Qiao, P.; Liu, Z. Development and properties of fish gelatin/oxidized starch double network film catalyzed by thermal treatment and schiff' base reaction. *Polymers* **2019**, *11*, 2065. [CrossRef] [PubMed]

32. Lin, W.H.; Yu, J.; Chen, G.; Tsai, W.B. Fabrication of multi-biofunctional gelatin-based electrospun fibrous scaffolds for enhancement of osteogenesis of mesenchymal stem cells. *Colloids Surf. B* **2016**, *138*, 26–31. [CrossRef] [PubMed]

33. Wang, J.L.; Chen, Q.; Du, B.B.; Cao, L.; Lin, H.; Fan, Z.Y.; Dong, J. Enhanced bone regeneration composite scaffolds of PLLA/β-TCP matrix grafted with gelatin and HAp. *Mater. Sci. Eng. C Mater. Biol. Appl.* **2018**, *87*, 60–69. [CrossRef] [PubMed]

34. Hosseini, S.F.; Javidi, Z.; Rezaei, M. Efficient gas barrier properties of multi-layer films based on poly(lactic acid) and fish gelatin. *Int. J. Biol. Macromol.* **2016**, *92*, 1205–1214. [CrossRef] [PubMed]

35. Qu, T.; Liu, X. Nano-structured gelatin/bioactive glass hybrid scaffolds for the enhancement of odontogenic differentiation of human dental pulp stem cells. *J. Mater. Chem. B* **2013**, *1*, 4764–4772. [CrossRef] [PubMed]

36. Etmimi, H.M.; Sanderson, R.D. New approach to the synthesis of exfoliated polymer/graphite nanocomposites by miniemulsion polymerization using functionalized graphene. *Macromolecules* **2011**, *44*, 8504–8515. [CrossRef]

37. Zhao, D.; Huang, J.; Zhong, Y.; Li, K.; Zhang, L.; Cai, J. High-strength and high-toughness double-cross-linked cellulose hydrogels: A new strategy using sequential chemical and physical cross-linking. *Adv. Funct. Mater.* **2016**, *26*, 6279–6287. [CrossRef]

38. Rizwan, M.; Peh, G.S.L.; Ang, H.-P.; Lwin, N.C.; Adnan, K.; Mehta, J.S.; Tan, W.S.; Yim, E.K.F. Sequentially-crosslinked bioactive hydrogels as nano-patterned substrates with customizable stiffness and degradation for corneal tissue engineering applications. *Biomaterials* **2017**, *120*, 139–154. [CrossRef]

39. Kokkinos, P.A.; Koutsoukos, P.G.; Deligianni, D.D. Detachment strength of human osteoblasts cultured on hydroxyapatite with various surface roughness. Contribution of integrin subunits. *J. Mater. Sci. Mater. Med.* **2012**, *23*, 1489–1498. [CrossRef]

40. Yao, S.; Xu, Y.; Zhou, Y.; Shao, C.; Liu, Z.; Jin, B.; Zhao, R.; Cao, H.; Pan, H.; Tang, R. Calcium phosphate nanocluster-loaded injectable hydrogel for bone regeneration. *ACS Appl. Bio Mater.* **2019**, *2*, 4408–4417. [CrossRef]

41. Yoon, H.J.; Shin, S.R.; Cha, J.M.; Lee, S.H.; Kim, J.H.; Do, J.T.; Song, H.; Bae, H. Cold water fish gelatin methacryloyl hydrogel for tissue engineering application. *PLoS ONE* **2016**, *11*, 0163902. [CrossRef] [PubMed]

42. Wang, C.; Shen, H.; Tian, Y.; Xie, Y.; Li, A.; Ji, L.; Niu, Z.; Wu, D.; Qiu, D. Bioactive nanoparticle-gelatin composite scaffold with mechanical performance comparable to cancellous bones. *ACS Appl. Mater. Interfaces* **2014**, *6*, 13061–13068. [CrossRef] [PubMed]

43. Shie, M.Y.; Ding, S.J.; Chang, H.C. The role of silicon in osteoblast-like cell proliferation and apoptosis. *Acta Biomater.* **2011**, *7*, 2604–2614. [CrossRef] [PubMed]

44. Zhai, D.; Xu, M.; Liu, L.; Chang, J.; Wu, C. Silicate-based bioceramics regulating osteoblast differentiation through a BMP2 signalling pathway. *J. Mater. Chem. B* **2017**, *5*, 7297–7306. [CrossRef]

45. Roohani-Esfahani, S.I.; Wong, K.Y.; Lu, Z.; Juan Chen, Y.; Li, J.J.; Gronthos, S.; Menicanin, D.; Shi, J.; Dunstan, C.; Zreiqat, H. Fabrication of a novel triphasic and bioactive ceramic and evaluation of its in vitro and in vivo cytocompatibility and osteogenesis. *J. Mater. Chem. B* **2014**, *2*, 1866–1878. [CrossRef]

46. Yang, F.; Yang, D.; Tu, J.; Zheng, Q.; Cai, L.; Wang, L. Strontium enhances osteogenic differentiation of mesenchymal stem cells and in vivo bone formation by activating Wnt/catenin signaling. *Stem Cells* **2011**, *29*, 981–991. [CrossRef]

47. Wu, C.; Zhou, Y.Z.; Lin, C.; Chang, J.; Xiao, Y. Strontium-containing mesoporous bioactive glass scaffolds with improved osteogenic/cementogenic differentiation of periodontal ligament cells for periodontal tissue engineering. *Acta Biomater.* **2012**, *8*, 3805–3815. [CrossRef]

48. Lei, Y.; Xu, Z.; Ke, Q.; Yin, W.; Chen, Y.; Zhang, C.; Guo, Y. Strontium hydroxyapatite/chitosan nanohybrid scaffolds with enhanced osteoinductivity for bone tissue engineering. *Mater. Sci. Eng. C Mater. Biol. Appl.* **2017**, *72*, 134–142. [CrossRef]

49. Shaltooki, M.; Dini, G.; Mehdikhani, M. Fabrication of chitosan-coated porous polycaprolactone/strontium-substituted bioactive glass nanocomposite scaffold for bone tissue engineering. *Mater. Sci. Eng C Mater. Biol. Appl.* **2019**, *105*, 110138. [CrossRef]

50. Prabha, R.D.; Nair, B.P.; Ditzel, N.; Kjems, J.; Nair, P.D.; Kassem, M. Strontium functionalized scaffold for bone tissue engineering. *Mater. Sci. Eng. C Mater. Biol. Appl.* **2019**, *94*, 509–515. [CrossRef]
51. Kazemi, M.; Dehghan, M.M.; Azami, M. Biological evaluation of porous nanocomposite scaffolds based on strontium substituted β-TCP and bioactive glass: An in vitro and in vivo study. *Mater. Sci. Eng. C Mater. Biol. Appl.* **2019**, *105*, 110071. [CrossRef] [PubMed]
52. Xie, H.; Gu, Z.; He, Y.; Xu, J.; Xu, C.; Li, L.; Ye, Q. Microenvironment construction of strontium–calcium-based biomaterials for bone tissue regeneration: The equilibrium effect of calcium to strontium. *J. Mater. Chem. B* **2018**, *6*, 2332–2339. [CrossRef]

 *applied sciences*

Article

Anti-Frothing Effect of Poultry Feathers in Bio-Based, Polycondensation-Type Thermoset Composites

Markus Brenner [1], Crisan Popescu [2] and Oliver Weichold [1],*

[1]   Institute of Building Materials Research, Schinkelstraße 3, 52072 Aachen, Germany; brenner@ibac.rwth-aachen.de

[2]   Kao European Research Laboratories, KAO Germany GmbH, Pfungstädter Str. 98–100, D-64297 Darmstadt, Germany; crisan717@yahoo.co.uk

*   Correspondence: weichold@ibac.rwth-aachen.de

Received: 21 February 2020; Accepted: 16 March 2020; Published: 21 March 2020

Featured Application: Potential green alternative to petrochemical thermoset resins in the fields of adhesives and fibre-reinforced composites.

Abstract: The formation of polycondensation-type thermoset resins from natural reactants such as citric and glutaric acid, as well as 1,3-propanediol and glycerol, was studied. Monitoring the mass loss by thermogravimetric analysis (TGA) allowed the rate constants of the esterification to be calculated, which were in the order of $7 \cdot 10^{-5}$ s^{-1} for glutaric acid and approximately twice as high for citric acid. However, the combination citric acid/glycerol was previously reported to froth up at high conversions, giving rise to foams, which makes the preparation of compact engineering composites challenging. In light of this, we observed that shredded poultry feathers not only increased the conversion and the reaction rate of the combination citric acid/glycerol, but increasing the amount of feathers continuously decreased the number of visible bubbles. The addition of 20 wt% of feathers completely prevented the previously reported frothing and gave rise to compact materials that were macroscopically free of defects. Besides this, the addition of feathers also improved the fire-retardant properties. The tensile properties of the first specimens are still rather low (σ = 11.6 N/mm^2, E = 750 N/mm^2), but the addition of poultry feathers opens a new path for green thermoset resins.

Keywords: thermoset; polyester; bio-based; poultry feathers; composite

---

1. Introduction

Thermoset resins are an important class of materials for adhesives [1] and load-bearing matrices in fibre-reinforced composites [2]. Typical examples are phenolic and epoxy resins, polyurethanes, unsaturated polyesters, or silicone rubbers [3]. However, the main source for these polymers is petrochemical feedstock and only very few attempts at a partial substitution by renewable materials, such as epoxidised castor and soybean oils [4], or bio-based polyols [5], have received wider attention. On the other hand, the growing environmental awareness and recent governmental regulations have increased the pressure on industry and academia to focus more on renewable and sustainable materials, closed material cycles, and materials with low or, ideally, zero carbon footprints. One reason for the difficulties in replacing petrochemical resources with natural ones, when it comes to thermoset resins, is the required high reactivity. Nature usually circumvents highly reactive functional groups, such as isocyanate and epoxy, by using highly specialised enzymes as catalysts in condensation reactions. As a result, bio-based feedstock requires in vitro reaction conditions not directly compatible with the requirements for thermoset resins. Thus, it seems that natural building blocks would not be suitable for making thermoset resins. However, the vast amount of natural or bio-based polyacids and polyols merits a closer look: by esterification, which is generally a rather slow, only slightly exothermic reaction type generating a lot of water as by-product, bio-based seminal thermoset materials were successfully generated, as presented in this paper.

Quite a common example of such an esterification is poly(butylene succinate), which can be synthesised from succinic acid and 1,4-butanediol via polycondensation and goes back to the work of Carothers [6]. Ever since, the polycondensation reaction has been refined and optimised, in particular to include new monomers and more active catalysts [7]. A recent example is the synthesis of a thermoset resin from lactic acid and glycerol, which was used to impregnate regenerated viscose fibre and compression-moulded to form a thermoset composite [8]. Naturally available di- and multifunctional carboxylic acids have been used to cross-link epoxidized sucrose soyate in order to produce a thermoset polyester that rapidly degrades in dilute sodium hydroxide [9]. A promising material is the combination citric acid and glycerol. Polymerised in solution at molar ratios from 0.75 to 1.5, the material obtained was initially an insoluble amorphous solid, that dissolves in water within 8 to 10 days and can be bio-degraded by *Aspergillius niger* and *E. coli* [10]. A material produced from citric acid and glycerol without a catalyst was used as a degradable thermoset polymer for drug delivery applications [11]. However, when prepared in bulk, the material turns out liquid, gel-like, or as an inhomogeneous foam due to the evaporation of substantial amounts of water as steam [12]. To the best of our knowledge, a catalysed method to produce compact solids from this mixture is not known.

Another aspect of substituting petrochemical feedstock with renewable resources is the exploitation of natural waste materials. Among these, poultry feathers are an interesting class, since they have one of the highest fracture toughnesses in nature [13] and make up approximately 10% of the weight of poultry [14]. This is a problem in poultry farms, because huge amounts of feather waste are generated, which are taken to landfills for disposal. The estimated global annual amount of feathers produced in poultry farms is in the order of 4 million tons [15]. Consequently, several research groups have studied the re-use of poultry feathers, e.g., in composite materials. Aranberri reported lightweight, biodegradable composites of polylactic acid or poly(butyrate adipate terephthalate) with up to 60 wt% poultry feathers, which were proposed as an insulation material [16]. Barone investigated the process parameters for compounding polyethylene and shredded feathers and found that feathers only act as reinforcements in polyethylene of low crystallinity [17,18].

Here, we present the first results of our work to bring these two topics together. It is our firm belief that polycondensation resins based on the above or similar natural or bio-based reactants can, at least in part, be an environmentally friendly alternative to petrochemical thermoset resins, provided that methods to circumvent the difficulties with evaporating condensation products are found. Citric and glutaric acid, as well as 1,3-propanediol and glycerol, are used as model ester resins, which liberate water during the reaction. Shredded poultry feathers are incorporated, because their dendritic structure and moisture sorption properties are expected to help in solving the problem caused by the generation of large amounts of water as by-product. Additionally, we expect the feathers to also improve the thermal and mechanical properties of the composites.

2. Materials and Methods

Citric acid, 1,3-propandiol, glutaric acid, and tetra-n-butyl ortotinanate were used as received. Glycerol was dried at 170 °C for three hours. Figure 1 shows the structures and abbreviations of the materials used. Goose feathers were provided washed and dried by a local manufacturer and shredded using a Retsch SM300 cutting mill with a 0.5 mm sieve. Prior to use, the feather shreds were dried at 120 °C for one hour.

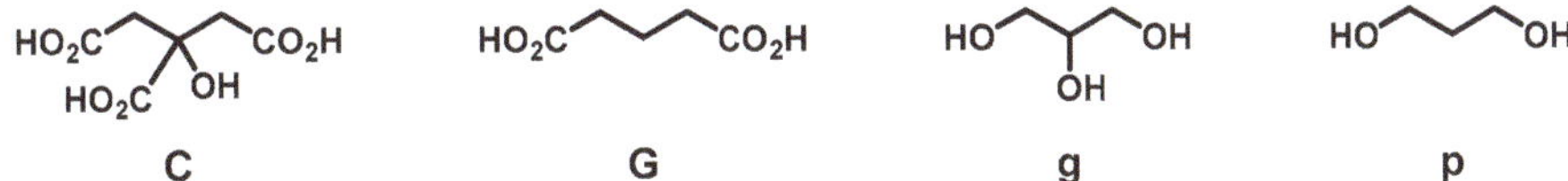

Figure 1. Structures and abbreviations of the compounds used to prepare the polyesters (C = citric acid; G = glutaric acid; g = glycerol; p = 1,3-propanediol). For example, the reaction of citric acid and 1,3-propanediol in a 1:1 molar ratio as well as the resulting product are labelled Cp1-1.

2.1. Polyester Preparation

The preparation of all mixtures followed a general procedure, which is illustrated using Cg1-1 (equimolar mixture of citric acid and glycerol) as an example: 7.04 mmol (1.53 g) citric acid was mixed with approximately half of the equimolar amount of glycerol (equimolar amount: 7.04 mmol, 0.65 g) and slowly heated to 110 °C while stirring. Meanwhile, 106 µmol (35.9 mg) Ti(O*n*Bu)$_4$—equivalent to 0.5 mol% per OH group—was added to the remaining glycerol. When the mixture of citric acid and glycerol was clear and homogeneous, the glycerol containing the catalyst was added and the combined mixture was stirred for 1 min at 110 °C to ensure complete mixing. For Cp2-3 and Gg3-2, the molar amounts of acid and alcohol were adjusted accordingly. Glutaric acid could be melted without addition of the alcohol.

At this point, the mixtures could be used for differential scanning calorimetry (DSC) and thermogravimetric analysis (TGA) analyses, cast directly into dog-bone shaped silicone moulds, or mixed with feather shreds prior to casting.

2.2. Feather Composite Preparation

The desired amount of feather shreds was weighed into a flask and the freshly prepared hot polyester mixture described above was added under continuous stirring. Mixing was continued for five minutes to ensure a homogeneous distribution of the shreds. During this time, the temperature needed to be kept above 60 °C to prevent solidification of the polyester. The mixture was then poured into 60 °C warm, dog-bone shaped silicone moulds. For curing, the moulds were covered with anti-adhesive paper, followed by a 2 mm thick, water-permeable, non-woven fabric, and an 18 mm screen-printing plate. The complete set-up was loaded with 750 g per sample and placed in a vacuum oven at 120 °C and 20 mbar for 20 h (Figure 2). After cooling in air, the samples could be removed from the mould. Gg samples were then heated for another 48 h at 120 °C.

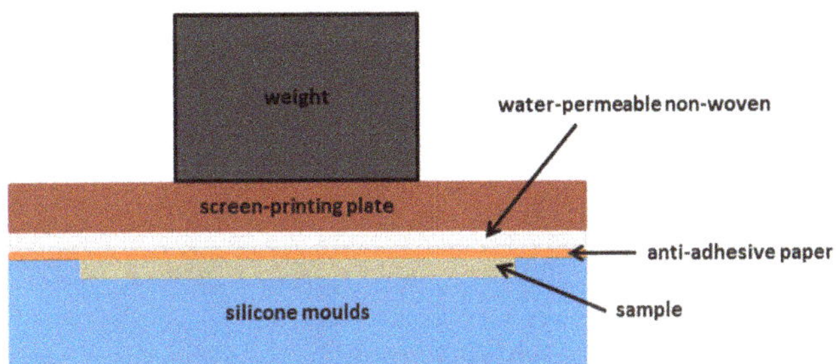

Figure 2. Experimental setup for composite preparation.

2.3. ATR-FTIR Spectroscopy

Infrared IR spectra were recorded using a Perkin Elmer Spectrum Two UATR FTIR spectrometer equipped with a diamond ATR (attenuated total reflection) window. All spectra were recorded in the spectral range of 4000–400 cm^{-1} with 12 scans at a spectral resolution of 4 cm^{-1}. Before each measurement, the diamond ATR crystal was cleaned with isopropanol.

2.4. DSC

The differential scanning calorimetry (DSC) measurements were performed on a Netsch DSC 204 F1 Phoenix. Approximately 40 mg of freshly prepared sample was placed in an aluminium sample pan and the heat flow was recorded between 5 and 170 °C. Each system was analysed at heating rates of 5, 10, 20, and 40 °C/minute. The measurements were carried out under nitrogen atmosphere with a flow rate 20 mL/minute.

2.5. Thermogravimetric Analysis (TGA and TGA-FTIR)

The mass loss during polycondensation was recorded using a Perkin Elmer TGA-4000. Approximately 40 mg of freshly prepared sample was placed in an aluminium oxide sample pan used for TGA. Each system was kept at 10 °C for 5 min, then heated to 120 °C at 10 °C/min, and finally kept at 120 °C for 14 h. The measurements were carried out under nitrogen atmosphere with a flow rate of 20 mL/min as sample purge. Coupled TGA/FTIR measurements were performed by using a Perkin Elmer TL8000 gas transfer line setup with Perkin Elmer Spectrum 2 FTIR spectrometer. A gas transfer line temperature and gas chamber temperature of 270 °C was used. The gas flow used for FTIR measurements was 37 mL/minute, which was two-thirds of the accumulated gas flow in the TGA sample chamber.

To evaluate the thermal stability of the composites, approximately 20 mg of the sample was put into the crucible, conditioned at 30 °C for 5 min, and then heated to 850 °C with a rate of 10 °C/min in a continuous stream of nitrogen or oxygen at a flow rate of 20 mL/min.

2.6. Mechanical Testing

Tensile tests were carried out on an Instron 5566 by using dog-bone shaped samples according to DIN EN ISO 527-2 type 1A and the sample holder shown in Figure 3. The tests were run strain-controlled using a strain rate of 1 mm/min. Recording started when the applied load reached 5 N.

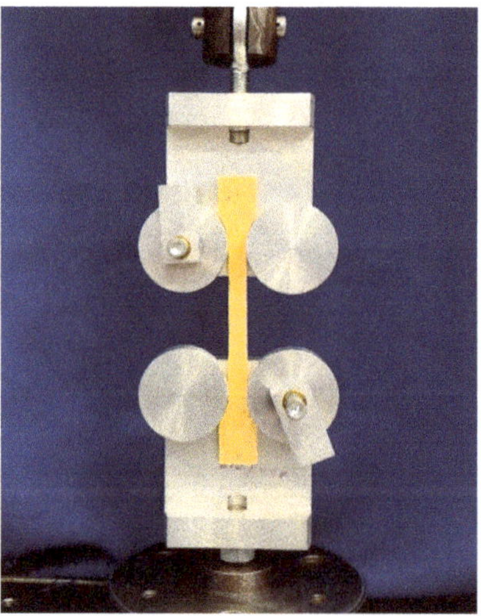

Figure 3. Set-up for mechanical testing.

3. Results and Discussion

The polycondensation was affected by titanium-catalysed esterification of the model compounds in bulk (Figure 4). The formation of ester bonds was confirmed by ATR-FTIR spectroscopy (Figure 5). For glutaric acid, the symmetric stretching vibration of the carbonyl group was observed at 1687 cm^{-1}, and shifted to higher wave numbers upon esterification (Figure 5 left). Citric acid exhibited two bands, at 1695 cm^{-1} and 1743 cm^{-1} (Figure 5 right), which merged into a single band at intermediate wave numbers of 1727 cm^{-1} for Cp1-1 and 1722 cm^{-1} for Cg1-1. This appears to be a common feature of citric acid ester, as triethyl citrate and tributyl citrate also showed only a single band at approximately 1740 cm^{-1} [19]. The carboxylic acid band was hardly visible in the products Gg1-1, Gp1-1, and Cp1-1, indicating high conversions [20]. In contrast, the ester band of Cg1-1 was comparatively broad and could hide some of the original acid bands.

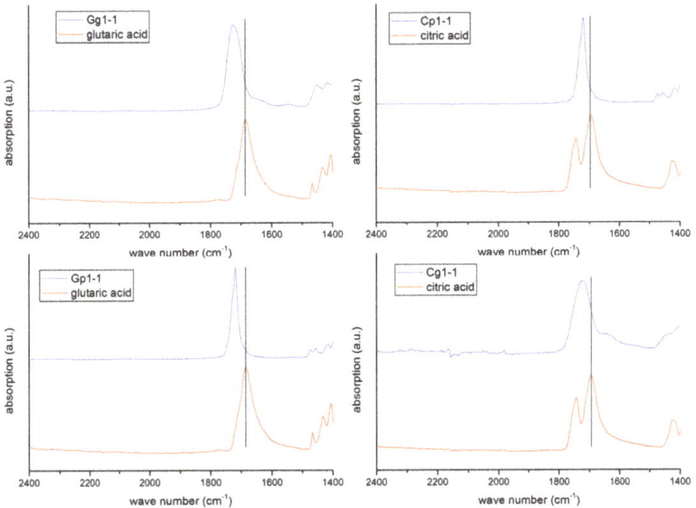

Figure 4. Schematic representation of the four reactions under investigation. x, y represent the stoichiometric coefficients in accordance with Equation (2). A star at the end of a bond indicates continuation of the polymer.

Figure 5. Comparison of the ATR-FTIR spectra of glutaric acid (left) and citric acid (right) with the four principal polycondensation products.

In order to assess the thermodynamic parameters for the polycondensation reactions in Figure 4, the reactions were monitored by DSC at four different heating rates (Figures S1–S4 in the Supplementary

Material). As expected for non-isothermal kinetics [21], the signal observed in the reactions Cp, Gp, and Gg shifted to higher temperatures when increasing the heating rate from 5 to 40 °C/min. However, the lack of effects in the DSC trace of the Cg mixture and the lack of any effects upon cooling and re-heating suggest that these might not relate to the polycondensation reaction shown to occur in all these systems. This is also supported by a comparative evaluation of DSC and TGA traces, exemplified for the Cp1-1 reaction in Figure 6.

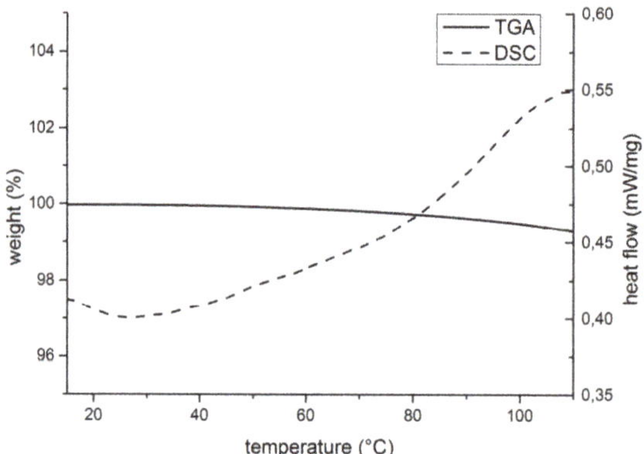

Figure 6. Comparison of DSC (dashed line) and TGA (solid line) traces of the Cp1-1 reaction up to the onset of the isothermal phase.

Clearly, the onset of the signal in the DSC trace at around 100 °C corresponded with the beginning of a mass loss observed by TGA that totalled approximately 2% and thus, could not be caused by the polycondensation reaction. This was further corroborated by monitoring the evaporating reaction products using a combined TGA-FTIR method (Figure 7). During the very early stages of the reaction, the IR spectrum showed, besides the prominent resonances of water vapour, others at 2946 cm^{-1} and 1046 cm^{-1}, indicating aliphatic C-H bonds (Figure 7a). These resonances disappeared not long after their first detection (Figure 7b). The most likely source of aliphatics is the catalyst Ti(O*n*Bu)$_4$, which reacted in the early stages of the polycondensation with either the alcohol or the carboxylic acid to displace n-butanol [22]. In addition, the weight loss of approximately 2% was in accordance with the amount of catalyst used.

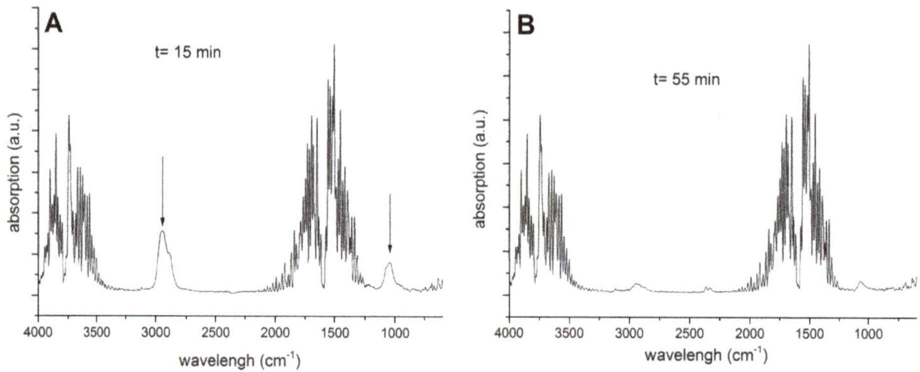

Figure 7. Gas-phase IR used to monitor the isothermal Cp reaction at 120 °C after 15 (**A**) and 55 (**B**) min. Arrows in A indicate the occurrence of *n*-butanol.

In consequence of the above findings, the kinetic parameters of the polycondensation reactions were evaluated using isothermal experiments monitored by TGA (Figure 8). The mass loss in the Gp1-1 reaction was considerably larger than in the other reactions. This is due to the joint evaporation of water and 1,3-propanediol—vapour pressure approximately 10 hPa at 100 °C as opposed to, e.g., 0.02 hPa for glycerol [23], which was confirmed by the combined TGA-FTIR method. On the other hand, 1,3-propanediol was not found to evaporate in the Cp reaction. The reason for this is not entirely clear, but could be caused by stronger hydrogen bonding in the Cp system. A potential alternative explanation is provided further below.

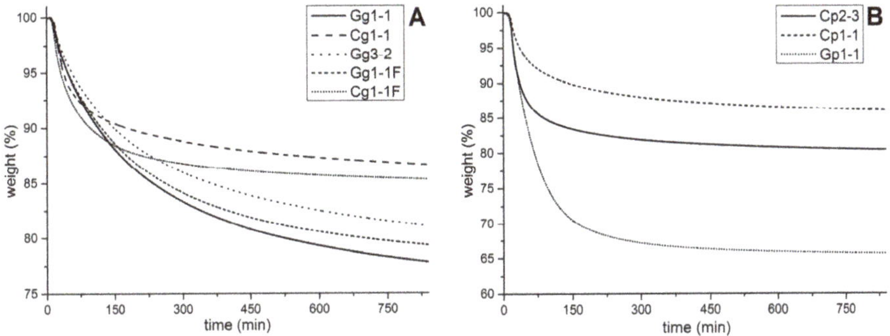

Figure 8. Mass loss of the polycondensation reactions of Figure 4 with glycerol (**A**) or 1,3-propanediol (**B**), as monitored by TGA.

All systems reached the equilibrium within approximately 14 h at 120 °C (Figure 8). The mass loss at equilibrium can be converted to the number of water molecules n liberated during the reaction by solving:

$$\text{TGA mass loss (\%)} = \frac{n_{\text{water molecules eliminated}} \cdot 18 \frac{g}{mol}}{M_{\text{acid}} + M_{\text{polyol}}} \times 100\%, \tag{1}$$

where M_{acid} and M_{polyol} are the molecular weights of the corresponding acid and alcohol. Based on the functionality of the reactants (it should be noted that in all calculations, the 2-hydroxy group of citric acid was considered to be unreactive) and the stoichiometry, the mixtures could liberate a maximum of either two or three water molecules. As can be seen from Table 1, some of the mixtures came close to the theoretical number (Table 1, entries 1, 2, 8), indicating that these reactions ran almost to completion. This is not too surprising for the system Cp1-1, for example, which, due to the excess of acid groups, should only form branched polymers with an average degree of polymerisation of 5. However, the systems Cp2-3 (Table 1, entry 2) and Gg3-2 (Table 1, entry 8) went to almost complete conversion, whereas Cg1-1 (Table 1, entry 3) appeared to stop at approximately 57%. This is in accordance with the results from IR spectroscopy (Figure 5). All three systems had a 1:1 ratio of functional groups and could give rise to highly cross-linked materials. Calculating the functionality factor f_{ar} according to:

$$f_{\text{ar}} = \frac{x \cdot p + y \cdot q}{x + y}, \tag{2}$$

where x, y are the stoichiometric coefficients and p, q the number of functional groups of acid and alcohol [24], afforded $f_{\text{ar}} = 2.4$ for Cp2-3 and Gg3-2, while for Cg1-1 $f_{\text{ar}} = 3$. According to the theory of non-linear step-growth reactions, gelation—i.e., the formation of a 3D network—occurs when the degree of polymerisation becomes infinite, and the corresponding limiting conversion p_G is given by $p_G = 2 \cdot f_{\text{ar}}^2$ [24]. This could be calculated as 83% for Cp2-3, Gg3-2 and 67% for Cg1-1. The latter is greater than the conversion determined by TGA (57%), indicating that the 3D network did not fully develop under the applied curing conditions. It can be easily observed during the reaction that the viscosity in the Cg1-1 system increased rather quickly. The reason for the low conversion in this reaction could, therefore, be the formation of a more rigid framework already at low conversions due to the dense packing of three functional groups in both reactants. This immobilises the chain ends and prevents them from meeting. Conversely, if one of the reactants is less crowded, such as 1,3-propanediol or glutaric acid in the Cp and Gg systems, the reaction reaches high conversions. It is not clear why for the Gg1-1 reaction the mass loss at equilibrium indicates the liberation of 2.5 water molecules, while theory calls for only 2. The joint evaporation of water and one of the reactants can be ruled out in the same way as wet reactants.

Table 1. Results of the kinetic evaluation using pseudo first order kinetics.

Entry	System [a]	Mass Loss at Equilibrium/%	n [b]	n_{theo} [c]	k/s^{-1}
1	Cp1-1	12	1.8	2	1.2×10^{-4}
2	Cp2-3	17	2.9	3	2.4×10^{-4}
3	Cg1-1	11	1.7	3	1.2×10^{-4}
4	Cg1-1F [d]	17 [e]	2.1	3	1.6×10^{-4}
5	Gp1-1	– [f]	– [f]	2	– [f]
6	Gg1-1	20	2.5	2	7.5×10^{-5}
7	Gg1-1F [d]	25 [e]	3.1	2	– [g]
8	Gg3-2	17	2.7	3	7.2×10^{-5}

[a] Abbreviations according to Figure 1; [b] number of water molecules calculated according to Equation (1); [c] theoretical amount of water molecules based on the stoichiometry; [d] mixture containing 20 wt% of feather shreds; [e] value calculated with regard to the feather content; [f] value not calculated due to the joint evaporation of water and diol; [g] value not calculated due to decomposition.

The mass loss over time in the isothermal regime followed an exponential decay, indicating first order kinetics. Although esterification is a classic example of a second order reaction, the use of only 0.5 mol% of the catalyst per OH group of the polyol resulted in a rather low concentration of active centres. This caused the mechanism to follow a pseudo first-order law, under the assumption that autocatalysed esterification was rather slow. Modelling the elimination of water according to:

$$\frac{dm}{dt} = -k \cdot m_0 \cdot t,$$ (3)

the TGA curves in Figure 6 can be fitted by rearranging Equation (1) to:

$$\text{TGA mass loss } (\%) = e^{-k \cdot t}.$$ (4)

Evidently, the reactions with citric acid were approximately twice as fast as those with glutaric acid. A potential reason for this could be the intramolecular formation of cyclic anhydrides of citric acid under the action of titanium catalysts such as $Ti(OnBu)_4$ [25]. The difference in reaction rates could also facilitate the evaporation of 1,3-propanediol from the Gp1-1 and impede evaporation from the Cp1-1 reaction mixture.

The addition of feathers caused an increase in both the reaction rate and conversion for the Cg1-1 system (Cg1-1F, Table 1, entry 4). The reason for this could be manifold, the simplest one being the vast number of different functional groups provided by the feather keratin, which could open up a number of alternative reaction pathways. Alternatively, the dendritic network of barbules might help in letting the water vapour escape the viscous mixtures and/or the feather keratin might absorb the moisture from the hot mixture. While the latter two would shift the equilibrium to the side of the products (increase of n), the former would increase the reaction rate. Thus, a combination of these alternatives is most likely.

Similar beneficial effects could not be observed when adding feather shreds to the Gg1-1 mixture (Gg1-1F, Table 1, entry 7). While the parent Gg1-1 mixture (Table 1, entry 6) was pale yellow and transparent, the addition of feathers caused the mixtures to turn reddish-brown (Figure S5 in the Supplementary Material). This apparent decomposition was also manifested in the rather large mass loss at equilibrium of 25%, which can be calculated as a total of three water molecules being liberated per molecule glutaric acid, while in theory it should only be two. Although citric acid is approximately one order of magnitude more acidic than glutaric acid, a similar decomposition was not observed in the Cg1-1F system and this could be due to the esterification occurring faster.

From the above mixtures, those based on citric acid appear more suitable for the preparation of engineering materials, particularly the Cg1-1 combination. However, the combination of citric acid and glycerol is known to froth massively, which prevents the formation of compact workpieces [12]. A plausible reason is that due to the rapid increase in viscosity already at low conversions, most of the water formed in the condensation reaction is trapped inside the viscous mixture in the form of bubbles. Based on the positive effects on the kinetics observed in the Cg1-1F system, the incorporation of shredded feathers was thought to enable the preparation of compact test specimens from citric acid and glycerol for the first time. Indeed, increasing the amount of feathers incorporated into Cg1-1 mixtures clearly reduced frothing and the number of visible bubbles, such that with 20 wt% of feathers, compact and macroscopically defect-free samples could be obtained (Figure 9). The incorporation of higher amounts of feathers appears possible and could potentially give rise to materials with improved properties, but requires more sophisticated mixing techniques.

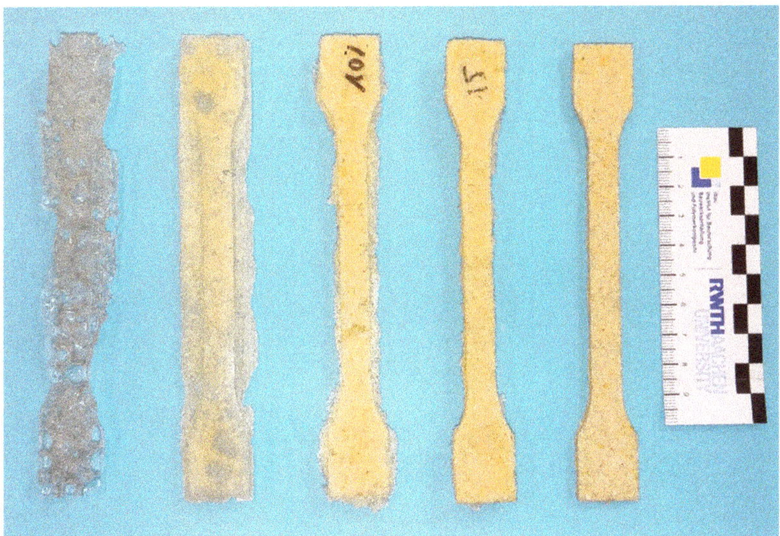

Figure 9. The Cg1-1 series containing an increasing amount of shredded feathers cured in dog-bone shaped moulds. From left to right: 0 (Cg1-1), 5, 10, 15, and 20 wt-% (Cg1-1F).

The Cg1-1F samples were obtained as solid and hard, but also rather brittle objects, which made testing in the set-up shown in Figure 3 somewhat difficult. It should be noted that all Cg1-1F samples failed close to the receiving rollers (cf. Figure S6 in the Supplementary Material), most likely because they did not fit precisely to the curvature of the specimen. This indicates that the properties shown in Table 2 and Figures S7–S10 do not represent the best performance of the tested materials. For comparison, the Gg1-1F and the Gg1-1 mixtures were also cured in dog-bone shaped moulds and tested. Due to the lower degree of cross-linking, both specimens were rubbery.

Table 2. Tensile properties of three samples from Table 1.

Sample	Number of Samples	Tensile Strength (N/mm^2)	Stiffness (N/mm^2)
Cg1-1F [a]	3	11.6 ± 2.06	730 ± 55.8
Gg1-1F [a]	3	1.20 ± 0.41	28.13 ± 10.9
Gg1-1	2	0.44 ± 0.13	2.58 ± 0.27

[a] Samples containing 20 wt% shredded feathers.

The thermal stability of the composites was investigated by TGA measurements under both nitrogen and oxygen atmosphere (Figure 10). The former represents rather a pyrolysis, the latter a combustion. Generally, the decomposition temperatures of the samples based on glutaric acid were approximately 100 °C higher than those based on citric acid. The reason for this might be the rather low decomposition temperature of citric acid of 175 °C, where it is transformed to aconitic acid by the elimination of water [26]. Under the assumption made above that the 2-hydroxy group does not participate in the polycondensation, a similar start to the decomposition appears likely for the citrate esters. Under oxygen, all samples were completely burnt at approximately 550 °C, while under nitrogen, a substantial residue remained. From this, the limiting oxygen index (LOI) can be calculated according to LOI = 17.5 + 0.4 · *cr*, where *cr* is the char residue at 850 °C in percentage (Table 3) [27].

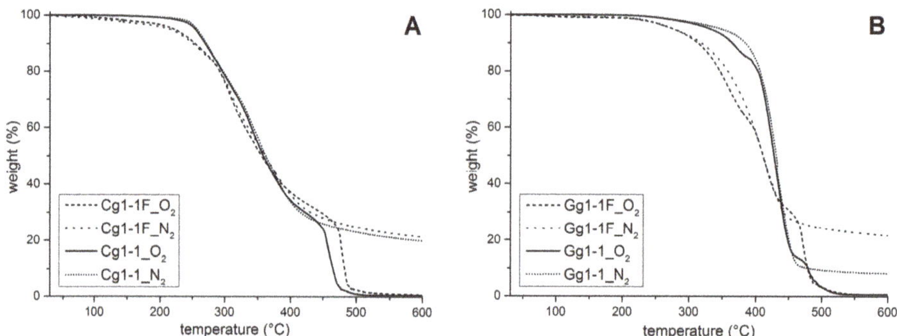

Figure 10. Thermogravimetric analysis of the products Cg1-1 (**A**) and Gg1-1 (**B**) with and without 20 wt-% feathers under nitrogen and oxygen atmosphere.

Table 3. TGA data and the calculated limiting oxygen index (LOI).

Sample	Char Residue at 850 °C	LOI/%
Cg1-1	12.304	22.4
Cg1-1F [a]	14.871	23.4
Gg1-1	6.761	20.2
Gg1-1F [a]	16.858	24.2

[a] Samples containing 20 wt% shredded feathers.

The addition of shredded feathers increased the LOI in all cases, and based on the observed values, Cg1-1F and Gg1-1F were considered to be hard to ignite and self-extinguishable in air. It should be noted that the decomposition of Gg1-1F under both nitrogen and oxygen started at considerably lower values than that of Gg1-1 (cf. Figures S11 and S12 in the Supplementary Material). As stated above, the Gg1-1 mixtures containing feathers turned reddish-brown during curing (Figure S5 in the Supplementary Material), which indicates undesired side-reactions that seem to affect the onset of decomposition, but not the fire-retarding properties.

4. Conclusions

In the quest for green alternatives to petrochemical feedstock, this article describes a significant step forward in the technologically important field of thermoset resins. The addition of shredded poultry feathers allowed, for the first time, the exploitation of natural or bio-based feedstock in the form of polyacids and polyols for the production of engineering composites from polycondensation-type thermoset resins. In the first results presented here, the feathers appeared to prevent the previously reported frothing of highly cross-linked combinations, such as citric acid/glycerol, and improved the thermal properties. However, the tensile properties of materials based on this mixture are so far rather low, and IR along with kinetic data indicate the conversion at equilibrium being below the conversion necessary to form a cross-linked 3D network. Improvements in processing and curing will allow the feathers to be activated as reinforcements, which will improve the mechanical properties.

Supplementary Materials: The following are available online at http://www.mdpi.com/2076-3417/10/6/2150/s1, From Figure S1 to Figure S12. DSC traces of the reactions Cg1-1, Cp1-1, Gg1-1, and Gp1-1, optical appearance of specimen Gg1-1 and Cg1-1 after fracture, stress-strain curves from which the results in Table 2 were calculated, and DTG curves for the mixtures Gg1-1 and Cg1-1.

Author Contributions: Conceptualization, O.W. and M.B.; formal analysis, C.P. and O.W.; investigation, M.B.; writing—original draft preparation, all authors; writing—review and editing, all authors. All authors have read and agreed to the published version of the manuscript.

Funding: The work received no external funding.

Acknowledgments: The authors thank Kaja Kensmann and Günther Wiwianka for technical assistance.

Conflicts of Interest: The authors declare no conflict of interest.

References

1. Engels, T. Chapter 10—Thermoset Adhesives. In *Thermosets*, 2nd ed.; Guo, Q., Ed.; Elsevier: Amsterdam, The Netherlands, 2018; pp. 341–368.
2. De, S.K.; White, J.R. *Short Fibre-Polymer Composites*; Woodhead Publishing: Cambridge, UK, 1996.
3. Hanna, D.; Goodman, S.H. *Handbook of Thermoset Plastics*; William Andrew: Amsterdam, The Netherlands, 2013.
4. Tan, S.G.; Chow, W.S. Biobased Epoxidized Vegetable Oils and Its Greener Epoxy Blends: A Review. *Polym.-Plast. Technol. Eng.* **2010**, *49*, 1581–1590. [CrossRef]
5. Li, Y.; Luo, X.; Hu, S. Introduction to Bio-Based Polyols and Polyurethanes. In *Bio-Based Polyols and Polyurethanes*; Springer: Berlin/Heidelberg, Germany; Cham, Switzerland, 2015; pp. 1–13.
6. Carothers, W.H. Polymerization. *Chem. Rev.* **1931**, *8*, 353–426. [CrossRef]
7. Xu, J.; Guo, B.H. Poly (Butylene Succinate) and Its Copolymers: Research, Development and Industrialization. *Biotechnol. J.* **2010**, *5*, 1149–1163. [CrossRef] [PubMed]
8. Esmaeili, N.; Bakare, F.O.; Skrifvars, M.; Afshar, S.J.; Åkesson, D. Mechanical Properties for Bio-Based Thermoset Composites Made from Lactic Acid, Glycerol and Viscose Fibers. *Cellulose* **2015**, *22*, 603–613. [CrossRef]
9. Ma, S.; Webster, D.C. Naturally Occurring Acids as Cross-Linkers to Yield Voc-Free, High-Performance, Fully Bio-Based, Degradable Thermosets. *Macromolecules* **2015**, *48*, 7127–7137. [CrossRef]
10. Pramanick, D.; Ray, T.T. Synthesis and Biodegradation of Copolyesters from Citric Acid and Glycerol. *Polym. Bull.* **1988**, *19*, 365–370. [CrossRef]
11. Halpern, J.M.; Urbanski, R.; Weinstock, A.K.; Iwig, D.F.; Mathers, R.T.; Von Recum, H.A. A Biodegradable Thermoset Polymer Made by Esterification of Citric Acid and Glycerol. *J. Biomed. Mater. Res. Part A* **2014**, *102*, 1467–1477. [CrossRef]
12. Tisserat, B.; O'kuru, R.H.; Hwang, H.; Mohamed, A.A.; Holser, R. Glycerol Citrate Polyesters Produced through Heating without Catalysis. *J. Appl. Polym. Sci.* **2012**, *125*, 3429–3437. [CrossRef]
13. Wegst, U.G.K.; Ashby, M.F. The Mechanical Efficiency of Natural Materials. *Philos. Mag.* **2004**, *84*, 2167–2186. [CrossRef]

14. Grazziotin, A.; Pimentel, F.A.; De Jong, E.V.; Brandelli, A. Nutritional Improvement of Feather Protein by Treatment with Microbial Keratinase. *Anim. Feed Sci. Technol.* **2006**, *126*, 135–144. [CrossRef]
15. Staroń, P.; Banach, M.; Kowalski, Z.; Staroń, A. Hydrolysis of Keratin Materials Derived from Poultry Industry. *Proc. ECOpole* **2014**, *8*, 443–448.
16. Aranberri, I.; Montes, S.; Azcune, I.; Rekondo, A.; Grande, H.J. Fully Biodegradable Biocomposites with High Chicken Feather Content. *Polymers* **2017**, *9*, 593. [CrossRef]
17. Barone, J.R.; Schmidt, W.F.; Liebner, C.F. Compounding and Molding of Polyethylene Composites Reinforced with Keratin Feather Fiber. *Compos. Sci. Technol.* **2005**, *65*, 683–692. [CrossRef]
18. Barone, J.R. Polyethylene/Keratin Fiber Composites with Varying Polyethylene Crystallinity. *Compos. Part A* **2005**, *36*, 1518–1524. [CrossRef]
19. Infrared Spectrum of Triethyl Citrate (Sdbs No. 2909). Spectral Database for Organic Compounds. Available online: https://sdbs.db.aist.go.jp/sdbs/cgi-bin/landingpage?sdbsno=2909 (accessed on 12 March 2020).
20. Gebhard, J.; Sellin, D.; Hilterhaus, L.; Liese, A. Online-Analyse Von Enzymatischen Polykondensationsreaktionen in Blasensäulenreaktoren Mittels Atr-Ftir-Spektroskopie. *Chem. Ing. Tech.* **2013**, *85*, 1016–1022. [CrossRef]
21. Popescu, C.; Segal, E. Critical Considerations on the Methods for Evaluating Kinetic Parameters from Nonisothermal Experiments. *Int. J. Chem. Kinet.* **1998**, *30*, 313–327. [CrossRef]
22. Otton, J.; Ratton, S.; Vasnev, V.A.; Markova, G.D.; Nametov, K.M.; Bakhmutov, V.I.; Komarova, L.I.; Vinogradova, S.V.; Korshak, V.V. Investigation of the Formation of Poly (Ethylene Terephthalate) with Model Molecules: Kinetics and Mechanisms of the Catalytic Esterification and Alcoholysis Reactions. Ii. Catalysis by Metallic Derivatives (Monofunctional Reactants). *J. Polym. Sci. Part A* **1988**, *26*, 2199–2224. [CrossRef]
23. Lide, D.R. *Crc Handbook of Chemistry and Physics, 2000-2001*; CRC Press: Boca Raton, FL, USA, 2000.
24. Carothers, W.H. Polymers and Polyfunctionality. *Trans. Faraday Soc.* **1936**, *32*, 39–49. [CrossRef]
25. Noordover, B.A.; Duchateau, R.; van Benthem, R.A.; Ming, W.; Koning, C.E. Enhancing the Functionality of Biobased Polyester Coating Resins through Modification with Citric Acid. *Biomacromolecules* **2007**, *8*, 3860–3870. [CrossRef]
26. Barbooti, M.M.; Al-Sammerrai, D.A. Thermal Decomposition of Citric Acid. *Thermochim. Acta* **1986**, *98*, 119–126. [CrossRef]
27. van Krevelen, D.W. Some Basic Aspects of Flame Resistance of Polymeric Materials. *Polymer* **1975**, *16*, 615–620. [CrossRef]

Article

FT-IR Characterization of Antimicrobial Hybrid Materials through Sol-Gel Synthesis

Michelina Catauro [1,*], **Simona Piccolella** [2] and **Cristina Leonelli** [3]

[1] Department of Engineering, University of Campania "Luigi Vanvitelli", Via Roma 29, 81031 Aversa, Italy
[2] Department of Environmental, Biological and Pharmaceutical Sciences and Technologies, University of Campania "Luigi Vanvitelli", Via Vivaldi 43, I-81100 Caserta, Italy; simona.piccolella@unicampania.it
[3] Department of Engineering "Enzo Ferrari", University of Modena and Reggio Emilia, Via P. Vivarelli n. 10, 41125 Modena, Italy; cristina.leonelli@unimore.it
* Correspondence: michelina.catauro@unicampania.it

Received: 5 January 2020; Accepted: 6 February 2020; Published: 10 February 2020

Abstract: Silica/polycaprolactone and titania/polycaprolactone hybrid organic/inorganic amorphous composites were prepared via a sol-gel method starting from a multi-element solution containing tetramethyl orthosilicate (TMOS) or titanium butoxide (TBT), polycaprolactone (PCL), water and methylethylketone (MEK). The molecular structure of the crosslinked network was based on the presence of the hydrogen bonds between organic/inorganic elements as confirmed by Fourier Transform Infra-Red (FT-IR) analysis. In particular, the structure of crosslinked network was realized by hydrogen bonds between the X-OH (X = Si or Ti) group (H donator) in the sol-gel intermediate species and ester groups (H-acceptors) in the repeating units of the polymer. The morphology of the hybrid materials; pore size distribution, elemental homogeneity and surface features, was studied by scanning electron microscopy/energy dispersive spectroscopy (SEM/EDS) and by atomic force microscopy (AFM). The bioactivity of the synthesized hybrid materials was confirmed by observing the formation of a layer of hydroxyapatite (HAP) on the surface of the samples soaked in a simulated body fluid. The antimicrobial behavior of synthetized hybrids was also assessed against *Escherichia coli* bacteria. In conclusion, the prepared hybrid materials are proposed for use as future bone implants.

Keywords: hybrid composite; PCL; TiO_2; SiO_2; FT-IR spectroscopy; antibacterial behavior

---

1. Introduction

Since the 1990s, organic/inorganic nanocomposites networks have become a growing field of investigation [1–3]. The sol-gel route represents an interesting approach, which can be effectively applied to the preparation of the inorganic phases within the inorganic/organic hybrid materials, due to the fact that it can occur in liquid solutions at room temperature. These materials are considered to be biphasic, where the organic and inorganic phases are mixed at the nanometric or submicronic scales. The smaller the scale of mixing, the more unexpected the properties are, since these materials may show properties that are not the sum of the individual contributions from both phases. Additionally, it has been proved that the role of the inner interfaces is predominant; in particular, the nature of the interface is the critical parameter used to distinguish two different classes [4]: class I collects those hybrid materials with weak bonds (hydrogen, van der Waals or ionic bonds), and class II gathers materials where the phases are linked together through strong chemical bonds (covalent or ionic-covalent bonds). In class I, materials, obtained combining the sol-gel process with polymer chemistry, can be enclosed with a large number of different applications [5–7], from non-linear optical materials [8] to mesoporous materials [9].

The basic chemical reactions of the sol-gel process are the hydrolyses of metal alkoxides in the proper solvent, usually an alcohol/water mixture, followed by their polycondensation. In the first stage, substituting the alkoxide groups (OR) with the hydroxyl groups (OH) in the metal alkoxide offers a large number of opportunities, due to the presence of such reactive OH groups, to match with organic monomers. Thus, a variety of organic polymers have been introduced into inorganic networks. The final material derives from the condensation of two OH groups with the elimination of a water molecule. The two adjacent OH groups may belong to one organic and one inorganic monomer to afford the formation of a hybrid or composite material with covalent bonds between the polymer and inorganic elements. The gel reaction might also give rise to weaker bonds, such as a hydrogen bond, in the organic/inorganic hybrid bulky materials. It is well known that many parameters, such as type and concentration of reactants, solvents, catalysts, reaction temperature and removal rate of by-products and solvents, influence the sol-gel process [10,11]. Hence, it is evident that the presence of organic elements modifies the morphology and physical properties of the final sol-gel products. The difference in products that are obtained through the base-catalyzed sol-gel reaction and through the carefully controlled acid-catalyzed one is evident, since from the first process, translucent or opaque products with visible organic/inorganic phase separation are obtained, while from the second, transparent and monolithic materials are realized. A key topic that is persistently challenging in these organic-altered materials is the degree of mixing of the organic/inorganic elements, i.e., the phase homogeneity. The high optical transparency to visible light suggests that the organic/inorganic phase separation, if any, occurs on a scale of $\leq$400 nm [12]. Two of the most-used techniques to investigate phase mixing in hybrid materials, at micrometric scale, are atomic force microscopy (AFM) [13] and scanning electron microscopy (SEM).

In this work, we propose a combination of bioactive organic and inorganic phases to obtain a single and homogeneous hybrid composite using the sol-gel route for the inorganic three-dimensional (3D) network. In particular, we chose SiO_2 and TiO_2-containing bioactive materials that are often produced in the amorphous state, typical of glasses, via the sol-gel route. Both of these oxides are known to bond to living bone [14]. Thus, a hybrid material based on smart combinations of biodegradable polymers, specifically polycaprolactone (PCL), and such bioactive inorganic compounds is of particular interest, since it exhibits tailored physical, biological and mechanical properties, as well as predictable degradation behavior.

The purpose of this manuscript is the microstructural characterization and bioactivity assessment of polycaprolactone/SiO_2 and polycaprolactone/TiO_2 hybrid materials prepared by the sol-gel process. These materials were tested in simulated body fluid (SBF) to prove their bioactivity in vitro [15] since the organic element, PCL, was preferred due to its biodegradable and biocompatible nature. Furthermore, as implant failure is often a consequence of microbial infections that occur in spite of antimicrobial prophylaxis and aseptic working conditions, the antimicrobial capability of the synthetized hybrids was investigated through a diffusion method against Gram-positive *Escherichia coli* after 24 h of incubation.

2. Materials and Methods

Organic/inorganic hybrid materials MO_2/PCL with M = Si or Ti were prepared, using the sol-gel process, from polycaprolactone of analytical reagent grade as the organic element, and tetramethyl orthosilicate (TMOS) as the source for SiO_2, or titanium butoxide as the source for TiO_2. The amount of the organic element was 6 wt.% in both PCL/SiO_2 and PCL/TiO_2 hybrid materials, as optimized in a previous work [15].

In Figure 1, the flow chart of hybrid synthesis by the applied sol-gel route is depicted.

The presence of hydrogen bonds between organic/inorganic elements of the hybrid materials was unraveled by acquiring their FT-IR spectra in the 400–4000 cm^{-1} region using a Prestige 21 (Shimadzu, Kyoto, Japan) system with a resolution of 4 cm^{-1} (45 scans). The spectrometer used a Deuterated Triglycine Sulfate (DTGS) detector with KBr windows. Disks with a diameter of 13 mm, a thickness of 2 mm, a weight of 200 mg and containing 1 wt.% of sample in KBr were obtained by pressing sample powders into a cylindrical holder using a Specac manual hydraulic press. The Prestige software (IR solution) was used to analyze the FT-IR spectra.

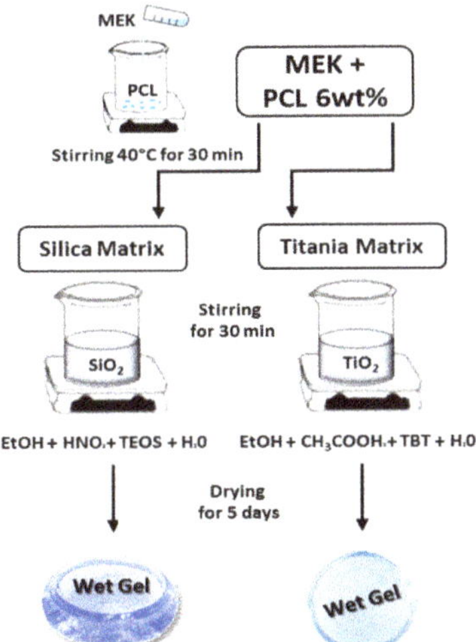

Figure 1. Flow chart of PCL/SiO$_2$ and PCL/TiO$_2$ gel synthesis.

The nature of MO$_2$ gel, PCL and PCL/MO$_2$ hybrid materials was confirmed by X-ray diffraction (XRD) analysis using a Philips diffractometer. Powder samples were scanned from $2\Theta = 5°$ to $60°$ using CuK$_\alpha$ radiation.

The surface morphology and phase distribution of the organic/inorganic hybrid materials were investigated by a scanning electron microscope (Cambridge model S-240) and by an atomic force microscope (Digital Instruments Multimode) in contact mode in air. For SEM, the specimens were previously coated with a thin Au film to obtain a conductive surface.

Bioactivity tests were performed on pieces of the hybrid materials soaked in a simulated body fluid (SBF) with the proper ion concentrations [16,17] nearly equal to those of human blood plasma. During soaking, the temperature was fixed at 37 °C and the ratio between the exposed surface and the volume solution was kept constant (50 mm^2 mL^{-1} of solution), due to its strong influence on the reactivity degree, as previously reported [17–19].

The ability of the hybrid materials to form an apatite layer during the SBF soaking test was ascertained by FT-IR analysis on SiO$_2$/PCL-reacted samples, where SEM and EDS analyses were performed on TiO$_2$/PCL-reacted samples. An SEM outfitted with an energy-dispersive X-ray fluorescence system (EDS, Link AN10000, Oxford Microanalysis) was used to investigate the morphology of the coated sample and to complete it with qualitative elemental analysis.

The antimicrobial activity of synthetized materials was investigated by a diffusion method. *Escherichia coli* (ATCC 25,922) bacteria culture was diluted in distilled water to achieve a 1.0×10^6 CFU/mL suspension. This latter was inoculated in Tryptone Bile Glucuronide Agar (TBX) medium (Liofilchem, Roseto degli Abruzzi (TE), Italy). Sample hybrids and a control disk were placed at the center of the Petri dishes using a pair of sterile tweezers. Incubation was carried out for 24 h in an incubator at 44 °C. The microbial growth was evaluated by observing the diameter of the inhibition halo (ID). Antimicrobial tests were performed in quadruplicate on different days to ensure reproducibility, and values are expressed as the mean values ± standard deviation (SD) of all the performed measurements.

3. Results and Discussion

Gelation is the result of hydrolysis and condensation according to the following reactions, where M = Si and R = CH_3 for PCL/SiO_2; and M = Ti and R = C_4H_9 for PCL/TiO_2.

$$M(OR)_4 + nH_2O \rightarrow M(OR)_{4\text{-}n}(OH)_n + nROH \tag{1}$$

$$\text{-MOH + RO-M-} \rightarrow \text{-M-O-M- + ROH} \tag{2}$$

$$\text{-M-OH + OH-M} \rightarrow \text{-M-O-M- + H}_2\text{O} \tag{3}$$

The reaction mechanism is not known in great detail; however, it is generally accepted that it proceeds through a second-order nucleophilic substitution [19]. The reaction (4) shows the formation of the organic/inorganic network in which a hydrogen bond between the ester C=O group of the organic polymer and the inorganic matrix was supposed to occur. Indeed, the carbonyl group of the polymer chain acts as an H-acceptor, whereas the O-H group of the inorganic matrix is the H-donor.

$$\tag{4}$$

In Figure 2, FT-IR spectra of pure PCL and SiO_2 are reported, together with the one recorded for $SiO_2/PCL_{6\,wt.\%}$ gel in order to highlight spectral differences due to the interaction of the two phases. The PCL spectrum (Figure 2a) shows the asymmetric and symmetric stretching of polymer $-CH_2-$ groups, corresponding to the bands at 2945 and 2866 cm^{-1}, respectively, and two peaks at 1470 cm^{-1} and 1360 cm^{-1} related to their bending modes. The sharp and strong band at 1730 cm^{-1} referred to the C=O stretching vibration. On the other hand, the hybrid's FT-IR spectrum (Figure 2b) showed a broad band in the range 3600–3200 cm^{-1}, the intensity of which almost completely covered the characteristic peaks of PCL methylene stretching. According to previous work, this referred to un-free O-H groups and could be explained considering the establishment of hydrogen bonds between the carbonyl groups of the polymer chains and the inorganic part [11]. The shift of the pure PCL signal at 1730 cm^{-1} to lower wavenumbers (1720 cm^{-1}) seemed to corroborate this hypothesis. Moreover, both in $SiO_2/PCL_{6\,wt.\%}$ and SiO_2 gel spectra (Figure 2b,c), the bands at 1080 and 460 cm^{-1}, corresponding to the stretching and bending modes of SiO_2 tetrahedra [20], were marked, as expected. This evidence indicates that the presence of the PLC in the hybrid composites does not inhibit the reticulation of the Si-O-Si 3D network.

In Figure 3, the infrared spectrum of PCL, as discussed above, was compared to those recorded for the TiO2 + $PCL_{6\,wt.\%}$ gel and the TiO2 gel. The signals detectable for the latter were the typical bands already described elsewhere [3]. In the hybrid's spectrum (Figure 3b), the broad and strong band in the region 3400–3200 cm^{-1}, due to –OH vibrations, obscured almost completely the polycaprolactone CH2 asymmetric and symmetric stretching modes (at 2945 and 2866 cm^{-1}, respectively). In addition, the band at 1715 cm^{-1}, attributable to the PCL ester bond, which shifted to low wavenumbers following the synthetic route, was not particularly evident. The scarce intensity could depend on the low amount of the polymer in the synthetized material. Furthermore, other signals typical of acetic-acid-containing titania sol-gel materials were detected. Indeed, the doublet at 1530 and 1445 cm^{-1} likely corresponded to the asymmetric and symmetric stretching vibrations of the carboxylic group coordinated to Ti as a bidentate ligand. The band around 800 cm^{-1} correlates with the vibrations of polyhedral TiOn with a coordination number less than 6.

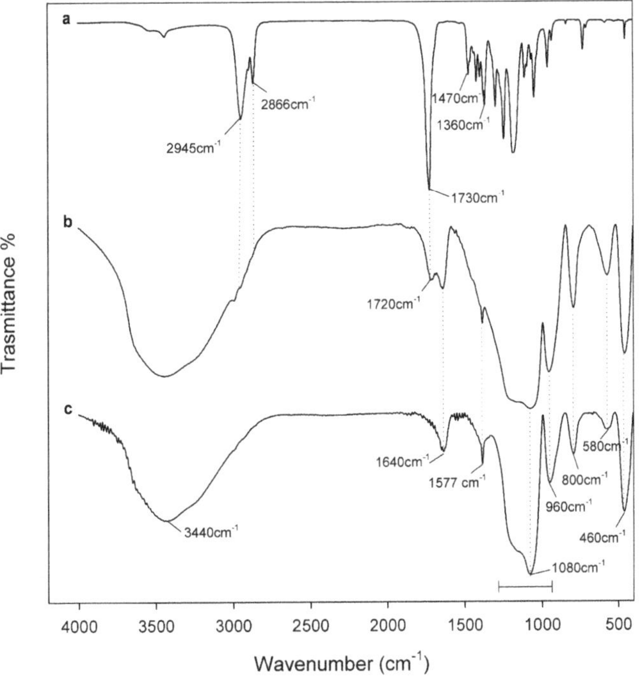

Figure 2. FT-IR spectra of (**a**) PCL, (**b**) SiO_2 + $PCL_{6\,wt.\%}$ gel and (**c**) SiO_2 gel.

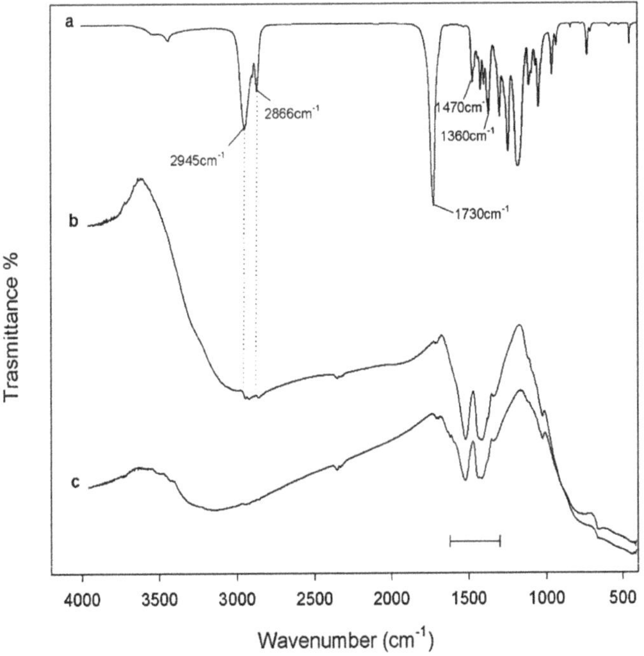

Figure 3. FT-IR spectra of (**a**) PCL, (**b**) TiO_2 + $PCL_{6\,wt.\%}$ gel and (**c**) TiO_2 gel.

The nature and microstructure of the SiO$_2$/PCL and TiO$_2$/PCL hybrid materials were investigated by X-ray diffraction (XRD) and scanning electron microscopy (SEM). In particular, based on the acquired diffractograms, hybrids exhibit amorphous features as their metal oxide precursors. In fact, taking into account Figure 4, it is observable that sharp peaks, typical of a crystalline material, can be detected on the diffractogram of pure polycaprolactone (Figure 4a), whereas the TiO$_2$ gel exhibits broad humps characteristic of amorphous materials (Figure 4b). On the other hand, the XRD spectrum of the TiO$_2$/PCL hybrid material gel was in line with an amorphous material, such as TiO$_2$ gel (Figure 4c).

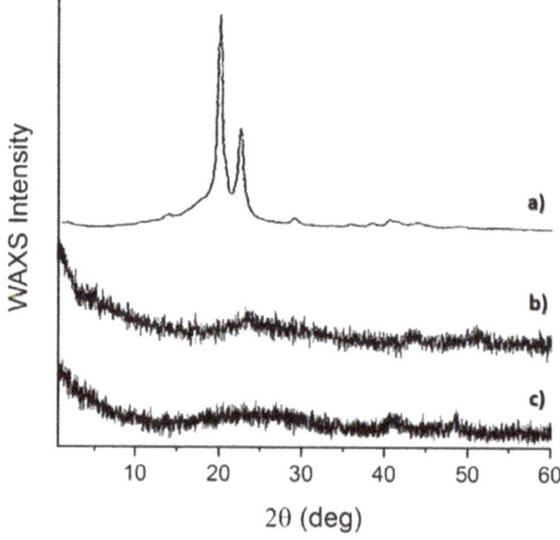

Figure 4. XRD spectra of (**a**) PCL, (**b**) TiO$_2$ gel and (**c**) TiO$_2$/PCL gel.

SEM micrographs of TiO$_2$ gel and TiO$_2$/PCL gel samples, reported in Figure 5, evidence a more homogenous morphology for the first one, as expected. With the introduction of the organic phase (Figure 5b), the unstructured TiO$_2$ gel presents a micrometric feature. Neither of the two morphologies, though, exhibit a regular geometric morphology indicating crystalline structures or phase separation. This result confirms the XRD evidence, i.e., the crystalline nature of the organic element was lost during the synthesis. A similar behavior was detected on XRD patterns and SEM micrographs of a SiO$_2$ gel sample and a SiO$_2$/PCL gel sample.

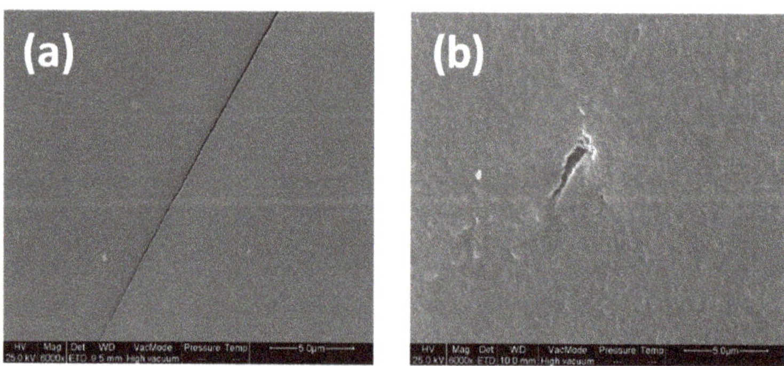

Figure 5. Scanning electron microscopy (SEM) micrograph of (**a**) TiO$_2$ gel and (**b**) TiO$_2$/PCL$_{6\,wt.\%}$ gel.

The degree of mixing of the elements in the hybrid material was investigated with Atomic Force microscopy (AFM). An AFM contact mode image can be measured in the height mode or in the force mode. In force images (z range in nN), differences appear sharper and richer and the contours of the nanostructure's essentials are clearer. In contrast, height images (z range in nm) provide a more exact reproduction of the height itself [21]. In this study, the height mode was adopted to estimate the degree of homogeneity of the hybrid materials. AFM topographic images of SiO_2/PCL and TiO_2/PCL gel samples are shown in Figure 6, where it can be observed that the average domain size is less than 130 and 40 nm. This answer confirms that the synthesized SiO_2/PCL and TiO_2/PCL gels can intrinsically be organic/inorganic hybrid materials [22].

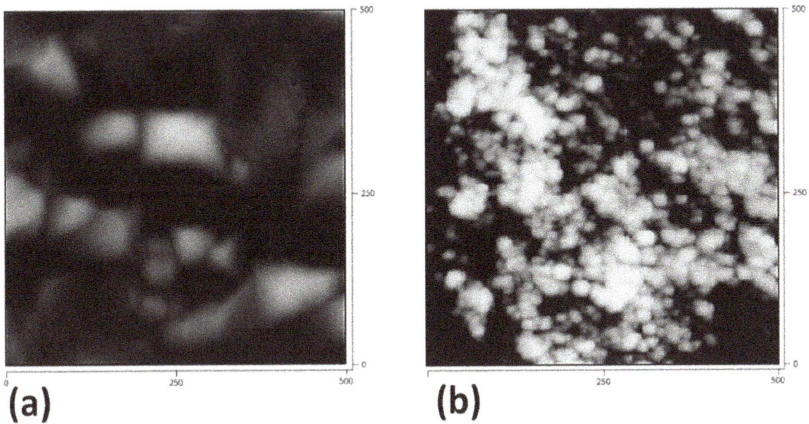

Figure 6. AFM images showing the microstructure of (**a**) SiO_2/PCL$_{6\ wt.\%}$ gel and (**b**) TiO_2/PCL$_{6\ wt.\%}$ gel. The numbers on the axis are in μm (micrometers).

The bioactivity of the SiO_2/PCL gel was ascertained by FT-IR measurements on a sample soaked in simulated body fluid for 7, 14 and 21 days as shown in Figure 7. The formation of a hydroxyapatite deposit was recognized by the appearance of the 587 and 550 cm^{-1} bands generally assigned to P-O stretching [23]. After soaking for 7 days, the splitting of the 580 cm^{-1} band into two others at 587 and at 550 cm^{-1} can be ascribed to the shape of crystalline hydroxyapatite [23]. These spectra modifications could be ascribable to the formation of the hydroxyapatite precipitate and, in particular, to the stretching of the hydroxyapatite -OH groups and the vibrations of PO_4^{3-} groups, respectively [24]. Moreover, a slight upshift of the Si-OH band (from 955 cm^{-1} to 960 cm^{-1}) suggested the interaction of the hydroxyapatite layer with the -OH groups of the silica matrix. Finally, the band at 800 cm^{-1} can be assigned to the Si-O-Si band vibration among two adjacent tetrahedra characteristic of silica gel [24]. The results obtained are in line with the mechanism of the formation of a hydroxy apatite deposit proposed in the literature [25].

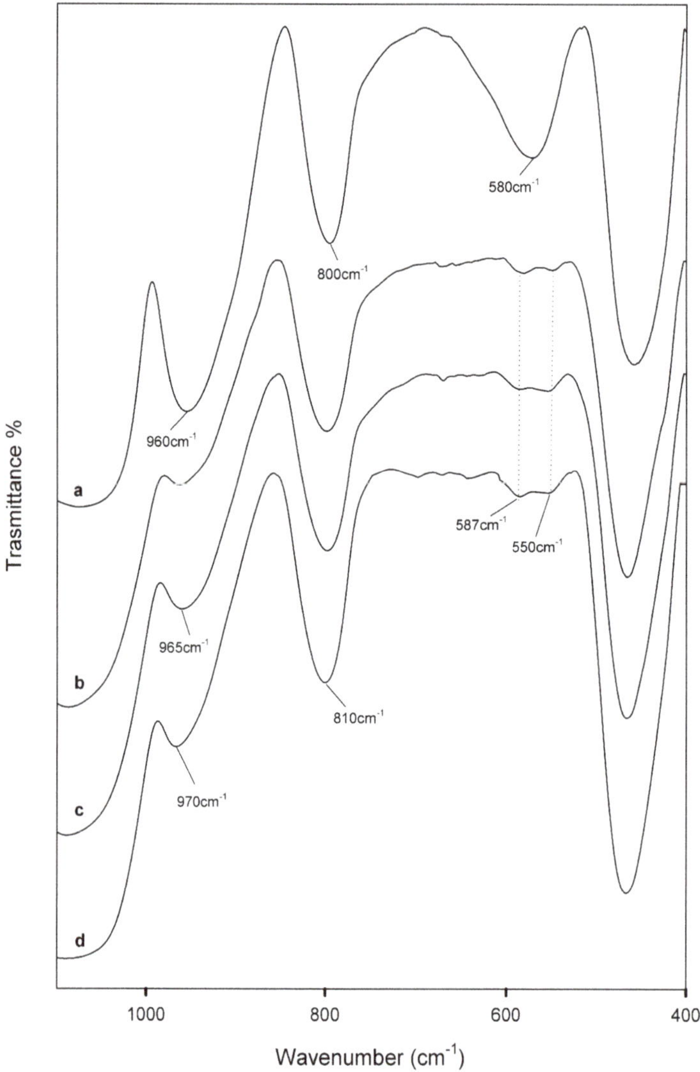

Figure 7. FT-IR spectra of SiO$_2$/PCL gel samples after exposure to simulated body fluid (SBF) for different times. (**a**) not exposed; (**b**) 7 days; (**c**) 14 days; (**d**) 21 days.

Figure 8a shows the SEM micrographs of a TiO$_2$/PCL gel sample soaked in SBF for 21 days. Figure 8b confirms that the surface layer observed in the SEM micrographs is composed of calcium phosphate.

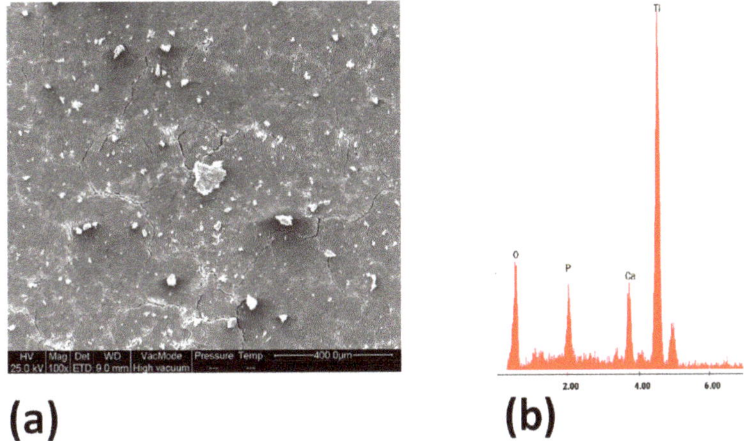

Figure 8. (**a**) SEM micrographs and (**b**) EDS of a TiO_2/PCL gel sample soaked in SBF for 21 days.

Finally, in order to propose the use of the synthetized materials in the biomedical field, as microbial infections can compromise the effectiveness and success of implants, preliminary data from a diffusion assay highlighted that $TiO_2/PCL_{6\ wt.\%}$ materials exerted a strong inhibitory effect against *Escherichia coli* growth (Figure 9). *E. coli*, a Gram-negative bacterium, being able to easily adapt to changing environmental conditions, is one of the primary causes of Gram-negative orthopedic implant infections. Its resistance to a great variety of antibiotics does not allow for its successful eradication even after treatment. Thus, antimicrobial materials could be advantageous in the biomedical field. Indeed, the significant antimicrobial effect of $TiO_2/PCL_{6\ wt.\%}$ could be substantially due to the TiO_2 element. In fact, previous findings state that TiO_2 nanoparticles found on pathogenic strains of *E. coli* were able to cause little pores to form in bacterial cell walls, leading to increased permeability and cell death [26].

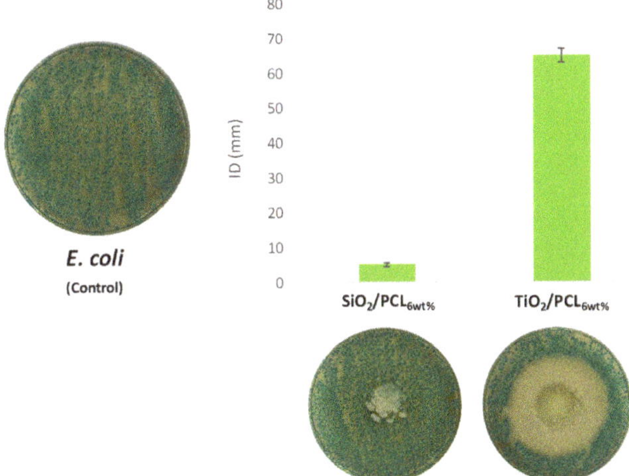

Figure 9. Representative image of *Escherichia coli* incubated with synthetized hybrid materials and antimicrobial activity reported as inhibitory zone diameter (mm).

4. Conclusions

MO$_2$/PCL materials, based on polycaprolactone and silica or titanium metal oxide (M = Si or Ti), prepared via the sol-gel process, were found to be organic/inorganic bioactive hybrid materials not altering the gelification and condensation route of the inorganic gel. The polymer (PCL) was crosslinked to the network by hydrogen bonds, as ascertained by FTIR spectra, between the ester groups of the organic polymer and the hydroxyl groups of inorganic matrices. Moreover, the AFM and SEM analyses confirmed that MO$_2$/PCL can be considered to be homogenous organic/inorganic hybrid materials because the average domains are less than 400 nm in size. Finally, the bioactivity of the PCL/MO$_2$ materials was ascertained by the formation of a layer of hydroxyapatite on the surface when samples were soaked in SBF, as shown in SEM micrographs and related EDS and detected in FTIR spectra. Although further investigations are required for evaluating the mechanical properties of the synthesized hybrid materials, these findings provide initial evidence of the strong antimicrobial effect of the TiO$_2$/PCL hybrid and its potential for favorable use.

Author Contributions: Conceptualization, M.C.; methodology, S.P. and M.C.; software, C.L.; validation, M.C., C.L. and formal analysis, S.P. and M.C.; investigation, M.C.; resources, C.L.; data curation, M.C.; writing—original draft preparation, M.C. and C.L.; writing—review and editing, M.C.; visualization, C.L.; supervision, M.C. All authors have read and agreed to the published version of the manuscript.

Funding: This research received no external funding.

Acknowledgments: The authors thank the Ecoricerche srl company for making the microbiology laboratories and materials used for experiments available.

Conflicts of Interest: The authors declare no conflicts of interest.

References

1. Joshua, D.Y.; Damron, M.; Tang, G.; Zheng, H.; Chu, C.-J.; Osborne, J.H. Inorganic/organic hybrid coatings for aircraft aluminum alloy substrates. *Prog. Org. Coat.* **2001**, *41*, 226–232. [CrossRef]

2. Zvonkina, I.J.; Soucek, M.D. Inorganic–organic hybrid coatings: Common and new approaches. *Curr. Opin. Chem. Eng.* **2016**, *11*, 123–127. [CrossRef]

3. Figueira, R.B.; Silva, C.J.R.; Petra, E.V. Organic–inorganic hybrid sol–gel coatings for metal corrosion protection: A review of recent progress. *J. Coat. Technol. Res.* **2015**, *12*, 1–35. [CrossRef]

4. Klukowska, A.; Posset, U.; Schottner, G.; Wis, M.L.; Salemi-Delvaux, C.; Malatesta, V. Photochromic hybrid sol-gel coatings: Preparation, properties, and applications. *Mater. Sci.* **2002**, *20*, 95–104.

5. Samuneva, B.; Djambaski, P.; Kashchieva, E.; Chernev, G.; Kabaivanova, L.; Emanuilova, E.; Salvado, I.M.M.; Fernandes, M.H.V.; Wu, A. Sol-gel synthesis and structure of silica hybrid biomaterials. *J. Non Cryst. Solids* **2008**, *354*, 733–740. [CrossRef]

6. Sanchez, C.; Ribot, F. Design of hybrid organic-inorganic materials synthesized via sol-gel chemistry. *New J. Chem.* **1994**, *18*, 1007–1047.

7. Catauro, M.; Bollino, F.; Papale, F.; Marciano, S.; Pacifico, S. TiO$_2$/PCL hybrid materials synthesized via sol–gel technique for biomedical applications. *Mater. Sci. Eng. C* **2015**, *47*, 135–141. [CrossRef] [PubMed]

8. Wei, Y.; Jin, D.; Brennan, D.J.; Rivera, D.N.; Zhuang, Q.; DiNardo, N.J.; Qiu, K. Atomic force microscopy study of organicâ´inorganic hybrid materials. *Chem. Mater.* **1998**, *10*, 769–772. [CrossRef]

9. Catauro, M.; Tranquillo, E.; Barrino, F.; Blanco, I.; Dal Poggetto, F.; Naviglio, D. Drug release of hybrid materials containing Fe (II) citrate synthesized by sol-gel technique. *Materials* **2018**, *11*, 2270. [CrossRef] [PubMed]

10. Brinker, C.J.; Scherer, G.W. *Sol-gel Science: The Physics and Chemistry of Sol-Gel Processing*; Elsevier, Inc.: Amsterdam, The Netherlands, 2013; ISBN 9780080571034.

11. Catauro, M.; Bollino, F.; Mozzati, M.C.; Ferrara, C.; Mustarelli, P. Structure and magnetic properties of SiO$_2$/PCL novel sol–gel organic–inorganic hybrid materials. *J. Solid State Chem.* **2013**, *203*, 92–99. [CrossRef]

12. Tranquillo, E.; Barrino, F.; Dal Poggetto, G.; Blanco, I. Sol–Gel Synthesis of Silica-Based Materials with Different Percentages of PEG or PCL and High Chlorogenic Acid Content. *Materials* **2019**, *12*, 155. [CrossRef] [PubMed]

13. David, I.A.; Scherer, G.W. An organic/inorganic single-phase composite. *Chem. Mater.* **1995**, *7*, 1957–1967. [CrossRef]

14. Kokubo, T.; Takadama, H. How useful is SBF in predicting in vivo bone bioactivity? *Biomaterials* **2006**, *27*, 2907–2915. [CrossRef] [PubMed]

15. Catauro, M.; Barrino, F.; Dal Poggetto, G.; Pacifico, F.; Piccolella, S.; Pacifico, S. Chlorogenic acid/PEG-based organic-inorganic hybrids: A versatile sol-gel synthesis route for new bioactive. *Mater. Sci. Eng. C* **2019**, *100*, 837–844. [CrossRef] [PubMed]

16. Catauro, M.; Tranquillo, E.; Dal Poggetto, G.; Pasquali, M.; Dell'Era, A.; Vecchio Ciprioti, S. Influence of the heat treatment on the particles size and on the crystalline phase of TiO_2 synthesized by the sol-gel method. *Materials* **2018**, *11*, 2364. [CrossRef]

17. Kokubo, T.; Matsushita, T.; Takadama, H.; Kizuki, T. Development of bioactive materials based on surface chemistry. *J. Eur. Ceram. Soc.* **2009**, *29*, 1267–1274. [CrossRef]

18. Yang, B.; Uchida, M.; Kim, H.M.; Zhang, X.; Kokubo, T. Preparation of bioactive titanium metal via anodic oxidation treatment. *Biomaterials* **2004**, *25*, 1003–1010. [CrossRef]

19. Catauro, M.; Barrino, F.; Dal Poggetto, G.; Crescente, G.; Piccolella, S.; Pacifico, S. Chlorogenic Acid Entrapped in Hybrid Materials with High PEG Content: A Strategy to Obtain Antioxidant Functionalized Biomaterials? *Materials* **2019**, *12*, 148. [CrossRef]

20. Blanco, I.; Abate, L.; Bottino, F.A.; Bottino, P. Synthesis, characterization and thermal stability of new dumbbell-shaped isobutyl-substituted POSSs linked by aromatic bridges. *J. Therm. Anal. Calorim.* **2014**, *117*, 243–250. [CrossRef]

21. Rädlein, E.; Frischat, G.H. Atomic force microscopy as a tool to correlate nanostructure to properties of glasses. *J. Non Cryst. Solids* **1997**, *222*, 69–82. [CrossRef]

22. Nguyena, K.; Garcia, A.; Sanic, M.-A.; Diaza, D.; Dubeyd, V.; Claytona, D.; Dal Poggetto, G.; Corneliuse, F.; Paynea, R.J.; Separovic, F.; et al. Interaction of N-terminal peptide analogues of the Na+,K+-ATPase with membranes. *BBA-Biomembr.* **2018**, *1860*, 1282–1291. [CrossRef] [PubMed]

23. Wang, D.S.; Pantano, C.G. Surface chemistry of multielement silicate gels. *J. Non Cryst. Solids* **1992**, *147*, 115–122. [CrossRef]

24. Kokubo, T.; Ito, S.; Huang, Z.T.; Hayashi, T.; Sakka, S.; Kitsugi, T.; Yamamuro, T. Ca, P-rich layer formed on high-strength bioactive glass-ceramic A-W. *J. Biomed. Mater. Res.* **1990**, *24*, 331–343. [CrossRef] [PubMed]

25. Kokubo, T.; Kushitani, H.; Sakka, S.; Kitsugi, T.; Yamamuro, T. Solutions able to reproduce in vivo surface-structure changes in bioactive glass-ceramic A-W3. *J. Biomed. Mater. Res.* **1990**, *24*, 721–734. [CrossRef] [PubMed]

26. Mantravadi, H.B. Effectivity of titanium oxide based nano particles on *E. coli* from clinical samples. *J. Clin. Diagn. Res.* **2017**, *11*, DC37–DC40. [CrossRef] [PubMed]

 *applied sciences*

Article

Thermo-Mechanical Properties of PLA/Short Flax Fiber Biocomposites

Laura Aliotta *, Vito Gigante, Maria-Beatrice Coltelli, Patrizia Cinelli, Andrea Lazzeri and Maurizia Seggiani *

Department of Civil and Industrial Engineering, University of Pisa, 56122 Pisa, Italy
* Correspondence: laura.aliotta@dici.unipi.it (L.A.); maurizia.seggiani@unipi.it (M.S.);
  Tel.: +39-05-0221-7854 (L.A.); +39-05-0221-7881 (M.S.)

Received: 7 August 2019; Accepted: 6 September 2019; Published: 11 September 2019

Abstract: In this work, biocomposites based on poly(lactic acid) (PLA) and short flax fibers (10–40 wt.%) were produced by extrusion and characterized in terms of thermal, mechanical, morphological, and thermo-mechanical properties. Analytical models were adopted to predict the tensile properties (stress at break and elastic modulus) of the composites, and to assess the matrix/fiber interface adhesion. The resulting composites were easily processable by extrusion and injection molding up to 40 wt.% of flax fibers. It was observed that despite any superficial treatment of fibers, the matrix/fiber adhesion was found to be sufficiently strong to ensure an efficient load transfer between the two components obtaining composites with good mechanical properties. The best mechanical performance, in terms of break stress (66 MPa), was obtained with 20 wt.% of flax fibers. The flax fiber acted also as nucleating agent for PLA, leading to an increment of the composite stiffness and, at 40 wt.% of flax fibers, improving the elastic modulus decay near the PLA glass transition temperature.

Keywords: poly(lactic acid); flax fibers; biocomposites; predictive analytical model; mechanical properties

1. Introduction

The development of biocomposites, produced using a biopolymer matrix and a reinforcement from renewable resources, is currently an extensive research area due to the promising mechanical properties, recyclability after service [1], and biodegradability [2] of these materials. These composites have potential applications in different fields such as automotive, packaging, and household goods [3]. Clearly, the application of natural fibers is being driven not only by environmental reasons, but also by economical ones. Natural fiber-reinforced composites, in fact, can be used as low-cost materials having at the same time different structural properties [4].

The use of natural fibers has many advantages. Natural fibers are renewable, biodegradable, and less abrasive to tooling. Furthermore, they can be formed into light composites leading to weight reductions and, especially in automotive field, to fuel saving [5]. Among natural fibers, lignocellulosic fibers are investigated as promising substitutes of the synthetic fibers in polymeric composites and several studies have examined drawbacks and advantages of the most significant lignocellulosic fibers and their related polymeric composites [6–9].

One of the strongest natural fibers is the flax fiber. Their structure is very complex and can be compared to a composite structure. In fact, the flax fibers are constituted by a series of polyhedron shape elementary fibers overlapped over a considerable length; they are held together by an interphase that mainly consists of hemicellulose and pectin. Each elementary fiber consists of a very thin primary cell wall, a secondary cell wall (dominating the cross section), and an open channel at the fiber center called "lumen" [10–12]. Typical dimeter for an elementary flax fibers are around 10–15 μm; on the other hand, technical flax fibers have a diameter that varies between 35–150 μm [11].

Different biopolymers such as poly(lactic acid) (PLA), polyhydroxyalkanoates (PHAs), and cellulose esters have been used as matrices coupled with natural fibers for various applications [13–15]. Among these biopolymers, PLA is currently one of the most used and promising biopolymers because of its lower production cost compared to that of other biopolymers such as PHAs, poly-butylene succinate (PBS), poly-butylene succinate-*co*-adipate (PBSA). PLA exhibits, at room temperature, very good mechanical properties: Young modulus of 3 GPa, tensile strength between 50 and 70 MPa, and impact resistance of 2.5–3 kJ/m^2 [16,17]. However, PLA is very stiff and brittle with poor thermal properties and a very slow crystallization rate; often it is necessary to use a plasticizer or introduce rubber particles to improve its break elongation and impact properties [18–20], while nucleating agents are added to increase its crystallinity and enhance its thermal stability [21].

Some of PLA drawbacks may be overcome by fiber-reinforcement and blending. Mechanical properties of natural fiber-reinforced PLA composites have been investigated by different researchers and different types of natural fibers were used in place of synthetic fibers [22]. However, the addition of natural fibers to polymeric matrices requires special attention in order to obtain biocomposites with good mechanical properties. In fact, compared to the synthetic composites, the composites containing natural fibers generally show lower mechanical properties, higher moisture absorption, and lower durability [23]. Many researchers have been working to address these issues, with particular attention to the surface treatments of the natural fibers and the improvement of the fiber/matrix adhesion. In fact, the fiber/matrix adhesion plays an important role in the final mechanical properties of the composites since a good adhesion guarantees an efficient transfer of the stress between the matrix and the fiber, obtaining the desired reinforcement. Natural fibers, containing a large amount of cellulose, hemicellulose, lignin, and pectin, tend to be active polar hydrophilic material while polymer matrices are not polar and exhibit hydrophobic properties. So there is typically a weak interfacial bonding between the highly polar natural fiber and non-polar organophilic matrix and this leads to a loss of the mechanical properties of the resultant composites [24]. Different approaches have been investigated to improve the interfacial compatibility and bond strength as the use of surface modification techniques [4,25]. The strength of the interfacial bond depends on different parameters as surface energy, chemistry, and fiber roughness that can be modified by different treatments [26,27]. In particular, chemical treatments with sodium hydroxide, peroxides, organic, and inorganic acids, silane, anhydrides, and acrylic monomers have been proposed to remove non-cellulosic components (in cellulosic fibers) and to add functional groups that provides better chemical bonding between the treated fibers and the polymeric matrix [28]. A weak interface adhesion decreases the composite tensile strength but increases the impact resistance, due to the fiber pull-out mechanism. On the other hand, a strong interface adhesion generally leads to a stiff composite having high tensile strength but brittle impact behavior (fiber pull-out is prevented) [29].

The fabrication methods of composites based on PLA and natural fiber such as flax fibers are the same as those used for the thermoplastics in which short fibers are added and traditional molding methods such as extrusion, compression, and injection molding are used [22]. The adopted processing conditions (screw speed, profile temperature, pressure, residence time, etc.) influence the final mechanical properties of PLA/fiber composites. The fibers damaging due to the long permanence at high temperature or to an excessive screw speed must to be considered. In particular, it was observed that compression molding technique does not damage the natural fiber during processing. However, this technique is time consuming and unsuitable for mass production. On the contrary, for high productions, the injection molding is used, showing easy processability that allows the prevention of fiber degradation during the process and generally it improves the fiber dispersion in the matrix [23]. In terms of fiber length, compression and hot press molding can be used for both long and short fibers, while injection molding can process only short fibers. Oksman et al. [30] manufactured PLA/flax fiber composites with a twin-screw extruder up to 40 wt.% of long fibers, followed by compression molding. Bax and Müssig [31] have produced PLA/flax fiber composites up to 30 wt.% fibers, using hot-pressing followed by pelletizing and injection molding.

In the present study, biocomposites based on PLA containing un-treated short flax fibers, up to 40 wt.% of fibers, were produced by melt extrusion followed by injection molding. The developed composites were characterized in terms of tensile properties, thermal properties by differential scanning calorimetry (DSC), and thermo-mechanical properties by dynamic mechanical thermal analysis (DMTA). Scanning electron microscopy (SEM) analysis was carried out to evaluate the fiber/matrix interface and the dispersion/structure of the fibers in the PLA matrix. Analytical models were used to predict the tensile properties of the developed composites and to assess the interfacial matrix/fiber adhesion.

2. Materials and Methods

2.1. Materials

PLA used in this study was PLA 2003D from NatureWorks LLC (Minnetonka, MN, USA), grade for thermoforming and extrusion processes with melt flow index (MFI) of 6 g/10 min (210 °C, 2.16 kg), having a nominal average molar mass of 200,000 g/mol, density of 1.24 g/cm^3, content of D-lactic acid units of 3–6%).

Chopped flax fibers with average length of 1 mm and aspect ratio of 8 were used as natural fibers. To evaluate the fiber volume fraction in the composites a density of 1.5 g/cm^3 [12] was considered.

2.2. Composite Preparation

All materials were dried in a circulating air oven at 60 °C for at least 24 h before the composite preparation. Formulations with 10, 20, and 40 wt % flax fibres, with respect to the total weight, were produced by melt blending using a Thermo Scientific™ HAAKE™ MiniLab II micro-compounder, a co-rotating conical twin-screw extruder with a screw rate of 110 rpm/min, a cycle time of 60 s at 180 °C; at this temperature PLA is at the molten state and the flax fibers are thermally stable. Sample names with the relative compositions in weight and volume are reported in Table 1.

Table 1. Composites and relative compositions.

Sample	PLA/Fiber (wt./wt.%/%)	PLA/Fiber (vol./vol.%/%)
PLA	100/0	100/0
PF10	90/10	92/8
PF20	80/20	83/17
PF40	60/40	64/36

After extrusion, the molten material was transferred, through a preheated cylinder, to a Thermo Scientific Haake MiniJet II mini injection press (mold temperature set at 70 °C) for the production of Haake type III dog-bone tensile specimens (size: 25 × 5 × 1.5 mm). The injection molding was carried out at 180 °C, at pressure of 680 bar and with a residence time in the mold of 15 s for all the tested formulations. In fact, as observed by Bos et al. [10], the thermal degradation of flax fibers starts at around 200 °C and it is not significant in the first minutes. Before thermal and mechanical tests, the specimens were stored in a climate chamber at room temperature and controlled relative humidity of 50%.

2.3. Tensile Tests

Tensile tests were carried out at room temperature and at a crosshead speed of 10 mm/min by an Instron 5500 R universal testing machine (Canton, MA, USA) equipped with a 10 kN load cell and interfaced with a computer running MERLIN software (version 4.42S/N-014733H). The samples were tested after 24 h from injection molding process. At least five specimens for each composite were tested and the average values were reported.

2.4. Morphological Analysis

The flax fibers and the cryogenic facturated cross-sections of the composites were analyzed by SEM (JEOL JSM-5600LV, Tokyo, Japan). In addition, SEM analysis was carried out also on the fractured sections of the dog-bone specimens obtained after doing tensile tests in order to evaluate the fiber-matrix adhesion. Prior to the morphological analysis, all the samples were coated with a gold layer.

2.5. Calorimetric Analysis

The composite sample used for the DSC analysis was obtained by cutting the corresponding tensile specimen and about 10–15 mg of sample was placed in sealed aluminum pan. The eventual crystallization of PLA that can occur during the injection molding process was evaluated using the first heating run from 25 to 200 °C at 10 °C/min. On the other hand, the second heating was also performed to eliminate the thermal history of the sample. After the first heating, the material was kept at 200 °C for 5 min to ensure the complete melting of the crystals and, then, it was cooled at 10 °C/min from 200 to 40 °C. Finally, the second heating was performed at 10 °C/min from 40 to 200 °C.

The cold crystallization temperature (T_{cc}) and the melting temperature (T_m) were evaluated in correspondence of the minimum cold crystallization peak and the maximum melting peak, respectively. Whereas, the enthalpies of cold crystallization (ΔH_{cc}) and melting (ΔH_m) were calculated as the areas of the corresponding peaks.

The degree of crystallinity (X_{cc}) of pure *PLA* and composites was determined using the following equation:

$$X_{cc} = \frac{\Delta H_m - \Delta H_{cc}}{\frac{\Delta H_m^\circ}{wt.\ \%\ of\ PLA}} \tag{1}$$

where ΔH_m° is the theoretical melting heat of 100% crystalline *PLA*. Because of the materials were processed in the range of the disorder-to-order phase transition [32] a double peak, due to the existence of α' and α crystalline phases, is present. As a consequence, considering the ΔH_m° values of α' and α phases [33], a mean value between the ΔH_m° of α' (107 J/g) and ΔH_m° of α (143 J/g) was adopted (125 J/g) according to other studies [17].

2.6. Thermo-Mechanical Analysis

Dynamic mechanical thermal analysis (DMTA) was carried out using a Gabo Eplexor® DMTA (Gabo Qualimeter, Ahiden, Germany) with a 100N load cell. At least two specimens were tested for each composite. Test bars (size: 20 × 5 × 1.5 mm), obtained by cutting the tensile specimens, were mounted on the machine in tensile configuration. The used temperature range varied from 35 to 120 °C with heating rate of 1.5 °C/min and at a constant frequency of 1 Hz.

3. Predictive Models of Mechanical Properties

In this work, different analytical models were adopted to better understand the results of the mechanical tests and to predict the mechanical properties of the composites as function of the fiber content.

3.1. Break Strength Prediction

The break strength is strongly affected by the adhesion between the fibers and the matrix [34]. Different analytical models, based on modified Kelly–Tyson equation, were proposed to predict the interfacial shear stress (IFSS) of composite taking into account the properties of the fibers and the matrix [35–37]. However, to apply them, it is necessary to know the critical length and the length distribution of the fibers into the composite. Since the procedure to have an IFSS estimation is laborious, one of the most used simple models to predict the yield/break strength and the fiber/matrix interaction is the Pukánszky's model [38,39] based on the Equation (2):

$$ln\,\sigma_{c,red} = ln\frac{\sigma_c\left(1 + 2.5V_f\right)}{\sigma_m\left(1 - V_f\right)} = BV_f \qquad (2)$$

where $\sigma_{c,red}$ is the predicted reduced stress (adimensional) of the composite that takes into account the reduced bearing section of the composites due to the presence of the filler, σ_c and σ_m are the stress at yield/break of the composite and matrix, respectively, V_f is the filler volume fraction, and B is the Pukánszky's interaction parameter. Plotting the natural logarithm of the reduced stress $\sigma_{c,red}$ as a function of V_f, the B parameter can be calculated from the slope of the linear correlation obtained. Consequently, applying the dependence of the strength on composition, an estimate of the fiber-matrix interaction can be obtained.

Although the B parameter has no a direct physical meaning, it is obviously connected with the interfacial properties of the system and it also depends on the matrix mechanical properties. The B parameter is related to the relative load-bearing capacity of the fiber, which depends by the interfacial interactions. The relationship between the B parameter and the interfacial properties is reported in the following equation (Equation (3)) [34]:

$$B = \left(1 + A_f\rho_f l\right)ln\frac{\sigma_i}{\sigma_m} \qquad (3)$$

where A_f, ρ_f, l and σ_i are the specific filler surface area, the filler density, the thickness of the interface and the strength of the interface, respectively.

3.2. Elastic Modulus Prediction

The composites analyzed are constituted by short flax fibers randomly oriented in the PLA matrix. The analytical models for short fibers were derived from the elasticity theory, starting from the models developed for long fiber composites. To better fit the experimental data and to take into account the differences between long and short fiber composites, a consistent numbers of geometrical, topological, and mechanical parameters [6,34] must to be taken into account. Different efforts to fit analytical models with experimental data can be found [40].

However, due to the low aspect ratio, it can be expected that the mechanical behavior and the elastic response of the produced composites is similar to the behavior of particulate filled composites. Consequently, some analytical models suitable for particulate filled composites [41–43] were applied to the PLA/short flax fibers composites. The analytical models adopted in this work to predict the Young's modulus of the composites are reported in Table 2.

Table 2. Analytical models used for the prediction of the composite Young's modulus.

Model	$E_{composite}$
Einstein [43]	$E_c = E_m\left(1 + 2.5V_f\right)$
Van Dyck [42]	$E_c = E_m\left(1 + \frac{1.25 \cdot V_f}{1 - 1.2 \cdot V_f}\right)$
Cox [40]	$E_c = \left(1 - V_f\right)E_m + V_fE_f\cdot\left(1 - \frac{\tanh(n\cdot a_r)}{(n\cdot a_r)}\right)$
Halpin-Tsai [44]	$E_c = \frac{3}{8}E_l + \frac{5}{8}E_t$

In the models of Table 2, E_c, E_m, and E_f represent the elastic modulus of the composite, matrix and fibers, respectively, V_f is the fiber volume fraction, a_r is the aspect ratio of the fibers. The adimensional parameter n of the Cox model is defined as (Equation (4)) [45]:

$$n = \frac{2E_m}{E_f(1+v)ln\left(\frac{P}{V_f}\right)} \tag{4}$$

where v is the Poisson ratio of the matrix (≈ 0.4) and P is the packing factor of the fibers, considered equal to $2\pi/\sqrt{3}$.

Concerning the Halpin–Tsai model [44], the two terms E_l and E_t are, respectively, the longitudinal and the tangential modulus; these values can be calculated by the following equations:

$$E_l = E_m \cdot \frac{1 + 2 \cdot a_r \cdot \left(\frac{\frac{E_f}{E_m}-1}{\frac{E_f}{E_m}+2a_r}\right) \cdot V_f}{1 - \left(\frac{\frac{E_f}{E_m}-1}{\frac{E_f}{E_m}+2a_r}\right) \cdot V_f} \tag{5}$$

$$E_t = E_m \cdot \frac{1 + 2 \cdot a_r \cdot \left(\frac{\frac{E_f}{E_m}-1}{\frac{E_f}{E_m}+2}\right) \cdot V_f}{1 - \left(\frac{\frac{E_f}{E_m}-1}{\frac{E_f}{E_m}+2}\right) \cdot V_f} \tag{6}$$

4. Results and Discussion

4.1. Processability

The possibility to use conventional melt processes is a very important for industrial use of these bio-composites. In this case PLA/flax fiber composites did not show any difficulties during the extrusion followed by injection molding. Under the same operating conditions adopted for the extrusion and injection molding, the mold filling was always complete, obtaining well-formed specimens also with 40 wt.% of fibers. These results are in accordance with the findings of Oksman et al. [30] who did not experience any difficulties in the extrusion and compression molding processes of PLA/long flax fiber composites up to 40 wt.% fibers.

4.2. Mechanical Properties and Modeling

Figure 1 shows the typical stress-strain curves of the PLA/flax fiber composites. As shown, PLA and all the composites show a fragile break without yielding. The main tensile properties are summarized in Figure 2.

As expected, the stiffness increased with increasing amount of flax fibers. This behavior is very common to other similar bio-composite systems encountered in literature [6]. The stiffness increase was reflected in a decrease in elongation at break. The tensile strength slightly increased with the fiber content up to 20 wt.% of fibers and, then, at higher fiber loading it resulted similar to that of pure PLA. This trend can be attributed to the well-known reinforcement effect of the flax fibers [30] up to 20 wt.%, while at 40 wt.% poor wetting of the fibers with the matrix may be present leading to reduced stress transfer across the fiber–matrix interface with consequent reduction of the tensile strength.

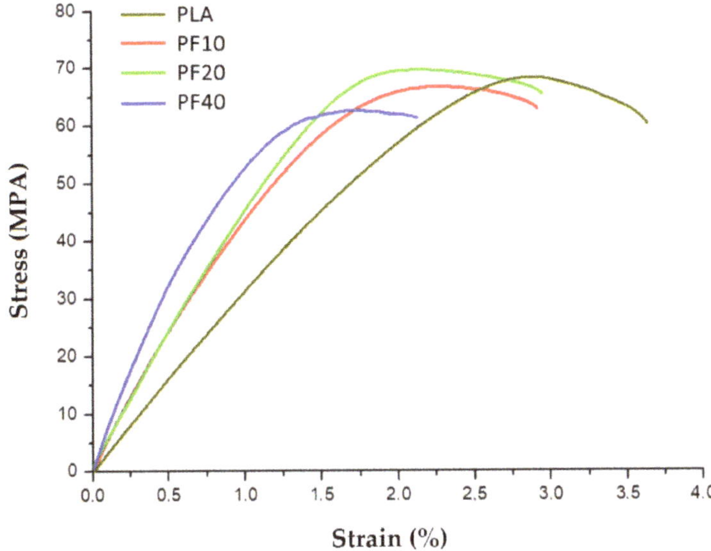

Figure 1. Stress-strain representative curves for pure poly(lactic acid) (PLA) and for PLA/flax fiber composites.

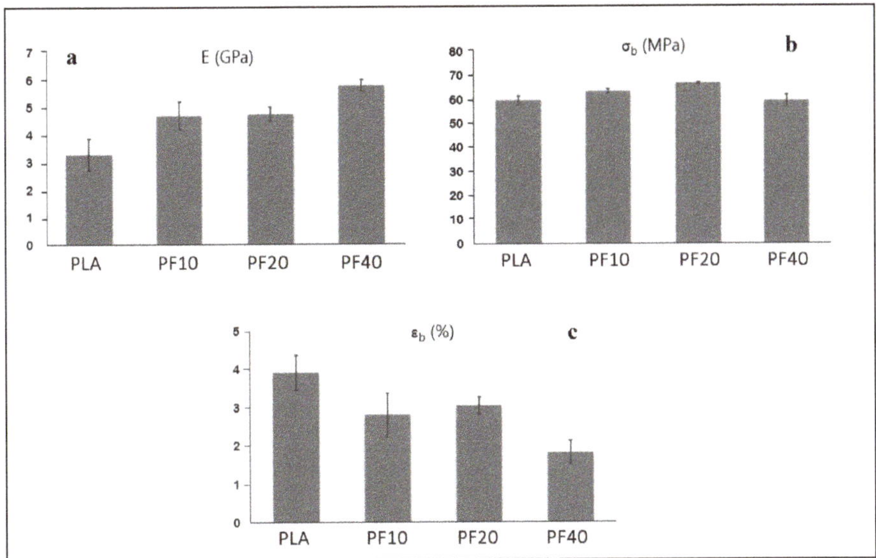

Figure 2. Mechanical properties for pure PLA and PLA/flax composites: (**a**) Elastic modulus; (**b**) Stress at break; (**c**) Elongation at break.

The values of the tensile strength are in agreement with those reported by Oksman et al. [30] and Bax and Müssig [31] that obtained tensile strengths of 53–54 MPa with 30 wt.% of fibers without fiber treatments and additives. In addition, the value for the Young's moduli of the PF20 is close to that found by Bax and Müssig [31] with 20 wt.% fibers (5.1 GPa).

Using the mechanical properties of the matrix (neat PLA) and the composites, the Pukánszky's plot can be obtained (Figure 3) and the B parameter can be evaluated.

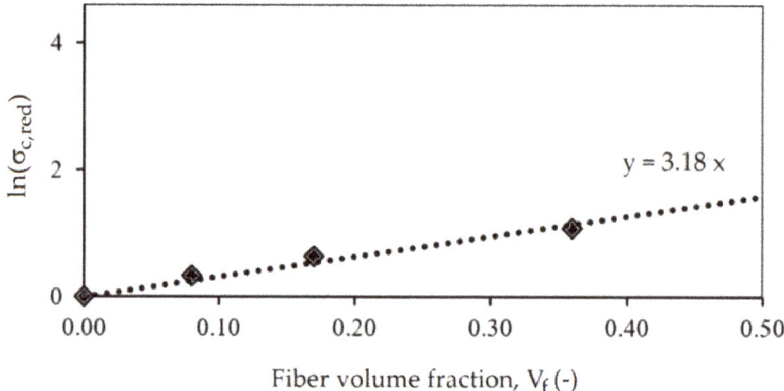

Figure 3. Pukánszky's plot for PLA/flax fiber composites.

The interaction B parameter value of about 3.2 indicates a good adhesion between the fibers and the matrix. In fact, in the case of weak interfacial bonding, the fibers do not carry the load and, consequently, B is almost zero [41]. This result is coherent with the general increase of the tensile strength of the composites indicating a good load transfer between the fibers and the matrix.

Interesting results were achieved using the analytical models for predicting the elastic modulus of the composites. In fact, as shown in Figure 4, the Cox and Halpin–Tsai models, typically used for aligned long fiber composites, did not fit the experimental data, especially at high fiber loading. On the other hand, the models for particulate filled systems (Einstein and Van Dyck models) provided a good fitting. This confirms that when the fibers are very short the mechanical behavior of their composites can be assimilated, as expected, to that of the particulate filled composites.

Furthermore, it can be confirmed the good adhesion between the fibers and the matrix since the Einstein model presupposes a perfect adhesion between the filler and the matrix and a good filler dispersion [46].

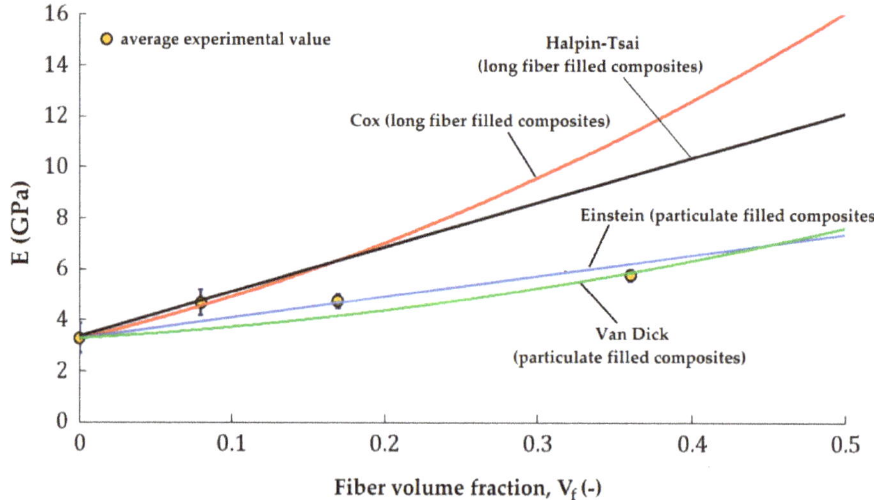

Figure 4. Comparison between experimental elastic modulus of the PLA/flax fiber composites and some predictive analytical models.

4.3. Morphological Analysis

SEM images of the chopped flax fibers and, as an example, the cross-sections of the composite with 10 wt.% flax fibers are reported in Figures 5 and 6, respectively. Further, Figure 7 shows the broken sections of the dog-bone specimens at different fiber loadings after doing tensile test.

Figure 5. Chopped flax fibers used in this work.

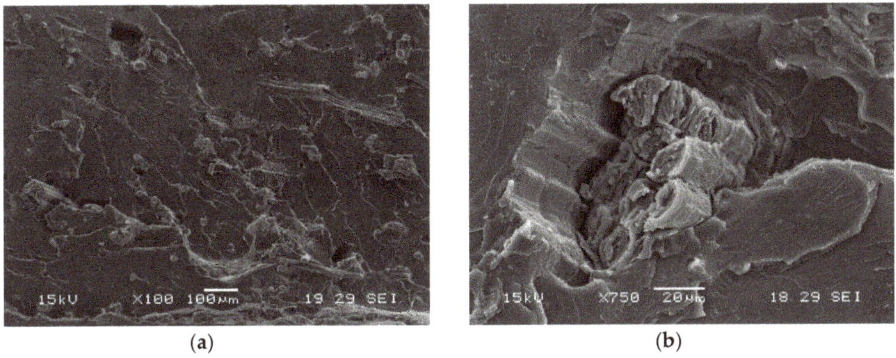

(a) (b)

Figure 6. SEM images of PLA/flax fibers composites with 10 wt.% flax fibers at (a) ×100 and (b) ×750 magnification.

It can be noticed that, due to their very low aspect ratio, no twisting was observed; however, a slight damage, due to the extrusion process, can be observed as many fibers became thinner and weaker.

As shown, the short flax fibers show a straight shape and appear like sticks. Figure 6b shows the typical polyhedron shape of flax fibers in which the elementary fibers are held pack together; the interphase is constituted by hemicellulose and pectin [5]. The adhesion between the fibers and the matrix is good; the fibers appear well adherent to the matrix and the very small space between the matrix and the fibers can be attributable to the different contraction/expansion coefficient of the fiber and matrix during the cryogenic fracture (Figure 6b). This supports the results of mechanical tests and analytical modelling. This good adhesion is confirmed also by the Figure 7 where there are many fibers that remain adhered to the matrix under the traction action and a moderate pull-out is observed. In addition, observing Figure 7, the short fibers are randomly oriented within the matrix and maintain their original straight shape without incurring evident deformation during the extrusion process.

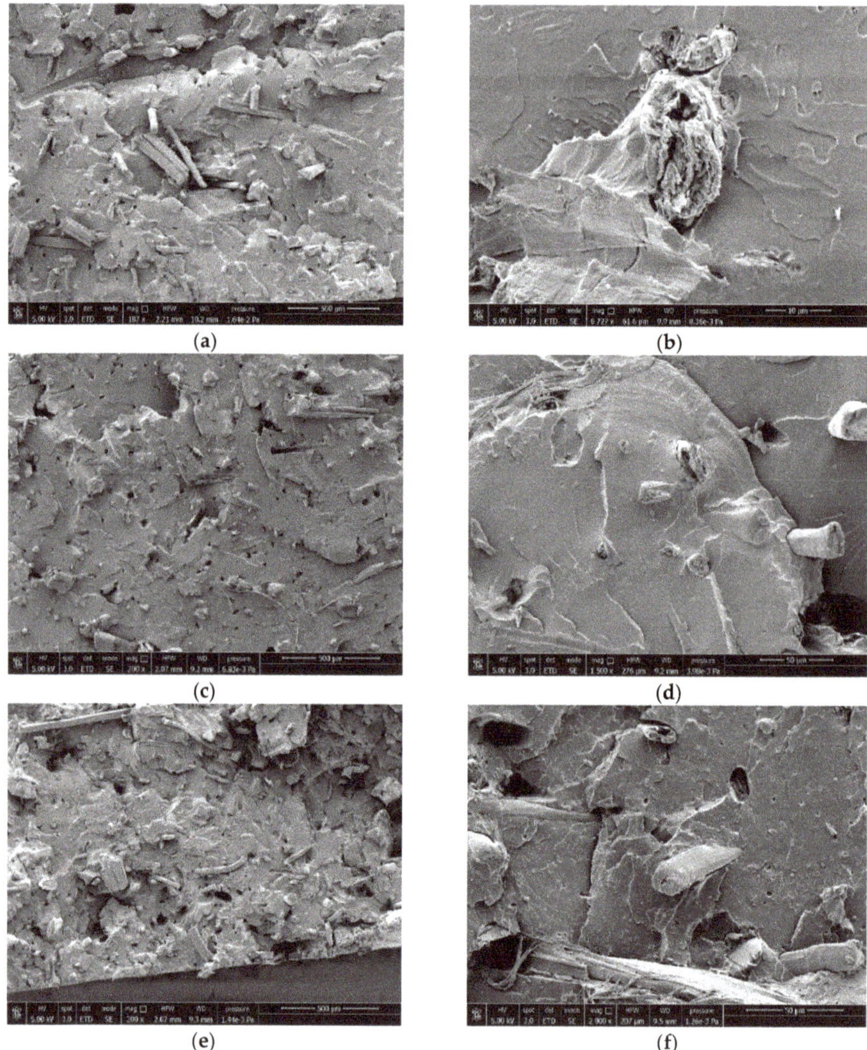

Figure 7. SEM images, at different magnifications, of PLA/flax fibers composites: (**a,b**) 10 wt.% of flax fibers, (**c,d**) 20 wt.% of flax fibers and (**e,f**) 40 wt. % of flax fibers.

Few works on the adhesion between PLA and flax fibers in biocomposites have been published. Most publications on the flax fiber composites concern the interaction between flax fibers and polymers including polypropylene (PP), polystyrene (PS), epoxy, bio-epoxy, and bio-phenolic resin, epoxy resins [47,48]. It is known that natural fibers have a poor adhesion to hydrophobic matrices like PP, because of their hydrophilic nature; consequently, to improve the interaction between these hydrophobic matrices and the flax fibers several chemical treatments were proposed such as mercerization, silane treatment, acylation, peroxide treatment and coatings [47,49,50]. Whereas, because of the hydrophilic nature of PLA a better interaction between PLA and flax fibers than between PP or epoxy resins and flax fibers can be expected as assumed by Heinemann and Fritz [51]. Nevertheless, Oksman et al. [30] and Bax and Müssig [31] observed a poor adhesion between long un-treated flax fibers and the PLA matrix and they concluded that adhesion needs to be improved to optimize the mechanical properties of the PLA/flax composites. In this case, very short flax fibers were used and the good adhesion, evidenced by SEM analyses and the analytical modelling, between the short un-treated flax fibers and the PLA matrix can be attributable to the short length of the fibers (low aspect ratio) which facilitated the dispersion of the fibers within the matrix and reduced the fiber agglomeration favoring the wetting of all the fibers by the matrix.

4.4. Calorimetric Analysis

Figure 8 shows the DSC thermograms of first and second heating run for neat PLA and its composites. The resultant thermal properties (T_g, T_{cc}, T_m, ΔH_m, ΔH_{cc}) obtained from the first and second heating run are summarized in Tables 3 and 4, respectively. The crystallinity percentage (X_{cc}) of PLA and its composites is also reported. Furthermore, only the temperature of the second peak was reported as PLA melting temperature.

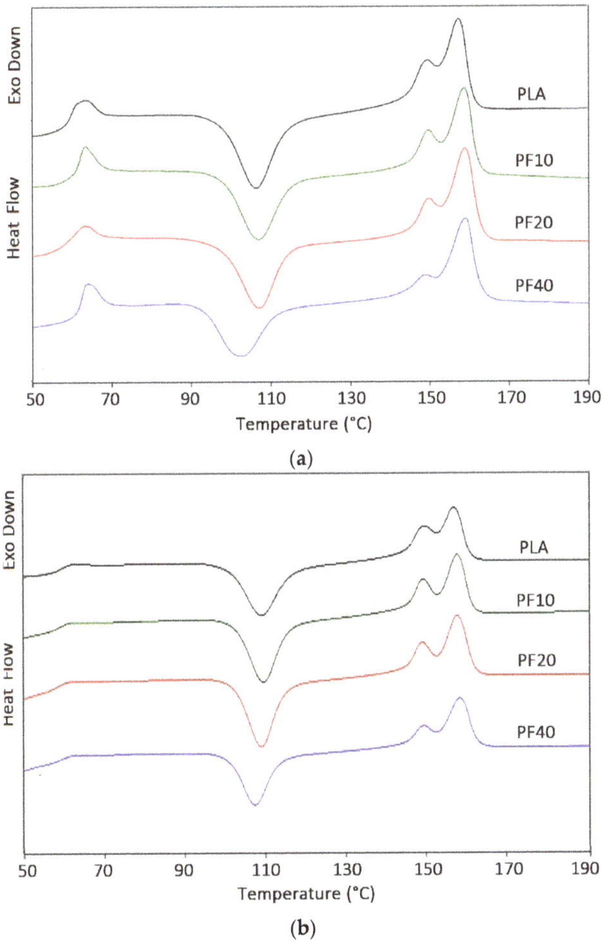

Figure 8. Differential scanning calorimetry (DSC) thermograms of neat PLA and its composites with flax fibers: (**a**) first and (**b**) second heating run.

Table 3. Thermal properties and crystallinity fraction (X_{cc}) of PLA and its composites (1st heating run).

PLA/Flax (wt./wt.%)	T_g (°C)	T_{cc} (°C)	T_m (°C)	ΔH_m (J/g)	ΔH_{cc} (J/g)	X_{cc} (%)
100/0	60.1	106.2	156.8	32.9	25.7	5.4
90/10	61.1	106.6	158.2	31.1	22.7	6.9
80/20	61.0	106.7	158.4	31.8	22.9	8.3
60/40	61.3	102.5	158.6	28.9	20.8	10.1

Table 4. Thermal properties and crystallinity fraction (X_{cc}) of PLA and its composites (2nd heating run).

PLA/Flax (wt./wt.%)	T_g (°C)	T_{cc} (°C)	T_m (°C)	ΔH_m (J/g)	ΔH_{cc} (J/g)	X_{cc} (%)
100/0	60.1	108.8	156.4	32.0	28.6	2.7
90/10	60.9	109.3	157.2	33.0	25.8	6.4
80/20	60.5	108.8	157.2	34.7	26.6	8.1
60/40	60.2	107.3	157.8	30.2	21.5	11.6

As shown, in the first heating, the glass transition of amorphous PLA, which typically occurs at 60 °C, is overlapped with an enthalpy recovery peak, due to permanence of the samples at room temperature before the tests. This effect disappeared in the second heating having deleted eliminated the sample thermal history.

The crystallinity slightly increased with the flax fiber content and at 40 wt.% fibers the crystallinity of PLA increased more than three times compared to pure PLA. A further trend that can be observed is the shifting of the cold crystallization peak to lower temperatures with the flax fiber amount. Consequently, the stiffness increment of the composites (Figure 2) with the increasing amount of fibers was due not only to the fiber stiffness but also to the PLA crystallinity increment.

The results of the DMTA, showed in Figure 9, confirm the nucleating effect of the flax fibers for the crystallization of PLA [52,53]. Both neat PLA and the composites start to soften after 60 °C (where the glass transition occurs); however, the decay of the elastic modulus is less marked for the composite with 40 wt.% of flax fibers. This result is in agreement with the DSC results in which the major crystallinity content is registered for this composite. More crystalline the polymer is, less marked is its elastic modulus decay near the T_g.

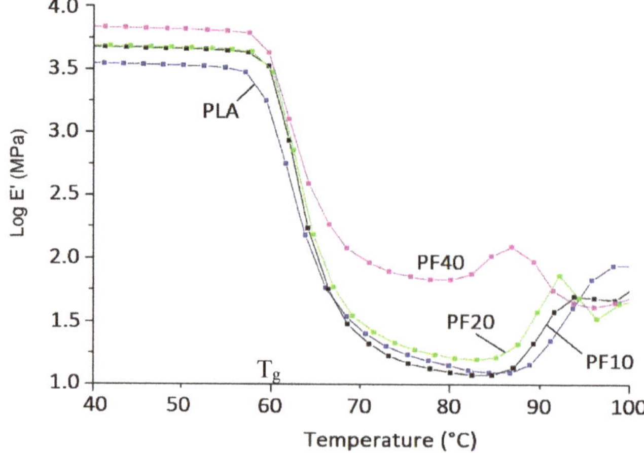

Figure 9. Dynamic modulus for PLA and PLA/flax fiber composites.

5. Conclusions

In this study PLA/short flax fiber composites with different content of fibers (10–40 wt.%) were produced by extrusion and characterized in terms of thermal, morphological, mechanical, and thermo-mechanical properties.

The flax fibers were not subjected to any superficial treatment in order to evaluate their intrinsic adhesion with the PLA matrix.

The processing of the composites by extrusion and injection molding, up to 40 wt.% of flax fiber content, did not create any difficulty. Consequently, they can be produced and processed using conventional melt processes and this is very important from the application point of view.

Although interfacial adhesion between natural fibers and polymeric matrices is generally weak, untreated flax fibers showed a good adhesion to the PLA matrix as confirmed by SEM images and also by the calculated Pukánszky's interaction B parameter (about 3.2) and the good fitting of the Einstein model to the experimental data for the prediction of Young modulus.

The well-known problem of PLA related to the low glass transition temperature was improved with the addition of significant amounts of flax fibers. In fact, short flax fibers acted both as reinforcement and nucleating agent for PLA, as confirmed by the mechanical and thermo-mechanical tests. The maximum value of stress at break (66 MPa) was obtained at 20 wt.% fibers and at 40 wt.% fibers an improvement in the elastic modulus decay near the glass transition temperature (60 °C) was observed.

Author Contributions: A.L. and M.S. supervised the study results and discussion; L.A. performed the experimental tests and modeling of the mechanical properties; L.A., V.G., and M.S. wrote the original draft; M.-B.C. and P.C. co-supervised the experimental activities.

Funding: This research received no external funding.

Acknowledgments: The authors acknowledge Luigi Torre of Perugia University that kindly provided the flax fibers.

Conflicts of Interest: The authors declare no conflict of interest.

References

1. Le Duigou, A.; Pillin, I.; Bourmaud, A.; Davies, P.; Baley, C. Effect of recycling on mechanical behaviour of biocompostable flax/poly(l-lactide) composites. *Compos. Part A Appl. Sci. Manuf.* **2008**, *39*, 1471–1478. [CrossRef]
2. Assarar, M.; Scida, D.; El Mahi, A.; Poilâne, C.; Ayad, R. Influence of water ageing on mechanical properties and damage events of two reinforced composite materials: Flax–fibres and glass–fibres. *Mater. Des.* **2011**, *32*, 788–795. [CrossRef]
3. Ibrahim, N.A.; Yunus, W.M.Z.W.; Othman, M.; Abdan, K.; Hadithon, K.A. Poly(Lactic Acid) (PLA)-reinforced kenaf bast fiber composites: The effect of triacetin. *J. Reinf. Plast. Compos.* **2010**, *29*, 1099–1111. [CrossRef]
4. Gurunathan, T.; Mohanty, S.; Nayak, S.K. A review of the recent developments in biocomposites based on natural fibres and their application perspectives. *Compos. Part A Appl. Sci. Manuf.* **2015**, *77*, 1–25. [CrossRef]
5. Bos, H.L.; Müssig, J.; Oever, M.J.V.D. Mechanical properties of short-flax-fibre reinforced compounds. *Compos. Part A Appl. Sci. Manuf.* **2006**, *37*, 1591–1604. [CrossRef]
6. Aliotta, L.; Gigante, V.; Coltelli, M.B.; Cinelli, P.; Lazzeri, A. Evaluation of Mechanical and Interfacial Properties of Bio-Composites Based on Poly(Lactic Acid) with Natural Cellulose Fibers. *Int. J. Mol. Sci.* **2019**, *20*, 960. [CrossRef] [PubMed]
7. Righetti, M.C.; Cinelli, P.; Mallegni, N.; Massa, C.A.; Aliotta, L.; Lazzeri, A. Properties of Poly (lactic acid) based Biocomposites with Potato Pulp Powder Treated with Waxes. *Materials* **2019**, *12*, 990. [CrossRef] [PubMed]
8. Bledzki, A. Composites reinforced with cellulose based fibres. *Prog. Polym. Sci.* **1999**, *24*, 221–274. [CrossRef]
9. Mohanty, A.K.; Misra, M.; Hinrichsen, G. Biofibres, biodegradable polymers and biocomposites: An overview. *Macromol. Mater. Eng.* **2000**, *276*, 1–24. [CrossRef]
10. Bos, H.L.; Oever, M.J.A.V.D.; Peters, O.C.J.J. Tensile and compressive properties of flax fibres for natural fibre reinforced composites. *J. Mater. Sci.* **2002**, *37*, 1683–1692. [CrossRef]

11. Ahmed, S.; Ulven, C.A. Dynamic In-Situ Observation on the Failure Mechanism of Flax Fiber through Scanning Electron Microscopy. *Fibers* **2018**, *6*, 17. [CrossRef]

12. Thuault, A.; Domengès, B.; Hervás, I.; Gomina, M. Investigation of the internal structure of flax fibre cell walls by transmission electron microscopy. *Cellulose* **2015**, *22*, 3521–3530. [CrossRef]

13. Cinelli, P.; Mallegni, N.; Gigante, V.; Montanari, A.; Seggiani, M.; Coltelli, B.; Bronco, S.; Lazzeri, A. Biocomposites Based on Polyhydroxyalkanoates and Natural Fibres from Renewable Byproducts. *Appl. Food Biotechnol.* **2019**, *6*, 35–43.

14. Seggiani, M.; Cinelli, P.; Balestri, E.; Mallegni, N.; Stefanelli, E.; Rossi, A.; Lardicci, C.; Lazzeri, A. Novel Sustainable Composites Based on Poly(hydroxybutyrate-co-hydroxyvalerate) and Seagrass Beach-CAST Fibers: Performance and Degradability in Marine Environments. *Materials* **2018**, *11*, 772. [CrossRef] [PubMed]

15. Seggiani, M.; Cinelli, P.; Mallegni, N.; Balestri, E.; Puccini, M.; Vitolo, S.; Lardicci, C.; Lazzeri, A. New Bio-Composites Based on Polyhydroxyalkanoates and Posidonia oceanica Fibres for Applications in a Marine Environment. *Materials* **2017**, *10*, 326. [CrossRef]

16. Raquez, J.-M.; Habibi, Y.; Murariu, M.; Dubois, P. Polylactide (PLA)-based nanocomposites. *Prog. Polym. Sci.* **2013**, *38*, 1504–1542. [CrossRef]

17. Aliotta, L.; Cinelli, P.; Coltelli, M.B.; Righetti, M.C.; Gazzano, M.; Lazzeri, A. Effect of nucleating agents on crystallinity and properties of poly (lactic acid) (PLA). *Eur. Polym. J.* **2017**, *93*, 822–832. [CrossRef]

18. Gigante, V.; Canesi, I.; Cinelli, P.; Coltelli, M.B.; Lazzeri, A. Rubber Toughening of Polylactic Acid (PLA) with Poly(butylene adipate-co-terephthalate) (PBAT): Mechanical Properties, Fracture Mechanics and Analysis of Ductile-to-Brittle Behavior while Varying Temperature and Test Speed. *Eur. Polym. J.* **2019**, *115*, 125–137. [CrossRef]

19. Aliotta, L.; Cinelli, P.; Beatrice Coltelli, M.; Lazzeri, A. Rigid filler toughening in PLA-Calcium Carbonate composites: Effect of particle surface treatment and matrix plasticization. *Eur. Polym. J.* **2018**. [CrossRef]

20. Argon, A.; Cohen, R. Toughenability of polymers. *Polymer* **2003**, *44*, 6013–6032. [CrossRef]

21. Byun, Y.; Rodríguez, K.; Han, J.H.; Kim, Y.T. Improved thermal stability of polylactic acid (PLA) composite film via PLA–β-cyclodextrin-inclusion complex systems. *Int. J. Boil. Macromol.* **2015**, *81*, 591–598. [CrossRef] [PubMed]

22. Thakur, V.K.; Thakur, M.K.; Gupta, R.K. Review: Raw Natural Fiber–Based Polymer Composites. *Int. J. Polym. Anal. Charact.* **2014**, *19*, 256–271. [CrossRef]

23. Dittenber, D.B.; GangaRao, H.V. Critical review of recent publications on use of natural composites in infrastructure. *Compos. Part A Appl. Sci. Manuf.* **2012**, *43*, 1419–1429. [CrossRef]

24. Jawaid, M.; Khalil, H.A. Cellulosic/synthetic fibre reinforced polymer hybrid composites: A review. *Carbohydr. Polym.* **2011**, *86*, 1–18. [CrossRef]

25. Biagiotti, J.; Puglia, D.; Torre, L.; Arbelaiz, A.; Cantero, G.; Marieta, C.; Llano-Ponte, R. A systematic investigation on the influence of the chemical treatment of natural fibers on the properties of their polymer matrix composites. *Polym. Compos.* **2004**, *25*, 470–479. [CrossRef]

26. Baley, C.; Busnel, F.; Grohens, Y.; Sire, O. Influence of chemical treatments on surface properties and adhesion of flax fibre–polyester resin. *Compos. Part A Appl. Sci. Manuf.* **2006**, *37*, 1626–1637. [CrossRef]

27. Arbelaiz, A.; Cantero, G.; Fernandez, B.; Mondragon, I.; Gañán, P.; Kenny, J.M. Flax fiber surface modifications: Effects on fiber physico mechanical and flax/polypropylene interface properties. *Polym. Compos.* **2005**, *26*, 324–332. [CrossRef]

28. Sawpan, M.A.; Pickering, K.L.; Fernyhough, A. Effect of fibre treatments on interfacial shear strength of hemp fibre reinforced polylactide and unsaturated polyester composites. *Compos. Part A Appl. Sci. Manuf.* **2011**, *42*, 1189–1196. [CrossRef]

29. Graupner, N.; Rößler, J.; Ziegmann, G.; Müssig, J. Fibre/matrix adhesion of cellulose fibres in PLA, PP and MAPP: A critical review of pull-out test, microbond test and single fibre fragmentation test results. *Compos. Part A Appl. Sci. Manuf.* **2014**, *63*, 133–148. [CrossRef]

30. Oksman, K.; Skrifvars, M.O.V.; Selin, J.-F. Natural fibres as reinforcement in polylactic acid (PLA) composites. *Compos. Sci. Technol.* **2003**, *63*, 1317–1324. [CrossRef]

31. Bax, B.; Müssig, J. Impact and tensile properties of PLA/Cordenka and PLA/flax composites. *Compos. Sci. Technol.* **2008**, *68*, 1601–1607. [CrossRef]

32. Zhang, J.; Duan, Y.; Sato, H.; Tsuji, H.; Noda, I.; Yan, S.; Ozaki, Y. Crystal Modifications and Thermal Behavior of Poly (L -lactic acid) Revealed by Infrared Spectroscopy. *Macromolecules* **2005**, *38*, 8012–8021. [CrossRef]
33. Righetti, M.C.; Gazzano, M.; Di Lorenzo, M.L.; Androsch, R. Enthalpy of melting of α′- and α-crystals of poly(l-lactic acid). *Eur. Polym. J.* **2015**, *70*, 215–220. [CrossRef]
34. Renner, K.; Kenyó, C.; Móczó, J.; Pukánszky, B. Micromechanical deformation processes in PP/wood composites: Particle characteristics, adhesion, mechanisms. *Compos. Part A Appl. Sci. Manuf.* **2010**, *41*, 1653–1661. [CrossRef]
35. Kelly, A.; Tyson, W. Tensile properties of fibre-reinforced metals: Copper/tungsten and copper/molybdenum. *J. Mech. Phys. Solids* **1965**, *13*, 329–350. [CrossRef]
36. Piggott, M.R. Short Fibre Polymer Composites: A Fracture-Based Theory of Fibre Reinforcement. *J. Compos. Mater.* **1994**, *28*, 588–606. [CrossRef]
37. Bader, M.; Bowyer, W. An improved method of production for high strength fibre-reinforced thermoplastics. *Composites* **1973**, *4*, 150–156. [CrossRef]
38. Pukánszky, B. Influence of interface interaction on the ultimate tensile properties of polymer composites. *Composites* **1990**, *21*, 255–262. [CrossRef]
39. Turcsányi, B.; Pukánszky, B.; Tüdös, F. Composition dependence of tensile yield stress in filled polymers. *J. Mater. Sci. Lett.* **1988**, *7*, 160–162. [CrossRef]
40. Kim, H.G.; Kwac, L.K. Evaluation of elastic modulus for unidirectionally aligned short fiber composites. *J. Mech. Sci. Technol.* **2009**, *23*, 54–63. [CrossRef]
41. Fu, S.-Y.; Feng, X.-Q.; Lauke, B.; Mai, Y.-W. Effects of particle size, particle/matrix interface adhesion and particle loading on mechanical properties of particulate–polymer composites. *Compos. Part B Eng.* **2008**, *39*, 933–961. [CrossRef]
42. Bourkas, G.; Prassianakis, I.; Kytopoulos, V.; Sideridis, E.; Younis, C. Estimation of Elastic Moduli of Particulate Composites by New Models and Comparison with Moduli Measured by Tension, Dynamic, and Ultrasonic Tests. *Adv. Mater. Sci. Eng.* **2010**, *2010*, 1–13. [CrossRef]
43. Einstein, A.; Fürth, R. *Investigations on the Theory of Brownian Movement*; Dover Publications: New York, NY, USA, 1956.
44. Affdl, J.C.H.; Kardos, J.L. The Halpin-Tsai equations: A review. *Polym. Eng. Sci.* **1976**, *16*, 344–352. [CrossRef]
45. Cox, H.L. The elasticity and strength of paper and other fibrous materials. *Br. J. Appl. Phys.* **1952**, *3*, 72–79. [CrossRef]
46. Fu, S.-Y. Effects of fiber length and fiber orientation distributions on the tensile strength of short-fiber-reinforced polymers. *Compos. Sci. Technol.* **1996**, *56*, 1179–1190. [CrossRef]
47. Zhu, J.; Zhu, H.; Njuguna, J.; Abhyankar, H. Recent Development of Flax Fibres and Their Reinforced Composites Based on Different Polymeric Matrices. *Materials* **2013**, *6*, 5171–5198. [CrossRef] [PubMed]
48. Yan, L.; Chouw, N.; Jayaraman, K. Flax fibre and its composites—A review. *Compos. Part B Eng.* **2014**, *56*, 296–317. [CrossRef]
49. Martin, N.; Mouret, N.; Davies, P.; Baley, C. Influence of the degree of retting of flax fibers on the tensile properties of single fibers and short fiber/polypropylene composites. *Ind. Crop. Prod.* **2013**, *49*, 755–767. [CrossRef]
50. Van De Weyenberg, I.; Truong, T.C.; Vangrimde, B.; Verpoest, I. Improving the properties of UD flax fibre reinforced composites by applying an alkaline fibre treatment. *Compos. Part A Appl. Sci. Manuf.* **2006**, *37*, 1368–1376. [CrossRef]
51. Heinemann, M.; Fritz, H.G. Polylactid-Struktur, Eigenschaften and Anwendungen. In *Stuttgarter Kunststoff-Kolloquium, 19*; Universität Stuttgart: Stuttgart, Germany, 2005.
52. Kumar, R.; Nair, K.; Thomas, S.; Schit, S.; Ramamurthy, K. Morphology and melt rheological behaviour of short-sisal-fibre-reinforced SBR composites. *Compos. Sci. Technol.* **2000**, *60*, 1737–1751. [CrossRef]
53. Yu, L.; Dean, K.; Li, L. Polymer blends and composites from renewable resources. *Prog. Polym. Sci.* **2006**, *31*, 576–602. [CrossRef]

Communication

Shape Memory Effect in Micro-Sized Shape Memory Polymer Composite Chains

Xin Lan [1,*], Weimin Huang [2] and Jinsong Leng [1,*]

[1]   Center for Composite Materials and Structures, Harbin Institute of Technology, Harbin 150001, China
[2]   School of Mechanical and Aerospace Engineering, Nanyang Technological University, Singapore 639798, Singapore
*   Correspondence: lanxin@hit.edu.cn (X.L.); lengjs@hit.edu.cn (J.L.); Fax: +86-451-86402328 (X.L. & J.L.)

Received: 6 June 2019; Accepted: 8 July 2019; Published: 22 July 2019

Featured Application: The potential applications of SMP/Ni vertical chain arrays may contain micro electro mechanical systems (MEMS) and biomedical devices.

Abstract: Since the shape memory effect (SME) has been confirmed in micron and submicron sized polyurethane (PU) shape memory polymer (SMP), it might be used in novel micro/nano devices even for surgery/operation inside a single cell. In this study, micron sized protrusive PU SMP composite chains are fabricated via mixing ferromagnetic nickel micro powders with PU SMP/dimethylformamide solution and then cured under a low magnetic field. Depending on the amount of nickel content, vertical protrusive chains with a diameter from 10 to 250 μm and height from 200 to 1500 μm are obtained. The SME in these chains is investigated to confirm the SME in SMP composites at microscale. An array of such protrusive chains may be utilized to obtain re-configurable surface patterns in a simple manner for applications, such as remarkable change in wetting and friction ability. Finally, its potential applications for micro electro mechanical systems (MEMS) and biomedical device are proposed.

Keywords: shape-memory polymer; nickel powder; protrusive chain; shape-memory effect; surface morphology

---

1. Introduction

As a smart material, shape-memory polymers (SMPs) attract much interest owing to the capability to fully recover a remarkable pre-strain (on an order of hundred percent) upon applying a variety of external stimulus [1,2], including Joule heating [3,4], magnetism-induced heating [5,6], moisture [7], light [8,9], and pH change [10]. The shape memory effect (SME) could be realized on the basis of the following mechanism and the associated shaping procedure in a SMP [11]. In a glass transition based thermo-responsive SMP with the hard/soft-segment system, the function of hard segments is to act as physical crosslinks, and the SME is driven by the entropic forces which tend to shape the stretched polymer chains in the soft elements back to their original coiled configuration [12,13]. During a typical shape memory procedure, the original shape of SMP is synthesized and solidified at below glass transition temperature (T_g), then it is deformed to a temporary and large-deformation shape above T_g. Then, it is cooled well below T_g and the frozen soft segment lock the large-deformation configuration. When it is heated above T_g again, the soft segment release the hard segment, and the SMP recover toward the original shape.

The SME has occurred in macroscale and microscale in recent years, and also demonstrated at nanoscale in SMPs [14,15]. Hence, the prospect of SME applying in microscale or even nanoscale patterning receives great attention, including adaptive programmable materials [15], switchable molecule-based materials [16], photoactive electroactive applications [7,8], soft electronics [17,18], biomedical applications (e.g., cell mechanobiology [19], biomimetic 4D printing [3], switchable dual pH-Responsiveness [10], robust microcarriers [20], soft microgrippers [21], and micropatterned containers [22]). At present, a number of SMPs are available in the market and many more are under developing [1]. Among these SMPs, some polyurethane SMPs are both thermo-responsive and moisture-responsive [9], which shows a prospective way to realize the operation and surgery in live cell at microscale and nanoscale. Some microscale polymeric machines have been developed [23,24] for operation or surgery at cell scale, and can be actuated outside the cells. But, it is still difficult to deliver those machines into cells and the following stimulus for the artificially controlled operation. Those polymeric machines may be fabricated with SMPs to deliver and operation inside the living cells on the basis of shape memory effect.

Considering the formation of chain structures of ferromagnetic particles of magnetorheological fluids and elastomers [25,26], the suspended ferromagnetic particles form chains, columns, or aggregates along the direction of magnetic field. To explain the above phenomenon, several research methods have been used, including macroscopic mechanical element combination method, macroscopic continuum mechanics, microscopic dipole mechanical method, and finite element method based on multi-field coupling [26–29]. Based on these methods, many models or theories have been proposed, and two typical theories among them are phase-nuclear theory and field-induced dipole moment theory [30]. Among those models, the main parameters include strength and distribution of magnetic field, properties of particles (volume fraction, size, and microstructures), viscosity of solvent, interface between particle and solvent, and time of action. However, the formation mechanism of particle chains under magnetic field is still unclear [25–30], and the testing results are not completely consistent with the theoretical predictions [31,32].

In this study, we prepared nickel (Ni) particles polyurethane-SMP chains in micron and submicron scale, and the shape memory effect and associated properties are demonstrated for those vertical protrusive chains upon thermal stimulus.

2. Sample Preparation and Formation of Shape Memory Polymer/Nickel (SMP/Ni) Vertical Chains

In this study, the polyurethane (PU) SMP solution (MS 5510) was bought from Diaplex Co. Ltd. (Tokyo, Japan), with 70 wt% dimethylformamide (DMF) solution and 30 wt% SMP solution. The nickel powders with average diameters 3–7 μm (99.8% purity, Good fellow, Delson, Canada), are electrically and ferromagnetic conductive. The array of microscale protrusive SMP-nickel composite chains in vertical direction was prepared in the following procedure. The nickel particles were blended into the SMP/DMF solution with the different volume fraction, namely 1%, 2%, 3%, and 5%. The well stirred nickel-particle/SMP/DMF mixture with a certain content of nickel particles was immediately injected into a petri dish (8 mm height, and 30 mm diameter) with a height of 4 mm. At this time, without applying the magnetic field, the nickel particles distributed uniformly and randomly in the well stirred Ni/SMP/DMF mixture. To avoid the deposition of nickel particles to the bottom of petri dish, the petri dish with SMP/DMF/Ni mixture was immediately placed in the middle of a pair of magnets as shown in Figure 1b, and the vertical chains immediately formed. The magnetic intensity was controlled at about 0.4T tested by a Gauss meter. The setup of magnets and petri dish was placed into an oven and kept at 80 °C for 48 h for solidification. The scanning electron microscope, SEM (JEOL 5600LV, MA, USA) and optical microscope was used to observe the morphology of SMP/Ni vertical chains. The differential scanning calorimetery, DSC (Modulated DSC, DSC-2920, TA Instruments, New Castle, DE, USA) was used to characterize the glass transition behavior of SMP/Ni composite.

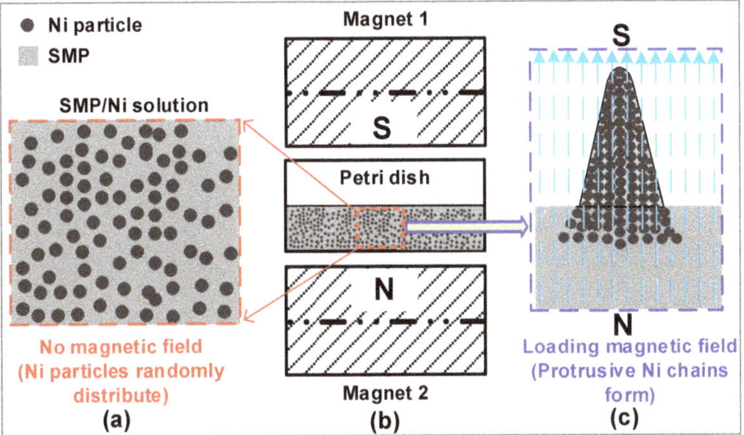

Figure 1. Illustration of preparation setup for vertical protrusive Shape Memory Polymer/Nickel (SMP/Ni) chains. (**a**) Random distribution of SMP/Ni solution without applying magnetic field; (**b**) Setup for preparing SMP/Ni vertical chain arrays; (**c**) Formation of SMP/Ni vertical chain arrays with applying magnetic field.

As revealed in Figure 1a, without loading a magnetic field, the ferromagnetic nickel particles were randomly distributed in the SMP/DMF solution. After loading a magnetic field, the ferromagnetic nickel particle aligned one by one in a line along the direction of magnetic field. In this way, the array of vertical protrusive nickel chains formed in the SMP/DMF solution as shown in Figure 1c. After evaporation of DMF from the SMP/DMF/Ni mixture, the aligned nickel particles were kept as shown in Figure 2; hence, the array of vertical protrusive SMP/Ni chains in solid state was obtained.

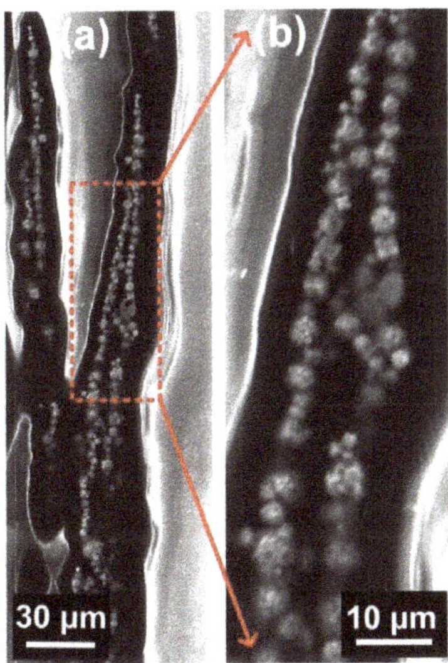

Figure 2. Nickel particles chains in Shape Memory Polymer (SMP) matrix. (**a**) Morphology of nickel particles chains; (**b**) Zoomed-in view of (**a**).

The typical vertical protrusive SMP/Ni chains and the morphology of separated nickel chains are exhibited as Figure 3. This protrusive SMP/Ni chain shows the dimension with approximate 1 mm height and 40–100 um diameter. As shown in the zoom-in image of Figure 3b, the tip of this SMP/Ni chain is a mixture of SMP and separated nickel chains aligned along the direction of magnetic field. In more detail, Figure 3c reveals the typical morphology of separated nickel chains distributed within the SMP matrix at micro/nano scale. Under the loading of magnetic field, the separated nickel particles moved and aligned one by one into regular array, and all the orientated lines are almost parallel in this field of view.

Figure 3. Typical chain structures of nickel particles in Shape Memory Polymer (SMP) matrix (3 vol% Ni). (**a**) Morphology of SMP/Ni vertical chain pillar; (**b**) Zoomed-in view of (a); (**c**) Typical morphology of nickel particles chains.

3. Morphology of Shape Memory Polymer/Nickel (SMP/Ni) Chain Arrays in Horizontal Plane

The distribution of nickel particles in SMP/DMF solution is essential for the formation of chain-like structures. After well stirring of the nickel-particle/SMP/DMF mixture, the nickel particles distribute uniformly and randomly at the initial time. As time flows, the nickel particles slowly deposit on the bottom of the petri dish. Hence, the uniform and random distribution of nickel particles in SMP/DMF solution is difficult to be observed by an optical microscope or SEM. Alternatively, to observe and study the formation characteristics of nickel chains, the magnetic field was applied in the direction parallel to the bottom of petri dish. The nickel chains formed, and finally solidified in horizontal plane with the volatilization of DMF solvent. Figure 4 indicates the distribution morphology of nickel particles under the magnetic field in horizontal plane. The straight nickel-chain structures along with the direction of magnetic field are obvious especial for the low volume fraction of nickel fillers, namely 0.25%, 0.5%, 1%, 2%, 3%, and 5%. The width of nickel chains gradually increases with the increasing of nickel contents. For the 0.25% sample, most of the cross sections of nickel chains only contain one nickel particle (3–7 μm width), and the length of nickel chains is general over 1000 μm, with an aspect ratio around 200. For the samples with higher nickel contents, they contain more nickel particles in one cross section, and the length limits of nickel chains approach 1500–3000 μm. In summary, for the varied contents of nickel fillers (0.25%, 0.5%, 1%, 2%, 3%, and 5%), the general widths of nickel chains are 3–7 μm, 10–50 μm, 20–80 μm, 20–100 μm, 100–300 μm, and 200–400 μm, respectively, and the general lengths are 1000–3000 μm.

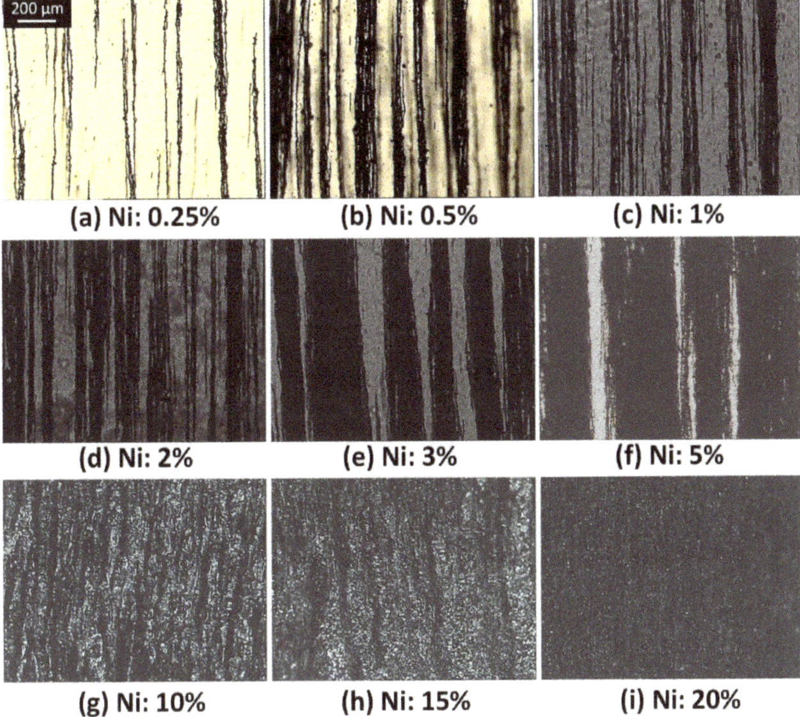

Figure 4. Distribution morphology of nickel particles in Shape Memory Polymer (SMP) matrix under the magnetic field in horizontal plane.

4. Differential Scanning Calorimetry (DSC) Test of Shape Memory Polymer/Nickel (SMP/Ni) Composite

In order to evaluate the fundamental material performance for pure polyurethane-based SMP and SMP/Ni composite, the DSC test was performed and the glass transition behavior was characterized (Figure 5). For SMP/Ni composite, the thin sample (around 100 μm) without applying magnetic field was prepared to maximally maintain the uniform distribution of nickel fillers in SMP matrix. The weight of sample for DSC test was 10–15 mg. The heating or cooling rate is 5 °C/min at the temperature range from −50 °C–150 °C. Figure 5 indicates the DSC curves in the cooling process, and the median point (half-height) was used to define the T_g. The glass transition temperatures of pure polyurethane-based SMP and SMP/Ni composite are 46 °C and 43 °C, which indicate that the T_g of pure SMP reduces a little bit when bending nickel powders into SMP. Furthermore, because the T_g is 6–9 °C higher than the temperature of human body, the SMP/Ni composites are suitable for thermal responsive actuation using for human (e.g., live cell operations).

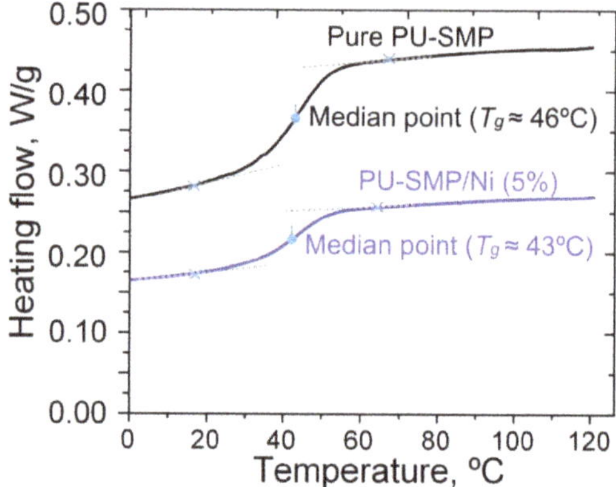

Figure 5. Differential Scanning Calorimetry (DSC) results of pure polyurethane-based Shape Memory Polymer (SMP) and SMP/Ni (Nickel) composite.

5. Morphology of Shape Memory Polymer/Nickel (SMP/Ni) Vertical Chain Arrays

The arrays of protrusive SMP nickel chains formed perpendicular to the SMP substrate. As shown in Figure 6, the morphology comparison of protrusive chains at four different volume fractions of nickel powders is presented. Statistical results of dimensions of protrusive chains, including diameter, height and separation were also obtained. Here, the diameter of a chain is defined as the measured diameter at the middle height of a chain. The interval distance is defined as the distance between the centrals of two adjacent chains. As shown in Table 1, it is clear that both the diameter (from 10 μm to 250 μm) and height (from 200 μm to 1500 μm) increase with the increasing of nickel contents, while the interval distance (from 100 μm to 700 μm) is relatively stable at five different nickel contents.

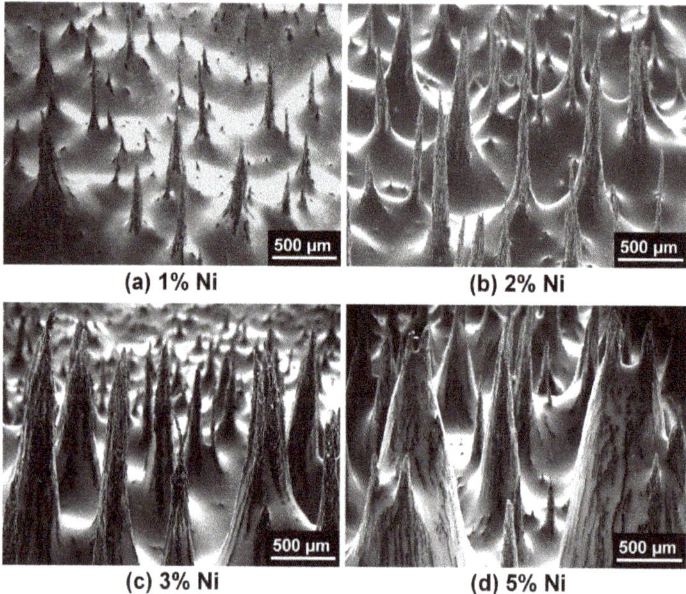

Figure 6. Typical array of vertical protrusive Shape Memory Polymer/Nickel (SMP/Ni) chains with varied volume fraction of nickel particles: 1%, 2%, 3%, and 5%.

Table 1. Dimensions (diameter, height, and interval distance) of vertical Shape Memory Polymer/Nickel (SMP/Ni) chains.

Filler Contents	Diameter (μm)	Height (μm)	Interval Distance (μm)
1%	25–45	270–310	100–400
2%	100–135	750–1000	300–550
3%	180–220	1100–1300	400–700
5%	200–240	1200–1400	450–600

Figure 7 reveals the patterns of the cross section cut from a vertical protrusive SMP/Ni chain with average 2% (Figure 7a) and 5% (Figure 7b) nickel particles in SMP matrix. The particles distributed randomly, which implies that the aligned nickel particles are also in random distributions in the horizontal cross-section plane (the direction of magnetic field is in the vertical direction). Furthermore, referred to the observed distributions of nickel particles, the volume fraction in this cross section is much higher than 2% (Figure 7a) and 5% (Figure 7b). Due to the attraction effect of magnetic field to the nickel particles, the nickel particles concentrate towards each vertical protrusive SMP/Ni chain, where the nickel content is much higher than those located far from the SMP/Ni chains. As shown in Figure 7a-2, the thin films of pure SMP are obvious at the exterior surface of the cross section, which ensure the strong banding of nickel particles on the surface. Furthermore, as shown in Figure 7b, a series of nickel chains are also embedded under the surface of a big vertical SMP/Ni chain, and the thin films of pure SMP also cover nickel particles. Moreover, on the tips of SMP/Ni vertical chains, the particles may be exposed to the air. However, due to high porosity of nickel particles, the nickel particles are still strong bonded with the SMP matrix.

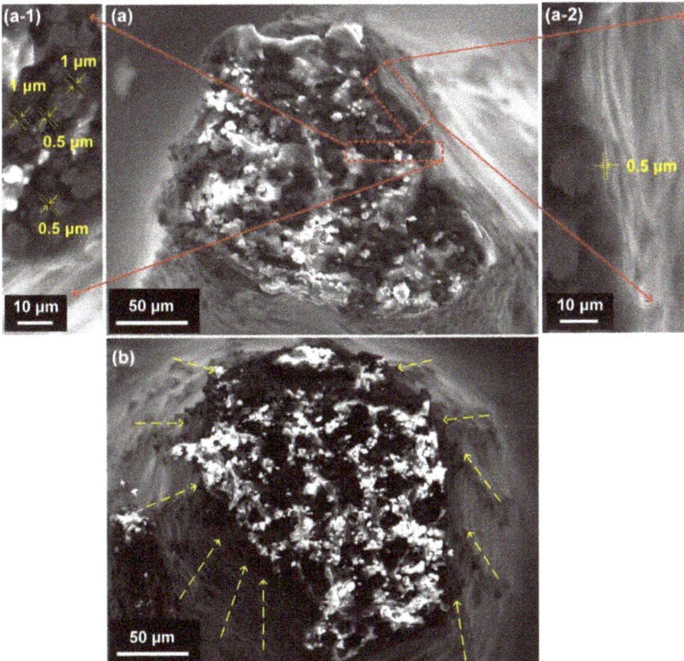

Figure 7. Cross section of the vertical protrusive Shape Memory Polymer/Nickel (SMP/Ni) chain ((**a**) 2 vol% Ni; (**b**) 5 vol% nickel).

6. Shape Recovery Properties of Shape Memory Polymer/Nickel (SMP/Ni) Vertical Chain Arrays

As illustrated in Figure 8, we heat initially vertical chains (Figure 8a) to 120 °C (well above T_g T_g = 43 °C–46 °C) for 20 min. After that, a piece of aluminum plate is placed atop the sample. Hence, the chain is flattened (Figure 8b). Subsequently, the SMP/Ni vertical chains are gradually cooled in the bending and flattened configuration while keeping the aluminum plate there. After the removal of the aluminum plate, the nickel protrusive chains are still in the bend-shape as shown in Figure 8c. Upon heating to 120 °C, if the chains can recover its initial shape (Figure 8d), the shape memory effect of SMP is demonstrated at microscale and nanoscale.

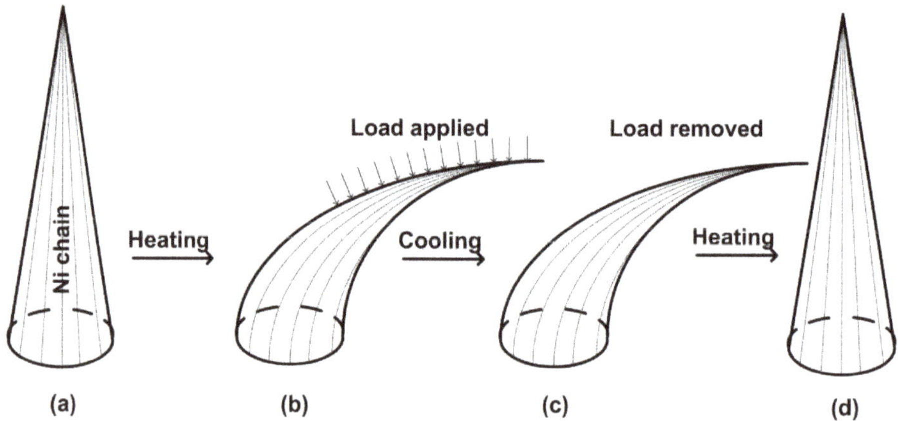

Figure 8. Illustration of a thermomechanical cycle to evaluate the shape memory effect of a vertical protrusive chain. (**a**) Initial shape; (**b**) pre-deformed shape; (**c**) flattened shape; (**d**) recovered shape.

The shape memory effect of vertical chains atop different samples with different nickel contents was tested following the procedure described above. Figure 9a shows one typical flattened chain of two samples with 2 vol% nickel. It is obvious that thin chains are easier to be flattened as compared with the thick chains. Hence, the chains obtained in samples with lower nickel content are preferred to demonstrate the SME. Furthermore, the local buckling phenomenon of SMP matrix can be found in the inner surface of the bending chain, which implies that the local compression strain is high. Figure 9b is the zoom-in view of chains of 2 vol% nickel content. Given that the size of nickel powders is around 3 to 7 μm, it reveals that the dimension of SMP (in particular, the thickness) is around 1 μm or less (i.e., at submicron scale), which is also indicated in Figure 7a. Hence, if the chains, although they are composite in nature, are demonstrated as with the SEM, we can conclude that the SMP/Ni chains structure does have the shape memory effect at around 1 micron and even submicron scale.

Figure 10 reveals the patterns of several SMP/Ni vertical raised chains in shape memory recycle. In the original shape Figure 10a, the SMP/Ni protrusive chains are strictly straight along with the direction of magnetic field. After heated above the T_g, the SMP matrix in the SMP/Ni chain became soft, and then the SMP/Ni protrusive chains were forced to be bent to approximate 180° to the flatten shape with a flat board. When cooled below T_g, after releasing the external constraint, this flatted pre-deformed configuration were obtained as shown in Figure 10b. The bending areas of vertical chains are mainly located at the root. When re-heating the SMP/Ni vertical chains, they gradually recovered toward the original vertical configuration. The macroscopic shape memory effect of pure SMP is the same as that at microscale, with shape recovery ratio 95–100% above T_g. However, with bending in solid and hard nickel particles, the shape recovery ratio of SMP/Ni vertical chains will reduce. The biggest unrecoverable angle in the recovered shape of SMP/Ni vertical chains is approximate 30° compared with the original straight shape, namely shape recovery ratio around 83%.

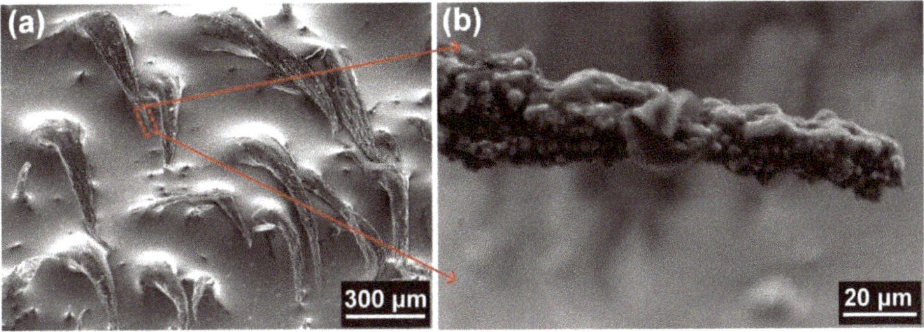

Figure 9. Typical configurations of flattened Shape Memory Polymer/Nickel (SMP/Ni) chain array with 2% nickel particles. (**a**) Morphology of flattened pillars; (**b**) Zoomed-in view of (a).

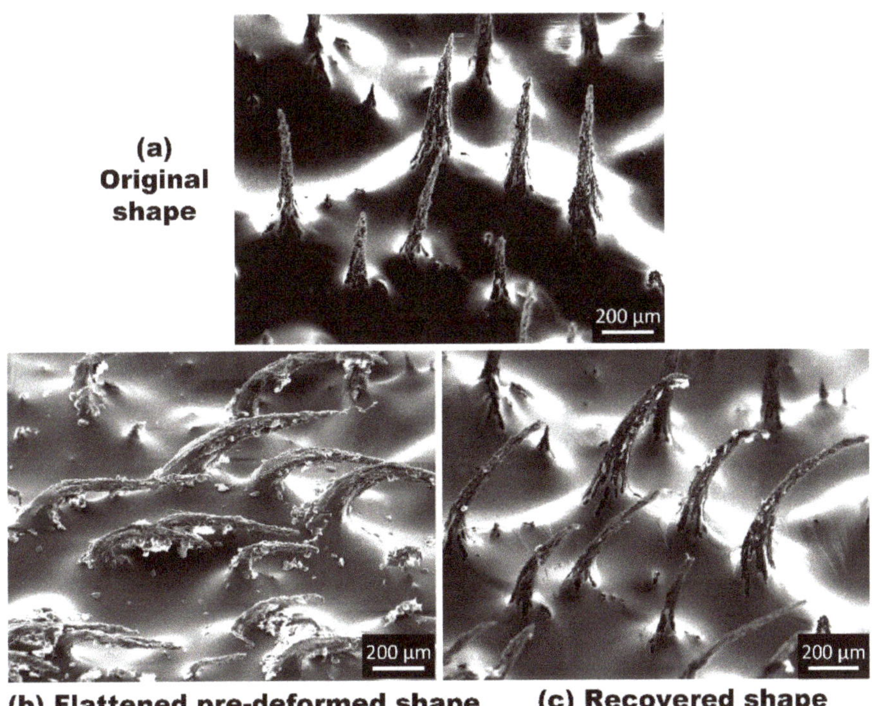

Figure 10. Typical shape recovery cycle of the vertical protrusive Shape Memory Polymer/Nickel (SMP/Ni) chains (2 vol% Ni). (**a**) Original shape; (**b**) Flattened pre-deformed shape; (**c**) Recovered shape.

7. Formation Mechanism and Dimensions Control Method of Shape Memory Polymer/Nickel (SMP/Ni) Vertical Chain Arrays

Considering the formation mechanism for nickel chains aligned along the magnetic field direction, the general explanation could be referenced to magnetomechanical coupling theory of the magnetorheological fluids [26]. Without loading a magnetic field, there are no permanent dipole moments in nickel particles; hence, all the nickel particles are in Brown motion and distributed randomly in the SMP/DMF solution prior to curing. Upon loading a permanent static magnetic field **H**, the permanent dipole moments in all ferromagnetic particles were induced and interacted. In local nonuniform magnetic field, they rotate, repel or attract between two adjacent nickel particles depending on the direction of induced dipole moments and their relative position [33–36]. Referenced to the aforementioned studies of magnetorheological fluids, the formation of nickel chain structures in SMP matrix should consider the following forces applied on nickel particles:

○ Magnetic force. SMPs are macromolecule and not ferromagnetic materials, and their magnetic susceptibility is negligible. The magnetic forces $\mathbf{F}_j^{mag}$ acting on a single particle j are induced by external magnetic field as well as induction field among nickel particles themselves [35]: $\mathbf{F}_j^{mag} = \mathbf{m} \cdot \nabla(\mathbf{B}_0 + \sum_{i \neq j} \mathbf{B}_i)$, where $\mathbf{m}$ indicates dipole moment, $\mathbf{B}_0$ reveals magnetic induction intensity of external magnetic field, $\mathbf{B}_i$ indicates magnetic induction intensity of particle j induced by particle i.

○ Repelling force. Repelling forces exist when a nickel particle j approaches other particles ($\mathbf{F}_j^{sphere}$). Moreover, repulsive forces of a nickel particle j from walls of container ($\mathbf{F}_j^{wall}$) are included.

○ Gravitation force. $\mathbf{F}_j^{gav}$. The gravitation of a nickel particle j.

○ Hydrodynamic force. The hydrodynamic force $\mathbf{F}_j^{hydro}$ of a moving nickel particle j for small Reynolds can be described by Oseen's equation $\mathbf{F}_j^{hydro} = -D\frac{dr_j}{dt}$, where $D = 6\pi a\eta$ reveals Stokes drag coefficient of a nickel sphere. η is the viscosity coefficient of the SMP/DMF solution. Note that, the viscosity coefficient increases during solidification process.

Hence, on the basis of references [33,34], the motion governing equations of a nickel particle j in the SMP/DMF solution can be expressed as

$$m\frac{d^2\mathbf{r}_i}{dt^2} + D\frac{d\mathbf{r}_i}{dt} = \mathbf{F}_i^{mag} + \mathbf{F}_i^{sphere} + \mathbf{F}_i^{wall} + \mathbf{F}_i^{gav}. \tag{1}$$

The properties of vertical protrusive chains (diameter, height, and interval between two adjacent chains) can be explained as follows according to Equation (1). For finitely long chains, the interaction between two adjacent chains may be attraction or repelling forces depending on their interval, and height of the two chains. Those interaction forces cause the local gather of adjacent nickel chains in local nonuniform magnetic field; hence, big and high nickel chains atop SMP substrate form. Naturally, with more nickel particle fillers, the diameter and height of nickel chains will be larger, and the interval will be smaller. After solidification of SMP/DMF/Ni mixture, the SMP/Ni chains form with different diameter, height, and interval. On the other hand, the forces among nickel chains are in short range, and the stable protrusive array forms. The field between neighboring protrusive chains is relatively stable and almost independent on the exact nickel content.

Because of the complexity and imperfection of the above theoretical models referenced to magnetomechanical coupling theory of the magnetorheological fluids [26,33–36], it is still difficult to quantitatively design the distributions or accurately control the dimensions (diameter, height, and interval distance). Alternatively, in order to form well distributions of SMP/Ni chain arrays as much as possible, the experimental methods are still employed. Base on the preliminary analysis of theoretical models (Equation (1)), the following key parameters in the preparation procedure of SMP/Ni vertical chain arrays were considered, including strength of magnetic field, volume fraction of nickel contents in SMP matrix, size of nickel particles, viscosity of SMP/DMF solution.

7.1. Strength of Magnetic Field

In this study, the strength of magnetic field was controlled within 0.35–0.4 T. When it was weaker than 0.35–0.4 T, the height of SMP/Ni chains was small, although the chain arrays are dense. If it was stronger than 0.35–0.4 T, the nickel particles aggregated into relative thick chains; hence, the uniform chain arrays were difficult to form.

7.2. Volume Fraction of Nickel Particles in Shape Memory Polymer (SMP) Matrix

The volume fraction of nickel particles was also controlled in a proper range, namely 1–5%. When it was lower than 1% (e.g., 0.5%), the diameter of vertical chains reduced to 3–20 μm with the interval distance 200–1000 μm (Figure 4a,b). Alternatively, when it was higher than 1% (e.g., 10%, 15% and 20%) the nickel particles aggregated into relative thick chains without forming regular chain arrays (Figure 4g–i). Therefore, for the fabrication of SMP/Ni vertical chains, the volume fractions 1%, 2%, 3%, and 5% were employed in this study.

7.3. Size of Nickel Particles

Considering the size of nickel particles, its diameters 3–7 μm are at microscale level, and the proper volume fractions for SMP/Ni mixture to form relative well vertical arrays are around 2–3%, as shown in Figures 3, 4, 6 and 7. Moreover, it is predicted that, when the size of nickel particles is smaller, the appropriate volume fractions as well as dimensions will also reduce.

7.4. Viscosity of Shape Memory Polymer/Dimethylformamide (SMP/DMF) Solution

Viscosity of SMP/DMF solution is essential for the formation of SMP/Ni chains. With a low viscosity of solvent, the nickel particles will quickly deposit at the bottom, and they will aggregate but not chains form when loading magnetic field. If the viscosity of solvent is too high, the movement of particles will be constrained in a small area, and therefore the nickel particles are difficult to align into chain structures along magnetic field.

8. Discussion, Future Work, and Potential Applications

The current investigation of SMP/Ni vertical protrusive chains in this paper is just a preliminary stage to study this type of ferromagnetic particle-filled composite under magnetic field. There are three purposes to prepare those pillar arrays with applying magnetic field. First, mechanism and experimental research are prospective to be studied to prepare designable microscale structures (e.g., aligned chains) for ferromagnetic particle-filled composite through the operation of magnetic field. The main research contents contain physical mechanism and associated theoretical model of chain formation of nickel particles under magnetic field, and the crucial parameters in the preparation and experimental study with the aim to control the distribution and dimensions of SMP/Ni chain arrays. Second, it is proposed to investigate special performance of SMP/Ni composite with nickel particle chain morphology, such as reinforcement of mechanical and electrical-conductivity along the direction of nickel chains. Third, the SMP/Ni pillar arrays are prospective to be fabricated into micro, sub-micro, and even nano size, and to realize the artificial and controllable operation at the corresponding scale levels. Hence, the nanoimprinting [37,38] and mold preparation method without applying magnetic field were not used in this study, although they are alternative methods to prepare more uniform pillar arrays for SMP/Ni composite. Based on the above three purposes, the further and more detailed study will be carried out in near future.

Additionally, the size of nickel particles is at microscale level (3–7 μm), and the diameters of SMP/Ni vertical chains are approximate at the magnitude of tens of micron for 1% or even lower filler contents, and hundreds of micron for 2%, 3%, and 5% filler contents. However, those dimensions are not suitable for the operations in sub-micro size. Alternatively, nanoscale nickel particles are a choice to be filled into SMP to prepare SMP/Ni vertical chains in sub-micro or even nano size. As a preliminary investigation, the nanoscale nickel particles with diameters of 100–500 nm have been used to prepare the nickel chains in horizontal plane with loading magnetic field. As shown in Figure 11, the nickel chains are in sub-micro, which provide a possibility to prepare SMP/Ni vertical chain arrays and for the corresponding operations in sub-micro size in future work.

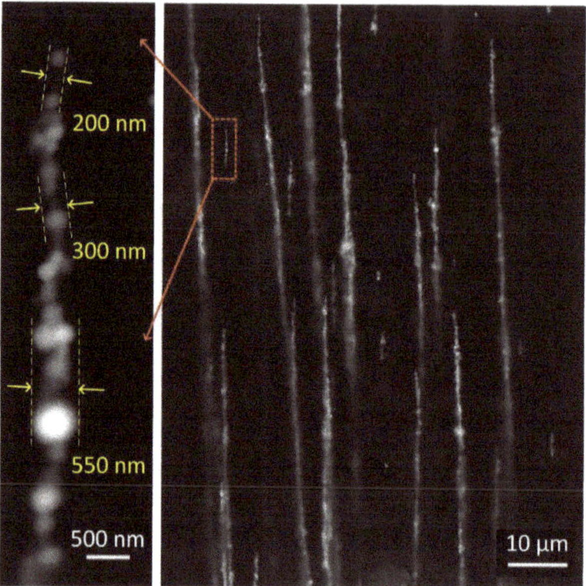

Figure 11. Typical array of Shape Memory Polymer/Nickel (SMP/Ni) chains with 1% volume fraction of sub-micro nickel particles.

Thermal-sensitive shape memory polymers have gained much attention because of their physical or chemical responses to quick changes by thermal-stimuli around the critical transition temperature. The vertical protrusive SMP chains show great prospects for the microscale and sub-micro actuation for biomedical materials and device, microscale patterns, and micro electro mechanical systems (MEMS), such as drug delivery, bioseparation, biomimetic 4D printing, switchable dual pH-Responsiveness, robust microcarriers, and soft microgrippers. For the microbrush or microgripper, the potential application may be described as follows: the chains of the microbrush or microgripper, which is flexible above transition temperature, can be used for finishing, cleaning, burnishing or gripping MEMS devices. The nature of the microbrush or microgripper is flexible to access the surface where the conventional tools are hardly applicable. After using, the protrusive chains can recover to the original shape by heating above the T_g. In this way, the shape memory microbrush or microgripper is recycling-used. For instance, Figure 12 illustrates the typical delivery and operation procedure of protrusive chains of microscale SMP brush machine. The array of protrusive SMP chains (Figure 12a) was compressed and bent to a flatten shape (Figure 12b) above T_g, and easily delivered into a live cancer cell in this compressed shape in a small volume. The bent SMP chains can recover to the original protrusive shape triggered by outside alternative magnetic field or moisture. In special, the T_g will reduce when the polyurethane SMP immersed into water or in moisture atmosphere. In this way, the pre-deformed polyurethane SMPs can be "actuated" by water or moisture, which is suitable to be used in human body.

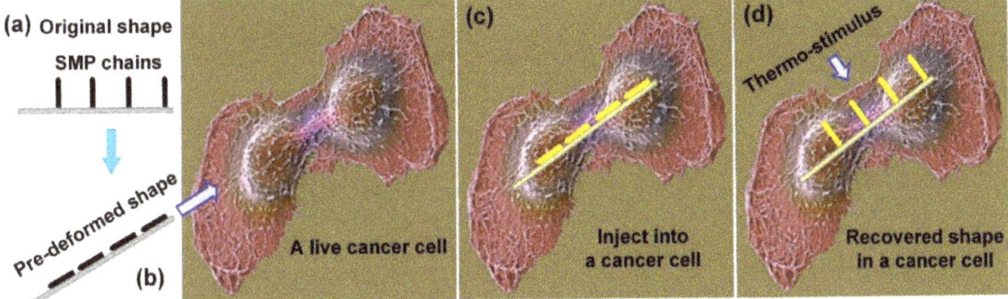

Figure 12. Schematic illustration of the delivery and operation procedure of microscale Shape Memory Polymer (SMP) brush machine in a live cancer cell. (The live cancer cell is referred to in Reference [39]). (**a**) Original shape of SMP chain pillars; (**b**) Pre-deformed SMP chain pillars; (**c**) Pre-deformed SMP chain pillars injected into a cancer cell; (**d**) Recovered SMP chain pillars in a cancer cell.

9. Conclusions

Arrays of SMP/Ni microscale chains were prepared along the direction of magnetic field. Depending on the volume fraction of nickel particles, protrusive chains were obtained with a diameter from 10 to 250 μm and height from 200 to 1000 μm. The shape memory effect in such vertical protrusive chains was demonstrated in microscale. As the SMP is thermo/moisture-responsive, it can be used for novel microscale devices. These micro protrusive chains may be utilized to significantly alter the surface morphology, and hence induce a dramatic potential for microscale actuation for biomedical materials and device, MEMS, such as microbrush or microgripper.

Author Contributions: X.L., W.H. and J.L. conceived and designed the experiments; X.L. performed the experiments; X.L. and W.H. analyzed the data; X.L. wrote the paper.

Funding: This work is supported by the National Natural Science Foundation of China (grant numbers: 11872020, 11632005, U1637207, and 11772109).

Conflicts of Interest: The authors declare no conflict of interest.

References

1. Leng, J.S.; Lan, X.; Liu, Y.J.; Du, S.Y. Shape memory polymers and their composites: Stimulus methods and applications. *Prog. Mater. Sci.* **2011**, *56*, 1077–1135. [CrossRef]
2. Xie, T. Tunable polymer multi-shape memory effect. *Nature* **2010**, *464*, 267–270. [CrossRef] [PubMed]
3. Sydney, G.A.; Matsumoto, E.A.; Nuzzo, R.G.; MahadevanL, J.A. Biomimetic 4D printing. *Nat. Mater.* **2016**, *15*, 413–418. [CrossRef] [PubMed]
4. Lan, X.; Liu, L.W.; Liu, Y.J.; Leng, J.S. Thermomechanical and electroactive behavior of a thermosetting styrene-based carbon black shape-memory composite. *J. Appl. Polym. Sci.* **2018**, *135*, 45978. [CrossRef]
5. Yu, K.; Westbrook, K.K.; Kao, P.H.; Leng, J.; Qi, H.J. Design considerations for shape memory polymer composites with magnetic particles. *J. Compos. Mater.* **2013**, *47*, 51–63. [CrossRef]
6. Marco, D.; Eckhouse, S. Bioerodible self-deployable intragastric implants. U.S. Patent 8,864,784 B2, 21 October 2014. Application granted.
7. Ayesha, K. Review on Technological Significance of Photoactive, Electroactive, pH-sensitive, Wateractive, and Thermoresponsive Polyurethane Materials. *Polym. Plast. Technol.* **2017**, *56*, 606–616.
8. Jin, B.J.; Song, H.J.; Jiang, R.Q. Programming a crystalline shape memory polymer network with thermo- and photo-reversible bonds toward a single-component soft robot. *Sci. Adv.* **2018**, *4*, 05195. [CrossRef]
9. Bertrand, O.; Gohy, J.F. Photo-responsive polymers: Synthesis and applications. *Polym. Chem.* **2017**, *8*, 52–73. [CrossRef]
10. Hu, Y.; Lu, C.H.; Guo, W.; Aleman-Garcia, M.A.; Ren, J.; Willner, I. A shape memory acrylamide/DNA hydrogel exhibiting switchable dual pH-responsiveness. *Adv. Funct. Mater.* **2015**, *25*, 6867–6874. [CrossRef]

11. Wang, X.; Sparkman, J.; Gou, J.H. Electrical actuation and shape memory behavior of polyurethane composites incorporated with printed carbon nanotube layers. *Compos. Sci. Technol.* **2017**, *141*, 8–15. [CrossRef]
12. Behl, M.; Lendlein, A. Shape-memory polymers. *Mater. Today* **2007**, *10*, 20. [CrossRef]
13. Lendlein, A.; Kelch, S. Shape-memory polymers. *Angew. Chem. Int.* **2002**, *41*, 2034. [CrossRef]
14. Wang, C.C.; Zhao, Y.; Purnawali, H. Chemically induced morphing in polyurethane shape memory polymer micro fibers/springs. *React. Funct. Polym.* **2012**, *72*, 7575–7764. [CrossRef]
15. Fan, X.S.; Chung, J.Y.; Lim, Y.X.; Li, Z.B.; Loh, X.J. Review of adaptive programmable materials and their bioapplications. *ACS Appl. Mater. Interfaces* **2016**, *8*, 33351–33370. [CrossRef]
16. Manrique-Juáreza, D.M.; Rata, S.; Salmona, L.; Molnára, G.; Quinterob, C.M.; Nicub, L.; Shepherdc, H.J.; Bousseksou, A. Switchable molecule-based materials for micro- and nanoscale actuating applications: Achievements and prospects. *Coordin. Chem. Rev.* **2016**, *308*, 395–408. [CrossRef]
17. Wang, Q.M.; Zhao, X.H. Beyond wrinkles: Multimodal surface instabilities for multifunctional patterning. *MRS Bull.* **2016**, *41*, 115–122. [CrossRef]
18. Lagrange, R.; LópezJiménez, F.; Terwagne, D.; Brojan, M.; Reis, P.M. From wrinkling to global buckling of a ring on a curved substrate. *J. Mech. Phys. Solids* **2016**, *89*, 77–95. [CrossRef]
19. Ebara, M. Shape-memory surfaces for cell mechanobiology. *Sci. Technol. Adv. Mater.* **2015**, *16*, 014804. [CrossRef]
20. Shim, T.S.; Kim, S.H.; Heo, C.J.; Jeon, H.C.; Yang, S.M. Controlled origami folding of hydrogel bilayers with sustained reversibility for robust microcarriers. *Angew. Chem. Int. Ed.* **2012**, *51*, 1420–1423. [CrossRef]
21. Breger, J.C.; Yoon, C.; Xiao, R.; Kwag, H.R.; Wang, M.O.; Fisher, J.P.; Gracias, D.H. Self-folding thermo-magnetically responsive soft microgrippers. *ACS Appl. Mater. Interfaces* **2015**, *7*, 3398–3405. [CrossRef]
22. Azam, A.; Laflin, K.E.; Jamal, M.; Fernandes, R.; Gracias, D.H. Self-folding micropatterned polymeric containers. *Biomed. Microdevices* **2011**, *13*, 51–58. [CrossRef] [PubMed]
23. Maruo, S.; Ikuta, K.; Korogi, H. Optically-driven nanomanipulators fabricated by two-photon microstereolithography. *Mat. Res. Soc. Symp. Proc.* **2003**, *739*, 269.
24. Ionov, L. Biomimetic 3D self-assembling biomicroconstructs by spontaneous deformation of thin polymer films. *J. Mater. Chem.* **2012**, *22*, 19366–19375. [CrossRef]
25. Sasena, P.; Hossain, M.; Steinmann, P. Nonlinear magneto viscoelasticity of transversally isotropic magneto active polymers. *Proc. Math. Phys. Eng. Sci.* **2014**, *470*, 20140082. [CrossRef] [PubMed]
26. Yuan, F.Y.; Wan, Q.; Zhang, C.Y.; Li, X. Advances in Magnetomechanical coupling constitutive relations of magnetorhelogical elastomers. *Mater. Rev. A* **2017**, *31*, 10.
27. Zhou, G.Y.; Jiang, Z.Y. Deformation in magnetorheological elastomer and elastomer-ferromagnetic composite driven by a magnetic field. *Smart Mater. Struct.* **2004**, *13*, 309. [CrossRef]
28. Ivaneyko, D.; Toshchevikov, V.; Borin, D.; Saphiannikova, M.; Heinrich, G. Mechanical properties of magneto-sensitive elastomers in a homogeneous magnetic field: Theory and experiment. *Macromol. Symp.* **2014**, *338*, 96. [CrossRef]
29. Sasena, P.; Hossain, M.; Steinmann, P. A theory of finite deformation magneto viscoelasticity. *Int. J. Solid Struct.* **2013**, *50*, 3886. [CrossRef]
30. Galipeau, E.; Rudykh, S.; Castaneda, P.P. Magnetoactive elastomers with periodic and random microstructures. *Int. J. Solid Struct.* **2014**, *51*, 3012. [CrossRef]
31. Itskov, M.; Khiem, V.N. A polyconvex anisotropic free energy function for electro and magneto rheological elastomers. *Math. Mech. Solid.* **2014**, *21*, 1126. [CrossRef]
32. Chen, S.W.; Li, R.; Zhang, Z.; Wang, X.J. Micromechanical analysis on tensile modulus of structured magneto rheological elastomer. *Smart Mater. Struct.* **2016**, *25*, 035001. [CrossRef]
33. Ly, H.V.; Reitich, F.; Jolly, M.R.; Banks, H.T.; Ito, K. Simulations of particle dynamics in magnetorheological fluids. *J. Comput. Phys.* **1999**, *155*, 160. [CrossRef]
34. Mohebi, M.; Jamasbi, N. Simulation of the formation of nonequilibrium structures in magnetorheological fluids subject to an external magnetic field. *Phys. Rev. E* **1996**, *54*, 5407. [CrossRef]
35. Gross, M.; Kiskamp, S. New long-range interaction between dipolar chains. *Phys. Rev. Lett.* **1997**, *79*, 2566. [CrossRef]
36. Zhou, L.; Wen, W.; Sheng, P. Ground states of magnetorheological fluids. *Phys. Rev. Lett.* **1998**, *81*, 1509. [CrossRef]

37. Dumond, J.J.; Low, H.Y. Recent developments and design challenges in continuous roller micro- and nanoimprinting. *J. Vac. Sci. Technol. B* **2012**, *30*, 010801. [CrossRef]
38. Sharstniou, A.; Niauzorau, S.; Ferreira, P.M. Electrochemical nanoimprinting of silicon. *PNAS* **2019**, *116*, 10264. [CrossRef]
39. University of California, San Diego Health Sciences. Live and Let-7: microRNA Plays Surprising Role in Cell Survival. *ScienceDaily*. 7 October 2014. Available online: https://phys.org/news/2014-10-let-microrna-role-cell-survival.html (accessed on 17 July 2019).

Article

Composite Films of Polydimethylsiloxane and Micro-Graphite with Tunable Optical Transmittance

Qi Wang, Bin Sheng *, He Wu, Yuanshen Huang, Dawei Zhang and Songlin Zhuang

Engineering Research Center of Optical Instruments and Systems, Ministry of Education and Shanghai Key Laboratory of Modern Optical Systems, University of Shanghai for Science and Technology, Shanghai 200093, China; w18817782027@163.com (Q.W.); 18818226501@163.com (H.W.); hyshyq@sina.com (Y.H.); dwzhang@usst.edu.cn (D.Z.); slzhuang@yahoo.com (S.Z.)
* Correspondence: bsheng@usst.edu.cn; Tel.: +86-189-6419-6659

Received: 24 May 2019; Accepted: 10 June 2019; Published: 13 June 2019

Abstract: In this paper we introduce a polydimethylsiloxane (PDMS) composite fabricated using a simple production process and demonstrate the optical transmittance properties of this composite in the 300–1000 nm wavelength region. We control the material's transmittance by varying the microcrystalline graphite powder concentration or the composite film's thickness. In addition, we tailor the specimens into various trapezoidal shapes and load these specimens by mechanically stretching them in the direction perpendicular to both their base lines and their top lines. The advantage of this method is that a wide range of transmittance properties can be obtained for a given specimen. Furthermore, samples with different trapezoidal shapes have different transmittance tuning capabilities.

Keywords: composite; polymer; crystalline graphite powder; transmittance property

---

1. Introduction

Composite optical materials are designed to have enhanced optical properties that are superior to those of their individual component materials [1–3]. At present, composites are used in a wide range of applications, including strain sensors, electronic displays, optical attenuators and smart windows [4–10]. Composites, which use carbon materials in high-performance applications, have been widely used owing to their strengthening effect, high stiffening and feasible preparation, such as graphene oxide-silica nanohybrid [11–16], graphene/polydimethylsiloxane [17,18] and polylactic acid (PLA) as a host polymer and different forms of carbon fillers [19,20]. Considerable progress has already been made in these fields. Composites have also been designed to have tunable transparency under specific electrical, magnetic, optical, thermal and chemical driving conditions but most industrially available devices of this type are costly and bulky [21]. There is thus a need for novel mechanisms to allow tunable composites to be realized.

Recently, many researchers have reported on their efforts to develop devices based on lightweight elastomers [22] such as polydimethylsiloxane (PDMS). PDMS is a macromolecule compound that is widely used in numerous fields because of its superb characteristics, which include good elasticity, optical properties, breathability, insulation properties and stability, along with a low glass transition temperature. PDMS-based devices thus have obvious advantages and increasing numbers of researchers are fabricating and integrating optical components using this flexible material [23–28]. In particular, a number of researchers have been considering the use of PDMS composites. By controlling the dopants that are added to the PDMS, they can control many of the properties of the resulting composites, including optical, electrical and mechanical properties [29–38]. However, these composites must be fabricated with high precision using vacuum deposition or evaporation or even by implantation via aerodynamic acceleration. While these methods have advantages, the fabrication techniques can be both complex and costly.

In this work, we propose a PDMS composite that can be fabricated using a simple production process and that offers flexibly tunable optical transmittance. We fabricated thin films of PDMS mixed with microcrystalline graphite powder, which can alter the transparency of the samples. By varying the concentration of the microcrystalline graphite powder and the sample thickness, we were able to control the transmittance of the resulting composite film. In addition, we tailored samples of the composite into various trapezoidal shapes and stretched them to tune their optical transmittance.

2. Experimental Section

2.1. Materials

Our experimental samples were made from PDMS polymer SYLGARD 184 SILICONE ELASTOMER, Dow Corning, Midland, TX, USA) consisting of two parts, the base polymer and the corresponding curing agent. The microcrystalline graphite powder (MCGP-1000, Sinopharm Chemical Reagent Co. Ltd. Shanghai, China) was about 1 micron in diameter, with the characteristics of excellent lubrication, conductivity and resistance for acid and alkali.

2.2. Preparation Routes

First, we mixed the base polymer together with its corresponding curing agent and the microcrystalline graphite powder and churned the mixture for 20 min to ensure that all materials were completely mixed. Specifically, the mass ratio of the base polymer to the curing agent was 15:1 and the microcrystalline graphite powder had a size of 1–3 microns. We weighed the materials using an electronic analytical balance that was accurate to 0.1 mg. Then the mixture was placed into a vacuum chamber to remove the gross air bubbles. Subsequently, we poured the mixture onto a glass substrate (7 × 7 cm) and degassed it in a vacuum chamber for a second time until no bubbles were observed. The liquid mixture was then solidified in an oven at 100 °C for 90 min to obtain a uniform thickness (h_0) of approximately 1 mm. By that stage, the composite film was completely solidified and could be separated from the substrate.

2.3. Characterizations

We used a spectrophotometer (Lambda 1050, PerkinElmer, Waltham, MA, USA), which is a two-beam scanning spectrophotometer with a double monochromator, to measure the samples' transmittance. The spectrophotometer performs the measurement of optical characteristics such as optical density, reflection and transmission coefficients of liquid and solid materials, including light-scattering inorganic, organic and biological objects within the wavelength range of 175–3300 nm. Metallographic Microscopes (10XB-PC, Shanghai Optical Instrument Factory, Shanghai, China) were used to observe the dispersion of the microcrystalline graphite powder in the composite films.

Figure 1a shows a pure PDMS sample placed on a piece of paper that has colored fonts on it. It is obvious that the pure PDMS has high transmittance. Figure 1b shows a composite PDMS sample placed in the same position. Figure 1b shows that the transmittance decreased considerably when compared with that of the pure PDMS sample. We also observed the formation of the composite film under an optical microscope, as shown in Figure 1c. When the sample was cured, it appeared to take the form of a nearly isotropic colloidal composite composed of microcrystalline graphite powder embedded in the PDMS.

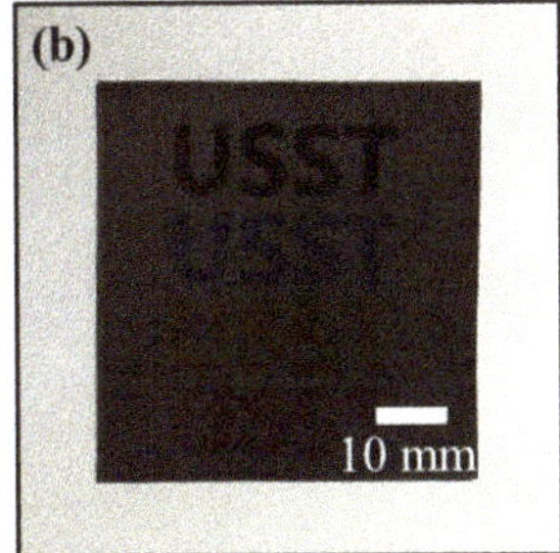
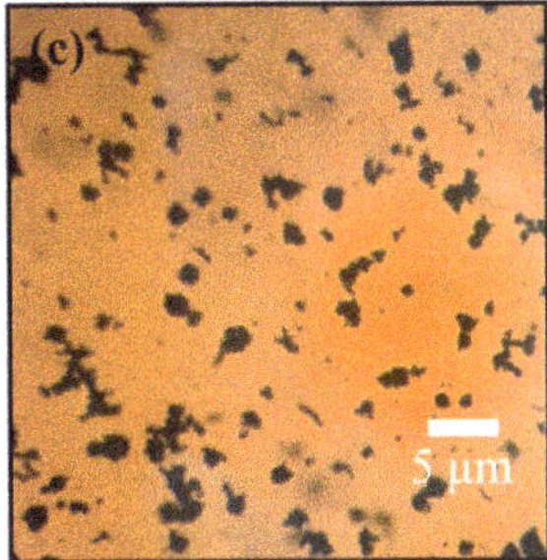

Figure 1. (**a**) Sample of pure polydimethylsiloxane (PDMS) (thickness $h_0 = 1$ mm), placed on top of paper with printed colored fonts. (**b**) Sample of the dyed PDMS (thickness $h_0 = 1$ mm), placed in the same position as that of the sample in (**a**). (**c**) Formation of the composite film observed under an optical microscope.

3. Results and Discussion

3.1. Transmittance Characterization for Different Concentrations

The transmittance distribution is an important property of each sample and this distribution was measured using a spectrophotometer (PerkinElmer Lambda 1050) over the 300–1000 nm range. As shown in Figure 2, the four transmission lines illustrate one pure PDMS film sample and three composite film samples of PDMS and microcrystalline graphite powder with different concentrations. For the different sample concentrations, which were 0%, 0.15%, 0.25% and 0.42%, the mean transmittance values over the 300–1000 nm range were $T_0 = 93.9 \pm 4.1\%$, $T_1 = 32.3 \pm 0.8\%$, $T_2 = 10.4 \pm 0.6\%$ and $T_3 = 1.4 \pm 0.2\%$, respectively. These results illustrate that each line shows an approximately constant level. The relative standard deviation (i.e., the ratio of the standard deviation to the corresponding transmissivity) of the sample is extremely low at approximately 0.01–0.07. This indicates that the transmittance shows almost no dependence on the wavelength of the incident light.

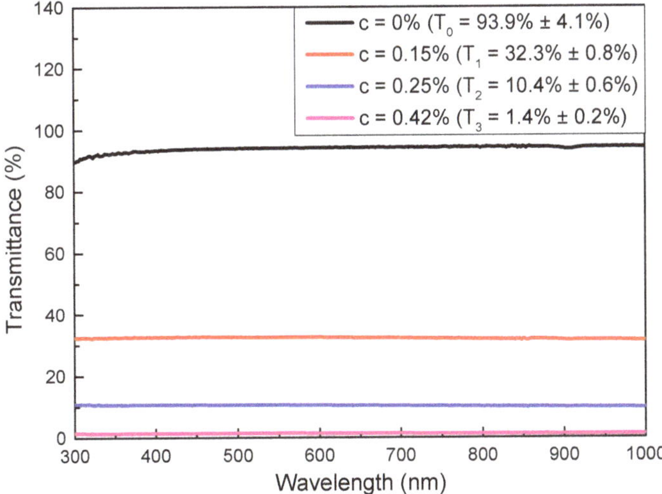

Figure 2. Four transmission lines for the pure PDMS sample and the three concentrations (0%, 0.15%, 0.25% and 0.42%) of microcrystalline graphite powder composites in the polydimethylsiloxane (PDMS) samples within the 300–1000 nm wavelength range.

In accordance with the Beer–Lambert law,

$$T = 10^{-kch_0} \tag{1}$$

where T is the transmittance of the sample, k is the absorptivity, c is the microcrystalline graphite powder concentration and h_0 is the sample thickness. If k and h_0 are both constants, then T and c will have an exponential relationship.

We fit a relationship curve between the transmittance and the powder concentration using the experimental data obtained, as shown in Figure 3, where $k = 3.24$ and $h_0 = 1$ mm. As the figure shows, the experimental data are in good agreement with the values that were theoretically predicted using the Beer–Lambert equation.

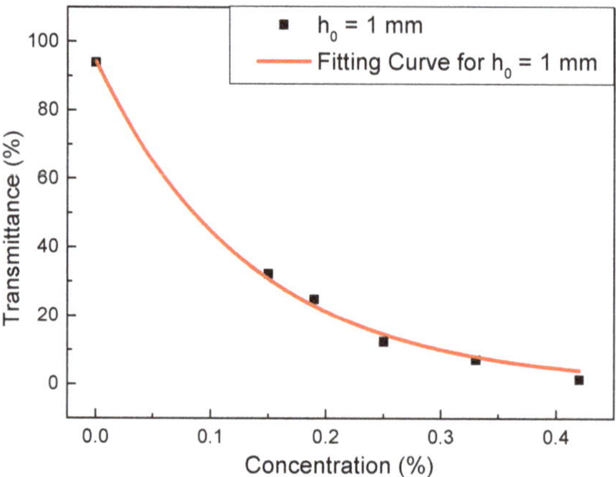

Figure 3. Transmittance characterization versus microcrystalline graphite powder concentration for samples with the same thickness ($h_0 = 1$ mm).

3.2. Transmittance Characterization of Wedge-Shaped Specimen

In addition to the microcrystalline graphite powder concentration, the sample thickness (i.e., the length of the optical path) also affects the transmittance. To demonstrate the regular pattern of the transmittance, we fabricated a cuneiform sample with a thickness that changed evenly from one side to the other while maintaining the same microcrystalline graphite powder concentration. Figure 4 shows a photograph of the cuneiform specimen. The transparency clearly shows a gradual change from the left side towards the right side.

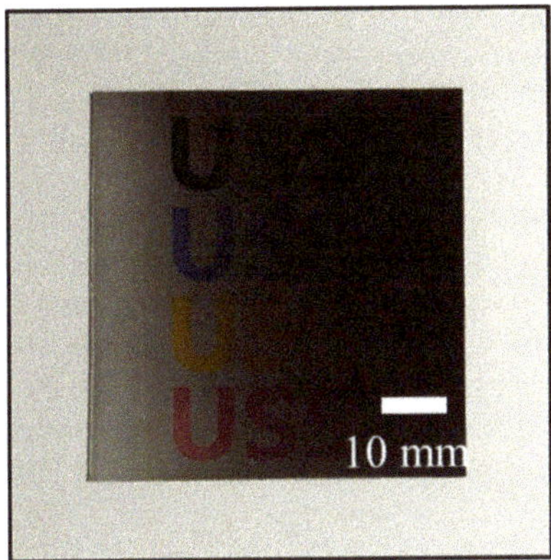

Figure 4. Cuneiform specimen (concentration $c = 0.2$ %; thickness, 0.3 mm $< h <$ 2.0 mm), placed on top of paper with printed colored fonts.

We took measurements at several different positions on the wedge-shaped sample along the direction of the change in thickness. In keeping with the Beer–Lambert law, when k and c are constant, the relationship between the transmittance and the thickness should be an exponential function. In Figure 5, we plotted lgT as a function of the specimen thickness (h), which can be changed along with the spatial position. The resulting relationship can then be expressed as follows:

$$lgT = -\varepsilon h \tag{2}$$

where the thickness has a value in the 0.3 mm $< h <$ 2.0 mm range and $\varepsilon = kc$ (where k is the specimen absorptivity, c is the microcrystalline graphite powder concentration and both values are constant) is calculated to be 0.93. This shows good agreement with the value calculated using the Beer–Lambert law.

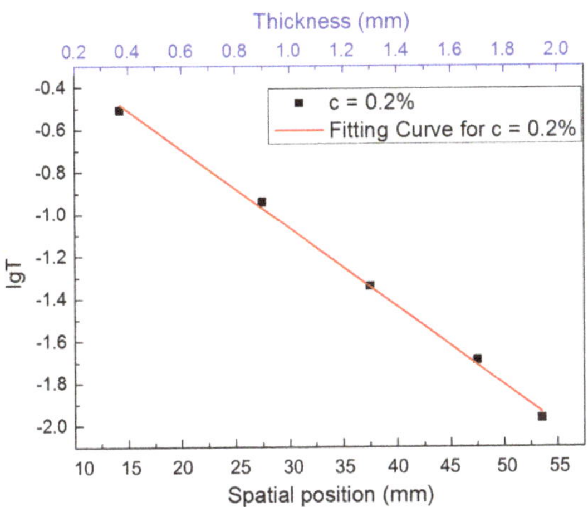

Figure 5. Characterization of transmittance versus sample thickness.

3.3. Mechanical Stretching of Trapezoid-Shaped Composite Films with Uniform Thickness

Mechanical loading can be used to control the transmittance of the sample by varying its thickness. Both uniaxial stretching and biaxial stretching can be used for this purpose. López [29] verified the feasibility of this method. However, because their sample had uniform thickness, it could only provide uniform transmittance on any given sample, even after being stretched.

To obtain a wide transmittance range on a specific sample, we tentatively proposed a more flexible method: first, we cut the uniform-thickness composite film into different trapezoidal shapes in plane and then stretched the resulting samples in the direction perpendicular to the base line and the top line. We tailored the samples to produce three different trapezoidal shapes as shown in Figure 6a. All three trapezoidal samples had the same base line length of 30 mm and the same height of 60 mm, but the lengths of their top lines were 10 mm, 15 mm and 20 mm. Henceforth, we called the three trapezoidal samples listed above A, B and C, respectively. We then fixed sample A using its base line and added a mechanical stretching load on its top line. This trapezoidal specimen was stretched out to an elongation of 20%; in other words, the height of the stretched specimen increased to 72 mm. The measurements were repeated and taken for the other two samples, B and C, in turn.

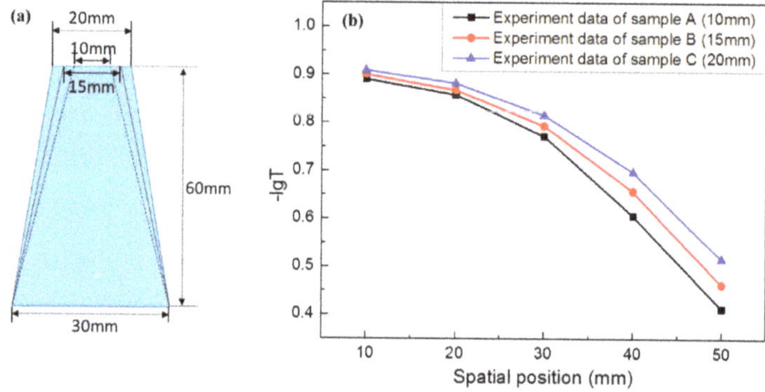

Figure 6. (**a**) Three different trapezoid-shaped samples with top line dimensions of (A) 10 mm, (B) 15 mm and (C) 20 mm; (**b**) transmittance at the central line for stretched composite films A, B and C along their stretching directions at intervals of 10 mm; the inset shows a diagram of the stretched sample.

For simplicity, we considered the case in which stretching occurred in the plane perpendicular to the direction of propagation of the light when loading was applied. When we stretched the sample, because of its incompressibility, the concentration of the microcrystalline graphite powder remained unaffected; however, the sample's thickness gradually decreased, i.e., the optical path was reduced by degrees. To be precise, the sample's thickness changed gradually from top to bottom because the forces that acted at the different locations were different. In this way, we acquired a type of composite film with a special feature in that its transmittance varied gradually from one side of the sample to the other. After stretching, we took the base line of the sample to be the initial position, i.e., x = 0, and then measured the transmittance values of the three trapezoidal samples at their central line along the stretching direction X at 10 mm intervals. Figure 6b shows the results.

If the stretched trapezoid-shaped samples maintain their normal trapezoidal shapes, then the thickness variation in the X direction will be linear. However, in the experiments, both the sides and the centers of the samples showed inhomogeneous shrinkage along the stretching direction. This led to inhomogeneous thickness gradients on all three stretched composite films, where the rate of variation in thickness increased as the measurement point became closer to the top side. The variation of transmittance shown in Figure 6b was consistent with the variation of thickness of the stretched composites. Furthermore, we found that samples with different trapezoidal shapes have different transmittance tuning capabilities. In addition, both the rate and the range of the variation in transmittance in sample A, which had the shortest top length, were maxima. In contrast, both the rate and the range of the variation in transmittance of sample C, which had the longest top length, were minimal. As a result, for samples A, B and C, the transmittance changed by 200.0%, 173.5% and 147.4% compared to their original transmittances, respectively.

This technique could enhance the optical transmittance tuning abilities of these composites. The novel opportunities that are offered by this composite material with its specific optical response can be used to provide system functionality with a wide range of potential applications.

4. Conclusions

In this work, we reported the fabrication of PDMS composite films and demonstrated the optical transmittance properties of these films in the 300–1000 nm wavelength region. By varying the microcrystalline graphite powder concentration or the composite film thickness, we could control the film's transmittance. It has been proved that the relationship among the transmittance, the concentration and the thickness follows the Beer–Lambert law. Furthermore, we focused on loading via mechanical stretching of the trapezoidal sample. Using this method, our system was able to tune the sample transmittance more flexibly. Specifically, the transmittance varied gradually from one side of a given sample to the other side. In addition, after stretching, samples with different trapezoidal shapes showed differing transmittance tuning abilities.

Author Contributions: Conceptualization, B.S., H.W. and Q.W.; writing—original draft preparation, H.W., B.S. and Q.W.; writing—review and editing, B.S., Y.H., D.Z. and S.Z.; supervision, B.S. and Y.H.; project administration, B.S. and Y.H.; funding acquisition, B.S. and Y.H.

Funding: This research was partially supported by the Natural Science Foundation of Shanghai (19ZR1436100), the National Natural Science Foundation of China (61775140, 61775141 and 11105149), the National Key Research and Development Program of China (2016YFB1102303) and the National Key Foundation for Exploring Scientific Instruments (2016YFF0101904).

Acknowledgments: We would like to express our gratitude to the editors and the reviewers for their constructive and helpful review comments.

Conflicts of Interest: The authors declare no conflict of interest.

References

1. Ma, J.; Hu, J.; Li, Z.; Nan, C.W. Recent progress in multiferroic magnetoelectric composites: From bulk to thin films. *Adv. Mater.* **2011**, *23*, 1062–1087. [CrossRef] [PubMed]
2. Qi, L.; Lee, B.I.; Chen, S.H.; Samuels, W.D.; Exarhos, G.J. High-dielectric-constant silver–epoxy composites as embedded dielectrics. *Adv. Mater.* **2005**, *17*, 1777–1781. [CrossRef]
3. Jones, R.M. *Mechanics of composite materials*, 2nd ed.; Taylor & Francis Group: Abingdon, UK, 1999; pp. 389–400.
4. Chen, J.; Zheng, J.; Gao, Q.; Zhang, J.; Zhang, J.; Omisore, O.M.; Wang, L.; Li, H. Polydimethylsiloxane (PDMS)-Based Flexible Resistive Strain Sensors for Wearable Applications. *Appl. Sci.* **2018**, *8*, 345. [CrossRef]
5. Gandhi, S.S.; Chien, L.C. High transmittance optical films based on quantum dot composites nanoscale polymer dispersed liquid crystals. *Opt. Mater.* **2016**, *54*, 300–305. [CrossRef]
6. Granqvist, C.G.; Avendaño, E.; Azens, A. Electrochromic coatings and devices: Survey of some recent advances. *Thin Solid Films* **2003**, *442*, 201–211. [CrossRef]
7. Dyer, A.L.; Grenier, C.R.G.; Reynolds, J.R.A. Poly(3,4-alkylenedioxythiophene) Electrochromic Variable Optical Attenuator with Near-Infrared Reflectivity Tuned Independently of the Visible Region. *Adv. Funct. Mater.* **2007**, *17*, 1480–1486. [CrossRef]
8. Sampanthar, J.T.; Neoh, K.G.; Ng, S.W.; Kang, E.T.; Tan, K.L. Flexible Smart Window via Surface Graft Copolymerization of Viologen on Polyethylene. *Adv. Mater.* **2000**, *12*, 1536–1539. [CrossRef]
9. Parkin, I.P.; Manning, T.D. Intelligent Thermochromic Windows. *J. Chem. Educ.* **2006**, *83*, 393–400. [CrossRef]
10. Hoi, S.K.; Chen, X.; Kumar, V.S.; Homhuan, S.; Sow, C.H.; Bettiol, A.A. Microfluidic Chip with Integrated Colloidal Crystal for Online Optical Analysis. *Adv. Funct. Mater.* **2011**, *21*, 2847–2853. [CrossRef]
11. Scaffaro, R.; Maio, A. Influence of Oxidation Level of Graphene Oxide on the Mechanical Performance and Photo-Oxidation Resistance of a Polyamide 6. *Polymers* **2019**, *11*, 857. [CrossRef]
12. Scaffaro, R.; Maio, A. A green method to prepare nanosilica modified graphene oxide to inhibit nanoparticles re-aggregation during melt processing. *Chem. Eng. J.* **2017**, *16*, 1034–1047. [CrossRef]

13. Lee, H.; Low, M.J.; Lim, C.H.J.; An, J.; Sandeep, C.S.; SRohith, T.M.; Rhee, H.; Murukeshan, V.M.; Kim, Y. Transferable ultra-thin multi-level micro-optics patterned by tunable photoreduction and photoablation for hybrid optics. *Carbon* **2019**, *149*, 572–581. [CrossRef]

14. Scaffaro, R.; Maio, A.; Re, G.L.; Parisi, A.; Busacca, A. Advanced piezoresistive sensor achieved by amphiphilic nanointerfaces of graphene oxide and biodegradable polymer blends. *Compos. Sci. Technol.* **2018**, *156*, 166–176. [CrossRef]

15. Zhang, Y.; Lu, Y.; Yan, X.; Gao, W.; Chen, H.; Chen, Q.; Bai, Y.X. Functional & Enhanced Graphene/Polyamide 6 composite fiber Constructed by A Facile and Universal Method. *Compos. Part A* **2018**, *57*, 10967–10976.

16. Wang, W.; Zhang, Y.; Han, B.; Ma, J.; Wang, J.; Han, D.; Ma, Z.; Sun, H. A complementary strategy for producing moisture and alkane dual-responsive actuators based on graphene oxide and PDMS bimorph. *Sens. Actuators B* **2019**, *290*, 133–139. [CrossRef]

17. Wang, Y.; Yang, R.; Shi, Z.; Zhang, L.; Shi, D.; Wang, E.; Zhang, G. Super-Elastic Graphene Ripples for Flexible Strain Sensors. *ACSNANO* **2011**, *5*, 3645–3650. [CrossRef]

18. Kou, H.; Zhang, L.; Tan, Q.; Liu, G.; Dong, H.; Zhang, W.; Xiong, J. Wireless wide-range pressure sensor based on graphene/PDMs sponge for tactile monitoring. *Sci. Rep.* **2019**, *9*, 3916. [CrossRef]

19. Scaffaro, R.; Maio, A. Integrated ternary bionanocomposites with superior mechanical performance via the synergistic role of graphene and plasma treated carbon nanotubes. *Compos. Part B* **2019**, *168*, 550–559. [CrossRef]

20. Scaffaro, R.; Maio, A.; Lopresti, F. Effect of graphene and fabrication technique on the release kinetics of carvacrol from polylactic acid. *Compos. Sci. Technol.* **2019**, *169*, 60–69. [CrossRef]

21. Baetens, R.; Jelle, B.P.; Gustavsen, A. Properties, requirements and possibilities of smart windows for dynamic daylight and solar energy control in buildings: A state-of-the-art review. *Sol. Energy Mater. Sol. Cells* **2010**, *94*, 87–105. [CrossRef]

22. Molberg, M.; Leterrier, Y.; Plummer, C.J.G.; Walder, C. Frequency dependent dielectric and mechanical behavior of elastomers for actuator applications. *J. Appl. Phys.* **2009**, *106*, 054112–054117. [CrossRef]

23. Sheng, B.; Luo, L.; Huang, Y.; Chen, G.; Zhou, H.; Zhang, D.; Zhuang, S. Tailorable Elastomeric Grating with Tunable Groove Density Gradient. *IEEE Photonics J.* **2017**, *9*, 2400406. [CrossRef]

24. Rosset, S.; Shea, H.R. Flexible and stretchable electrodes for dielectric elastomer actuators. *Appl. Phys. A* **2013**, *110*, 281–307. [CrossRef]

25. Zhang, Y.; Xu, S.; Fu, H.; Lee, J.; Su, J.; Hwang, K.C.; Rogers, J.A.; Huang, Y. Buckling in serpentine microstructures and applications in elastomer-supported ultra-stretchable electronics with high areal coverage. *Soft Matter* **2013**, *9*, 8062–8070. [CrossRef] [PubMed]

26. Adrega, T.; Lacour, S.P. Stretchable gold conductors embedded in PDMS and patterned by photolithography: Fabrication and electromechanical characterization. *J. Micromech. Microeng.* **2010**, *20*, 055025. [CrossRef]

27. Tooley, W.W.; Feghhi, S.; Han, S.J.; Wang, J.; Sniadecki, N.J. Thermal fracture of oxidized polydimethylsiloxane during soft lithography of nanopost arrays. *J. Micromech. Microeng.* **2011**, *21*, 54013–54019. [CrossRef]

28. Xin, Q.B.; Ookawa, K.; Wong, J.Y. Evaluation of polydimethylsiloxane scaffolds with physiologically-relevant elastic moduli: Interplay of substrate mechanics and surface chemistry effects on vascular smooth muscle cell response. *Biomaterials* **2005**, *26*, 3123–3129.

29. López Jiménez, F.; Kumar, S.; Reis, P.M. Soft Color Composites with Tunable Optical Transmittance. *Adv. Opt. Mater.* **2016**, *4*, 620–626. [CrossRef]

30. Minnai, C.; Di, V.M.; Milani, P. Mechanical-optical-electro modulation by stretching a polymer-metal nanocomposite. *Nanotechnology* **2017**, *28*, 355702. [CrossRef]

31. Mohamed-Noriega, N.; Hinojosa, M.; González, V.; Rodil, S.E. Polymer-based composite with outstanding mechanically tunable refractive index. *Opt. Mater.* **2016**, *58*, 18–23. [CrossRef]

32. Apostoleris, H.N.; Chiesa, M.; Stefancich, M. Improved transparency switching in paraffin-PDMS composites. *J. Mater. Chem. C* **2015**, *3*, 1371–1377. [CrossRef]

33. Prasse, T.; Flandin, L.; Schulte, K.; Bauhofer, W. In situ observation of electric field induced agglomeration of carbon black in epoxy resin. *Appl. Phys. Lett.* **1998**, *72*, 2903–2905. [CrossRef]

34. Cooper, C.A.; Ravich, D.; Lips, D.; Mayer, J.; Wagner, H.D. Distribution and alignment of carbon nanotubes and nanofibrils in a polymer matrix. *Compos. Sci. Technol.* **2002**, *62*, 1105–1112. [CrossRef]

35. Schwarz, M.K.; Bauhofer, W.; Schulte, K. Alternating electric field induced agglomeration of carbon black filled resins. *Polymer* **2002**, *43*, 3079–3082. [CrossRef]

36. Ajayan, P.M. Aligned carbon nanotubes in a thin polymer film. *Adv. Mater.* **1995**, *7*, 489–491. [CrossRef]

37. Kimura, T.; Ago, H.; Tobita, M.; Ohshima, S.; Kyotani, M.; Yumura, M. Polymer Composites of Carbon Nanotubes Aligned by a Magnetic Field. *Adv. Mater.* **2002**, *14*, 1380–1383. [CrossRef]

38. Choi, E.S.; Brooks, J.S.; Eaton, D.L.; Al-Haik, M.S. Enhancement of thermal and electrical properties of carbon nanotube polymer composites by magnetic field processing. *J. Appl. Phys.* **2003**, *94*, 6034–6039. [CrossRef]

 *applied sciences*

Article

Synthesis and Catalytic Activity of Activated Carbon Supported Sulfonated Cobalt Phthalocyanine in the Preparation of Dimethyl Disulfide

Zhiliang Cheng [1,2], Mingxing Dai [2], Xuejun Quan [1,2,*], Shuo Li [2,*], Daomin Zheng [1], Yaling Liu [1] and Rujie Yao [1]

[1] Chongqing Unis Chemical Company Ltd., Chongqing 401121, China; purper886@126.com (Z.C.); zhengdaomin@163.com (D.Z.); liuyaling886@126.com (Y.L.); ccyao@163.com (R.Y.)

[2] School of Chemistry and Chemical Engineering, Chongqing University of Technology, Chongqing 400054, China; dai_mingxing@yeah.net

* Correspondence: hengjunq@cqut.edu.cn (X.Q.); lishuo@cqut.edu.cn (S.L.); Tel.: +86-(0)23-6256-3180 (X.Q.); +86-(0)23-6256-3183 (S.L.)

Received: 29 November 2018; Accepted: 26 December 2018; Published: 31 December 2018

Featured Application: Dimethyl disulfide (DMDS) is an important fine chemical, which is widely used as soil fumigant, herbicide, food additive, etc. and exhibits broad market prospects. In this work, an activated carbon (AC)-supported sulfonated cobalt phthalocyanine (AC-CoPcS) catalyst through a chemical linkage of ethylenediamine was successfully synthesized and applied to the preparation of DMDS. The new catalyst shows better catalytic and reuse performance than the commercial one, and exhibits good potential for the industrial application of manufacturing DMDS.

Abstract: The Merox process was widely applied in the fine chemical industry to convert mercaptans into disulfides by oxidation with oxygen, including dimethyl disulfide (DMDS). In this paper, a new activated carbon (AC)-supported sulfonated cobalt phthalocyanine (AC-CoPcS) catalyst was prepared through the chemical linkage of ethylenediamine between them. UV−VIS, FT-IR, BET, and XPS were used to characterize the structure of the new catalyst. Then AC-CoPcS was applied to catalyze sodium methylmercaptide (SMM) oxidation for the preparation of DMDS. The effect of process parameters, such as reaction time, catalyst dosage, reaction temperature, and oxygen pressure on SMM conversion per pass (CPP_{SMM}), yield ($Yield_{DMDS}$), and purity of the DMDS ($Purity_{DMDS}$) product were investigated to evaluate the catalytic performance of AC-CoPcS. The new supported catalyst exhibits better catalytic performance than the commercial one and can be properly reused four times to obtain CPP_{SMM} and $Yield_{DMDS}$ higher than 90% and 70%. Under the optimum experimental conditions, the CPP_{SMM} and $Yield_{DMDS}$ could reach as high as 98.7% and 86.8%, respectively, and the purity of the DMDS product is as high as 99.8%. This new supported catalyst exhibits good industrial application prospects.

Keywords: Merox process; dimethyl disulfide; metal phthalocyanine; activated carbon; sodium methylmercaptide oxidation

1. Introduction

As an important fine chemical product, dimethyl disulfide (DMDS) is widely used as a soil fumigant [1–3], herbicide [4], food additive [5], sulfide hydrogenation catalyst, and chemical raw material [6]. Thus, the preparation method of DMDS is vital for all the chemical companies, which is commonly classified as four kinds of manufacturing processes: (1) the dimethyl sulfate process

(Equation (1)) [7]; (2) the methanol vulcanization process (Equation (2)) [8,9]; (3) the methanethiol vulcanization process (Equation (3)) [10,11]; and (4) the methyl mercaptan oxidation process which is also called Merox process (Equation (4)) [12]:

$$(CH_3O)_2SO_2 + Na_2S + S \rightarrow CH_3SSCH_3 + Na_2SO_4 \tag{1}$$

$$2CH_3OH + S + H_2S \xrightarrow{catalyst} CH_3SSCH_3 + 2H_2O \tag{2}$$

$$2CH_3SH + S \xrightarrow{catalyst} CH_3SSCH_3 + H_2S \tag{3}$$

$$4CH_3SH + O_2 \xrightarrow{catalyst} 2CH_3SSCH_3 + 2H_2O \tag{4}$$

Compared with other processes, the Merox process shows many advantages, such as high yield and purity of DMDS products, no usage of highly toxic raw materials, etc. However, the traditional Merox process is a gas-gas reaction, so it has a high risk of explosion and pollutes the environment because of the exhaust gas produced in the process. Additionally, the process has very high requirements for reactor quality, airtightness, and pressure endurance. In order to solve these problems, engineers refined the Merox process of the gas-gas reaction into a gas-liquid reaction using sodium methylmercaptide (SMM) as the reagent, as shown in Equations (5) and (6) [13–15]:

$$CH_3SH + NaOH \rightarrow CH_3SNa + H_2O \tag{5}$$

$$4CH_3SNa + O_2 + 2H_2O \xrightarrow{catalyst} 2CH_3SSCH_3 + 4NaOH \tag{6}$$

Unlike the catalysts used in the reaction shown in Equation (4) of MgO or Na$_2$O [12], the commonly-used catalysts in the reaction shown in Equation (6) are metal phthalocyanines and their derivatives, such as sulfonated cobalt phthalocyanine (CoPcS) [13,14], cobalt tetraaminophthalocyanine (CoTAPc) [16], etc. The former is the most widely-used catalyst because of its lower cost, high efficiency, and high selectivity of the catalytic performance. However, the free CoPcS catalyst is hydrosoluble, therefore, it very easily gets into the DMDS product, creating separation problems between the catalyst and the product. This leads to high cost of post-treatment of DMDS products, and the catalyst will not be able to be reused. In order to solve these problems, the researchers supported the metal phthalocyanines on solid supports, such as activated carbon (AC) [17,18], activated carbon fiber [19,20], viscose fibers [21], carbon nanotubes [22], graphene oxide [23], silicas [24], mineral carriers [25,26], TiO$_2$ [27,28], molecular sieves [29], etc. In contrast, AC is the best choice for industrial applications because of its special advantages of high surface area, well-developed pore structure, high absorptivity, and stability. However, the AC-supported sulfonated cobalt phthalocyanine (AC-CoPcS) catalyst is commonly prepared by the physical method of dipping [13,14,17] or dispersion [18]. There still exists the problem of loss of the active ingredient of the supported catalyst, resulting in dissatisfactory reuse performance. Thus, the CoPcS should be supported on AC by the chemical grafting method.

In this paper, we developed a proprietary technology of supported catalyst preparation method through a chemical linkage of ethylenediamine between AC and CoPc. Then the new AC-CoPcS catalyst was characterized by UV−VIS spectroscopy (UV–VIS), Fourier transformation infrared spectrum (FT-IR), BET surface area test, and X-ray photoelectron spectroscopy (XPS) before it was applied to prepare DMDS. During the catalytic experiments, the effect of operation parameters of reaction time (t), catalyst dosage (C_{ca}), reaction temperature (T_{re}), and oxygen pressure ($P(O_2)$) on SMM conversion per pass (CPP_{SMM}), yield ($Yield_{DMDS}$), and purity of the DMDS product ($Purity_{DMDS}$) were investigated. Moreover, the catalytic performance comparison between the newly prepared AC-CoPcS and commercial AC-CoPcS was also analyzed.

2. Materials and Methods

2.1. Materials

The SMM solution (18.0 wt%, free alkali 2.4 wt%), commercial CoPcS (99 wt%, Co $\geq$ 6 wt%), AC-CoPcS (CoPcS $\geq$ 5 wt%, BET 265.4 m^2/g, D_{50} = 55.5 μm, prepared by the dipping method) were provided by Chongqing Unis Chemical Co., Ltd. (Chongqing, China). The compressed oxygen (99.99 wt%, 45 kg) and AC (analytical purity, BET 704.8 m^2/g, D_{50} = 50.7 μm) were purchased from a local chemical reagent company (Chongqing Maoye Chemical Reagent Co., Ltd., Chongqing, China). The sulfoxide chloride (SOCl$_2$), ethylenediamine (EDA), N, N-dimethylformamide (DMF), absolute ethyl alcohol (CH$_3$CH$_2$OH), and nitric acid (HNO$_3$) are of analytical purity without further purification; methenyl trichloride (CHCl$_3$, HPLC grade); high purity helium (He, 99.999%) and all these experimental chemicals were purchased from Chengdu Aikeda Chemical Reagent Co., Ltd. (Chengdu, Sichuan, China).

2.2. Synthesis of AC-Supported CoPcS (AC-CoPcS)

The synthetic procedure of AC-CoPcS has applied for a Chinese patent [30], and mainly includes the following three steps:

2.2.1. Synthesis of Modified AC by Ethylene Diamine (AC-E)

Activated carbon (AC) was washed with deionized water and dried to have 10.0 g of AC which was added to a 250 mL beaker that contains 150 mL nitric acid solution ($V(\mathrm{HNO_3}){:}V(\mathrm{H_2O})$ = 1:2). The reaction mixture was stirred at 60 °C for 24 h. The product was then washed with deionized water to neutralize and dried in vacuum oven at 105 °C to obtain nitric acid-treated AC (AC-T). Then 2.0 g AC-T and 5 mL SOCl$_2$ were added into a flask. The reaction mixture was heated at 80 °C for 12 h. After the reaction, the mixture was heated up to 130 °C to evaporate residual SOCl$_2$ and cooled down to room temperature (25 °C). The product was added to 100 mL DMF which dissolved 5 mL EDA and reacted at 50 °C for 8 h. After cooling to room temperature, the product was washed with deionized water, DMF, and absolute ethyl alcohol, respectively, several times, and then dried in a vacuum oven at 105 °C to obtain AC-E. The synthetic process is illustrated in Scheme 1.

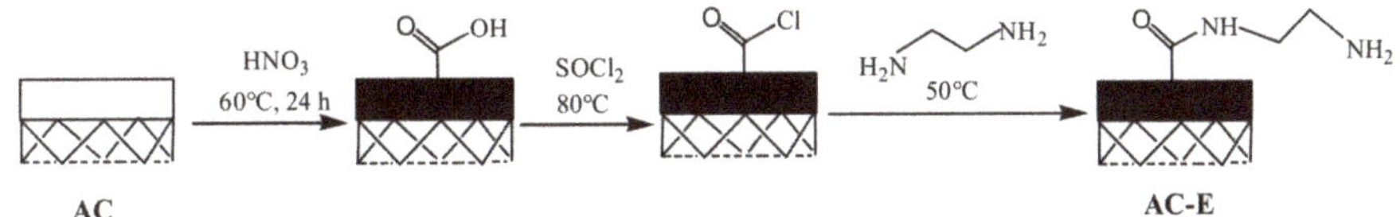

Scheme 1. Schematic representation of the preparation of modified AC by ethylene diamine (AC-E).

2.2.2. Synthesis of Modified CoPcS by Sulfoxide Chloride (CoPc(SO$_2$Cl)$_4$)

As CoPcS is a very common chemical product that can be purchased easily, so we do not synthesize it, but purchase it directly. Commercial CoPcS (0.10 g) was dissolved with 100 mL DMF in a flask, and five milliliters of SOCl$_2$ was added to the flask. Then the mixture was heated at 75 °C for 24 h. After the reaction, the product was cooled down to 50 °C and the residual unreacted SOCl$_2$ was removed through vacuum distillation to obtain the CoPc(SO$_2$Cl)$_4$ solution. The synthetic route is illustrated in Scheme 2.

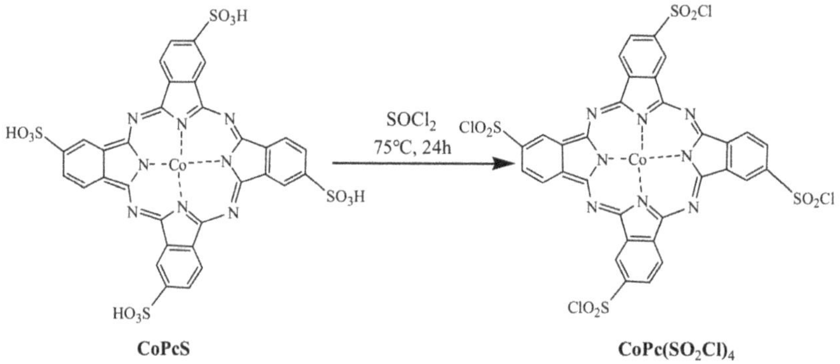

Scheme 2. Schematic representation of preparation of modified sulfonated cobalt phthalocyanine by sulfoxide chloride (CoPc(SO$_2$Cl)$_4$).

2.2.3. Synthesis of the AC-Supported CoPcS (AC-CoPcS)

Then 2.00 g of previously prepared AC-E was added into the CoPc(SO$_2$Cl)$_4$ solution to react at 45 °C for 12 h and then cooled to room temperature after the reaction. The product was washed by deionized water several times to remove the unreacted component and was dried in a vacuum oven at 60 °C to obtain the final product of AC-CoPcS. The synthetic route is shown in Scheme 3.

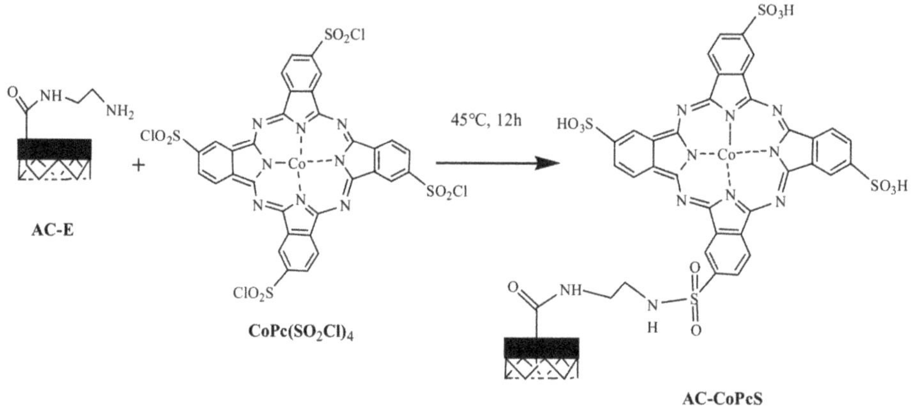

Scheme 3. Schematic representation showing the preparation of AC-supported sulfonated cobalt phthalocyanine (AC-CoPcS).

2.3. Characterization Methods

The UV–VIS absorption spectra were recorded with a UV–VIS absorption spectrometer (U3010, Hitachi, Loveland, CO, USA) by dissolving the CoPcS into DMF with the concentration of 1.0×10^{-6} mol/L. The FT-IR spectrum was characterized by a Spectrum One infrared spectrometer (Perkin Elmer, Waltham, MA, USA) using a KBr presser. X-ray photoelectron spectroscopy (XPS) was obtained on an AXIS Ultra DLD Thermo scientific spectroscope (Shimadzu, Kyoto, Japan) with the measurement condition of an X-ray source, a monochromatic aluminum target at 1486.6 eV, ion gun current of 3 mA, and electric voltage of 15 kV. The specific surface area based on nitrogen physisorption of the catalyst samples was measured by a JW-BK100C surface area instrument (JWGB, Sci.&Tech., Beijing, China).

2.4. Catalytic Experiment of AC-CoPcS Catalyst for the Preparation of DMDS

The catalytic experiments were carried out in a high-pressure reactor (HPR, $Volume_{max}$ = 1 L, $Pressure_{max}$ = 14.5 MPa, its structure is shown in Figure 1). All the catalytic experiments were carried out in a batch operation, and during each experimental circle, different amounts of AC-CoPcS catalyst was added into the reactor with 450 g SMM solution (18.0 wt%). Before the reaction, the air in the HPR should be replaced by oxygen. Then the control box of the HPR was turned on, and the stirring speed was fixed at 600 rpm. When the temperature in the reactor reached the desired value, the oxygen was inlet to start the catalytic reaction, which was labeled as the zero time point, and the reaction time was recorded accurately. After the reaction, heating electricity was turned off while cooling water was turned on until the temperature in HPR reached room temperature. Before opening the HPR, the exhaust valve was opened to get atmospheric pressure. The oily organic product of DMDS was separated by a separating funnel and was measured by an electronic balance (0.0001 g) to calculate the $Yield_{DMDS}$. The initial and final concentration of SMM solution was determined by the iodometric method to calculate the CPP_{SMM}. The principle of the iodometric method is that SMM was reacted by excessive iodine standard solution, and residual iodine was detected by sodium hyposulfite standard solution. The AC-CoPcS catalyst in the aqueous phase was filtered, washed by deionized water, and dried in a vacuum oven for reuse. All the catalytic experiments were conducted in triplicate and the statistical variance was found not to exceed five percent.

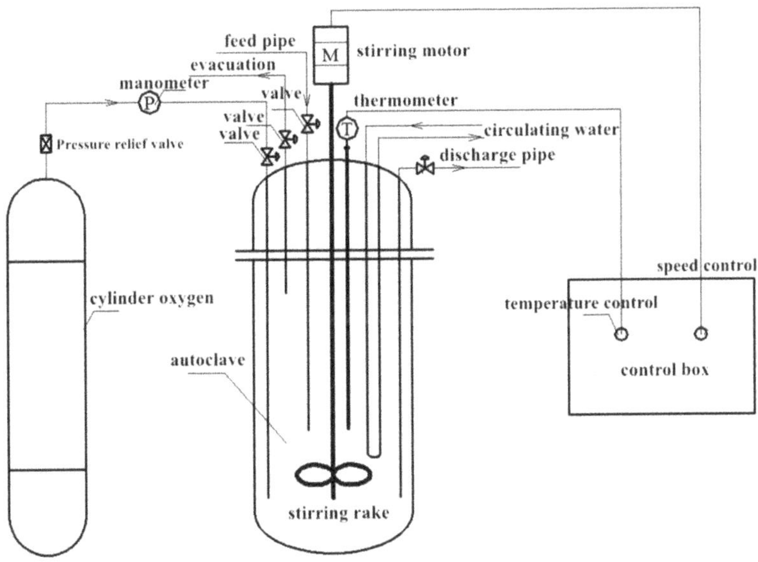

Figure 1. Structure schematic diagram of the reactor and reaction process for the preparation of DMDS.

2.5. The Determination of Purity of DMDS

The purity of DMDS was calculated by the percentage of different components of the organic product which was measured by a GC6890/MS5973 gas Chromatograph-Mass Spectrometer (GC-MS, Agilent, Santa Clara, CA, USA) with an OV1701 capillary column (30 m × 0.25 mm × 0.25 µm). During each GC-MS test, a test sample of DMDS was prepared by dissolving 1.00 g of DMDS product in 100 mL of methenyl trichloride, and the sample injection volume is 0.5 µL. The heating program of the GC-MS is 50 °C kept for 3 min, then heated at a rate of 10 °C/min to 150 °C and kept for 3 min.

2.6. Calculate Method of CPP$_{SMM}$ and Yield$_{DMDS}$

The conversion per pass of SMM and the yield of the DMDS were calculated as shown in the following equations:

$$CPP_{SMM} = \frac{consumed\ mol\ of\ SMM\ in\ the\ reaction}{nitial\ mol\ of\ SMM} \tag{7}$$

$$Yield_{DMDS} = \frac{quality\ of\ generated\ DMDS\ in\ the\ reaction}{nitial\ quality\ of\ the\ SMM} \tag{8}$$

3. Results and Discussion

3.1. Characterization of CoPcS and AC-CoPcS Catalysts

The UV−VIS and FT-IR were used to confirm the structure of the purchased CoPcS catalyst. The metal phthalocyanines and their derivatives have special absorption spectra in the ultraviolet and visible regions. That is the Q band of 600~800 nm and B band or Sorct band of 300–400 nm [31]. The UV−VIS spectra result of CoPcS catalyst is shown in Figure 2. The CoPcS have obvious absorption spectra in the B band and Q band with the maximum wavelengths located at 326 nm and 664 nm, which can be confirmed as the characteristic absorption peak of the CoPcS [32].

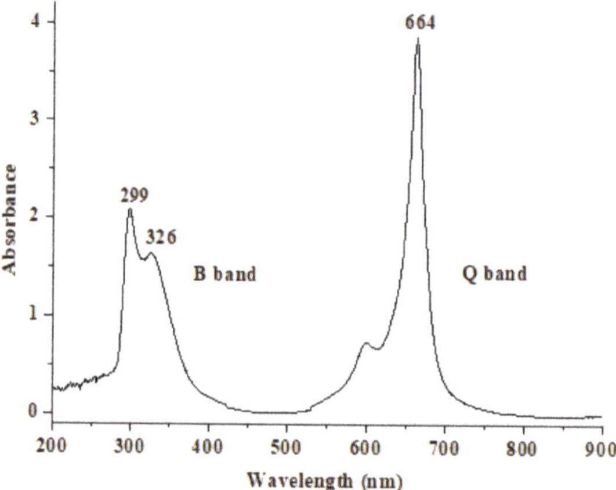

Figure 2. UV–VIS spectra of the free CoPcS catalyst (test sample of CoPcS DMF solution with the concentration of 1.0×10^{-6} mol/L).

FT-IR result of CoPcS is shown in Figure 3, the peaks of 1716, 1542, 1510, and 1400 cm^{-1} corresponded to C=N stretching vibration, C–H bending vibration, C=C skeleton vibration in the benzene ring, and C-C stretching, respectively. We also found the peaks of 1120, 1053, 914, 732, and 717 cm^{-1} of CoPc reported by Seoudi et al. [33], and two characteristic peaks at 623 and 484 cm^{-1} owing to the phthalocyanine skeletal and cobalt ligand vibrations reported by Zhang et al. [34] confirming the structure of the CoPcS.

XPS was used to analyze the surface element and its chemical states of the AC-CoPcS catalyst, and the results are shown in Figure 4. The surface element of the AC-CoPcS is mainly carbon (C), oxygen (O), nitrogen (N), cobalt (Co,) and sulfur (S). This result indicates the fact that CoPcS was successfully combined into the AC through a chemical linkage of ethylenediamine because the Co and S appear in the XPS pattern of the new material. The CoPcS is very easy to dissolve in water, while the AC-CoPcS catalyst is well washed by deionized water after being prepared, so CoPcS appearing in the new materials cannot be physically adsorbed by the AC, but is chemically connected with it.

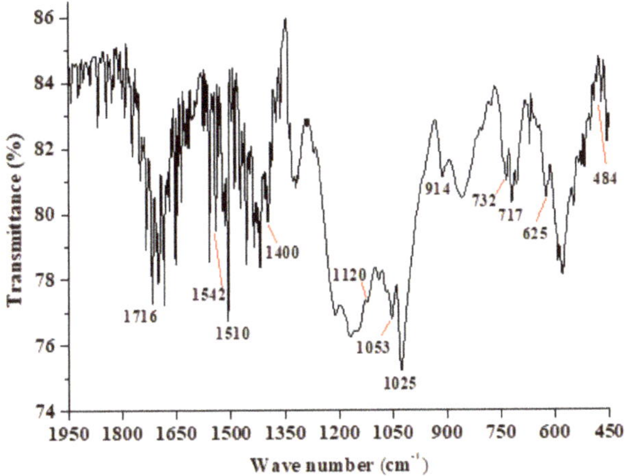

Figure 3. The FT-IR spectra of the free CoPcS catalyst (KBr tabletting treatment).

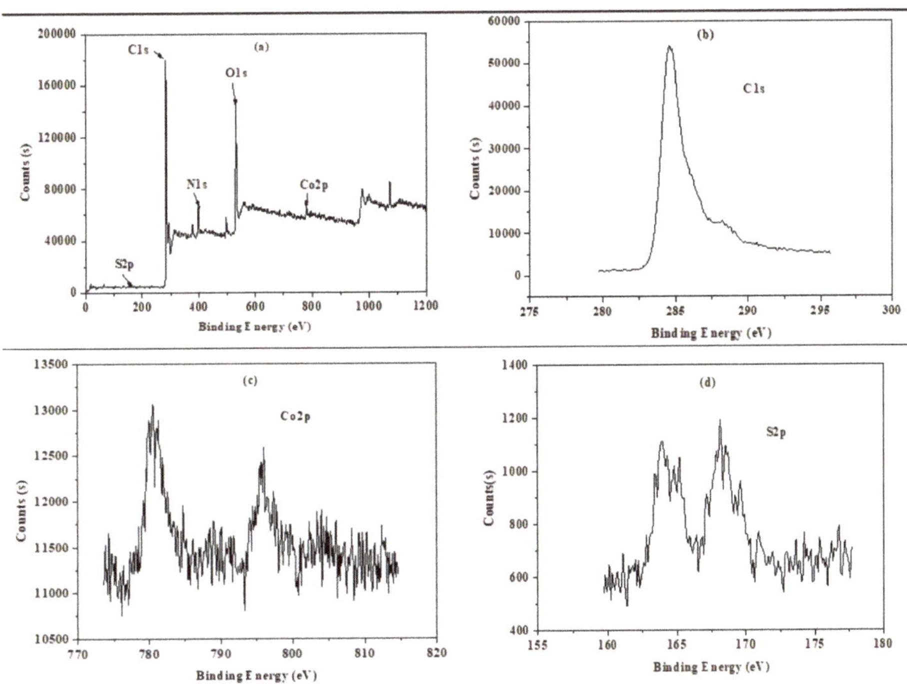

Figure 4. The XPS spectra of the AC-CoPcS (**a**), C1s (**b**), Co2p (**c**), and S2p (**d**).

The surface area results of the AC and AC-supported catalysts are listed in Table 1, the BET specific surface area of AC support, commercial, and newly prepared AC-CoPcS catalysts are 704.8 m^2/g, 265.4 m^2/g, and 301.3 m^2/g. Compared with the commercial AC-CoPcS, the new AC-CoPcS catalyst shows a slightly higher BET specific surface area, larger total pore volume, and smaller average pore size, which may benefit the adsorption of the reagents to accelerate the catalytic reaction, but this should be proved by the catalytic performance of the two catalysts.

Table 1. The surface area test results of the AC, commercial, and newly prepared AC-CoPcS catalysts.

Sample	BET Surface Area (m^2/g)	Total Pore Volume (cm^3/g)	Average Pore Size (nm)
AC	704.8	1.30	2.50
AC-CoPcS (commercial)	265.4	0.46	4.16
AC-CoPcS (new)	301.3	0.51	3.95

3.2. The Effect of Operation Parameters on AC-CoPcS Catalytic Performance

3.2.1. The Performance Comparison between Free CoPcS and AC-CoPcS Catalysts

In order to show the advantages of the supported catalyst, the catalytic performance of free CoPcS and AC-CoPcS catalysts were compared in Table 2 and Figure 5. As is shown in Figure 5, the free CoPcS catalyst is so highly soluble that it runs into not only the water layer but also the organic layer of DMDS product, resulting in post-treatment of DMDS by AC absorption and an inability to reuse the catalyst. However, the supported AC-CoPcS catalyst could be a good solution to these problems: as is shown in Figure 5b, the AC-CoPcS catalyst only stays in the water layer and can be reused through simple filtration and washing treatment. In addition, the AC-CoPcS also exhibits better catalytic performance than the free CoPcS catalyst: as is shown in Table 2, the $Yield_{DMDS}$ and $Purity_{DMDS}$ using the AC-CoPcS catalyst are higher than the one using the free CoPcS catalyst of 86.8% to 81.4% and 99.8% to 98.7%, respectively. The better catalytic performance of the AC-CoPcS catalyst could be attributed to the following reasons: (1) Large specific surface area and flourishing pore structure of AC support, which can not only absorb the reagents of SMM and oxygen to react on its surface, but also remove the byproduct of the reaction to improve the yield and purity of DMDS; or (2) AC itself can be used as the catalyst in the Merox process, which can also help to improve the catalytic performance of the new AC-CoPcS catalyst [35].

(a)   (b)

Figure 5. The unseparated DMDS product (upper layer) after the catalytic oxidation reaction using (**a**) free CoPcS and (**b**) AC-CoPcS catalysts. Experimental conditions: SMM (18.0 wt%) 450 g, RR = 600 r/min, free CoPcS dosage 160 ppm, AC-CoPcS dosage 888.9, t = 60 min, T_{re} = 65 °C, $P(O_2)$ = 0.9 MPa.

Table 2. The performance comparison between free CoPcS and AC-CoPcS catalysts [1].

Catalysts	C_{ca}	$Yield_{DMDS}$	$Purity_{DMDS}$	Post-Treatment of DMDS
CoPcS	160 ppm	81.4%	98.7%	AC adsorption
AC-CoPcS	888.9 ppm	86.8%	99.8%	None

[1] Other experimental conditions: SMM (18.0 wt%) 450 g, RR = 600 r/min, t = 60 min, T_{re} = 65 °C, $P(O_2)$ = 0.9 MPa.

3.2.2. The Effect of Reaction Time (t)

The reaction time is one of the most important process parameters that should be optimized firstly, so the effect of reaction time on CPP_{SMM}, $Yield_{DMDS}$, and $Purity_{DMDS}$ was investigated. The purity of the DMDS product was detected by GC-MS, and the spectra are shown in Figure 6. As can be seen, the retention time of DMDS is around 4 min, and there is no other peak of the byproduct in the GS-MS spectra, indicating the high quality of the DMDS is prepared by the new catalytic reaction system. As is shown in Figure 7, when the reaction time increases from 30 min to 60 min, the CPP_{SMM} and $Yield_{DMDS}$ increase gradually from 62.4% and 44.1% to 98.7% and 86.8%, respectively. However, when the reaction time continues to increase from 60 min to 90 min, the CPP_{SMM} decreases slightly to 84.1% at first and then remains nearly constant, while the $Yield_{DMDS}$ remains constant at the same time. Nevertheless, the reaction time seems have a little effect on the $Purity_{DMDS}$, when the reaction time increases from 30 min to 90 min, the $Purity_{DMDS}$ decreases slightly from 99.8% to 99.1%, which verifies the high quality of the DMDS product that was prepared in the new system. As a result, the reaction time of 60 min is optimal, when the reaction time extends to 75 min or 90 min, the DMDS product may react with each other to form the byproduct, causing the decrease of $Yield_{DMDS}$ and $Purity_{DMDS}$ [30].

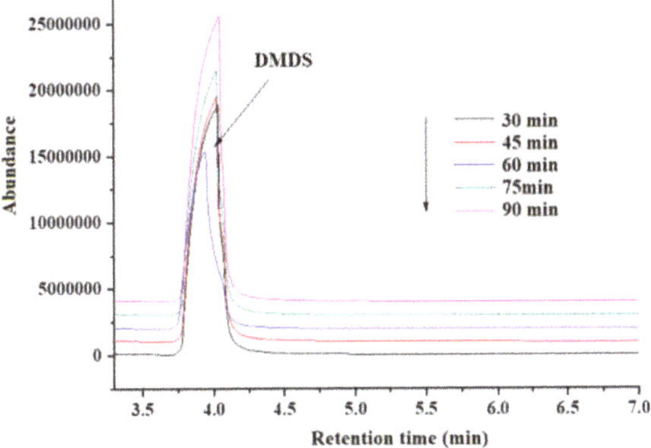

Figure 6. The GC-MS spectra of the DMDS product under different reaction time. Experimental conditions: SMM (18.0 wt%) 450 g, RR = 600 r/min, C_{ca} = 888.9 ppm, T_{re} = 65 °C, $P(O_2)$ = 0.9 MPa.

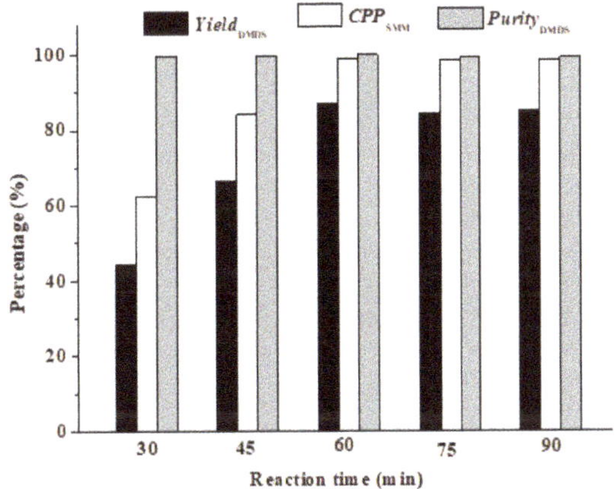

Figure 7. The effect of reaction time on CPP_{SMM}, $Yield_{DMDS}$, and $Purity_{DMDS}$ using the AC-CoPcS catalyst. Experimental conditions: SMM (18.0 wt%) 450 g, RR = 600 r/min, C_{ca} = 888.9 ppm, T_{re} = 65 °C, $P(O_2)$ = 0.9 MPa.

3.2.3. The Effect of AC-CoPcS Catalyst Dosage (C_{ca})

Figure 8 shows the effect of AC-CoPcS catalyst dosage (C_{ca}) on CPP_{SMM}, $Yield_{DMDS}$, and $Purity_{DMDS}$. The CPP_{SMM} and $Yield_{DMDS}$ increase gradually when the C_{ca} increases from 222.2 ppm (µg/g, 0.1 g/450 g) to 888.9 ppm and then begins to decrease continuously when C_{ca} continues to increase to 1555.6 ppm. The optimum AC-CoPcS catalyst dosage is 888.9 ppm where maximum CPP_{SMM} and $Yield_{DMDS}$ of 98.7% and 86.8% are achieved. When the C_{ca} increases, the more active spot of the catalyst and reaction surface involves, so the reaction rate of Equation (6) process accelerates, resulting in a sharp increase of the CPP_{SMM} and $Yield_{DMDS}$ at first. However, when it reaches, and even exceeds, the appropriate dosage, too much heterogeneous solid matter may hinder the gas-liquid mass transfer between SMM and oxygen, so the CPP_{SMM} and $Yield_{DMDS}$ decrease gradually. However, C_{ca} seems to have no effect on the purity of the DMDS product (Figure 8), and the $Purity_{DMDS}$ is nearly constant and its value of all the samples is above 99.5%.

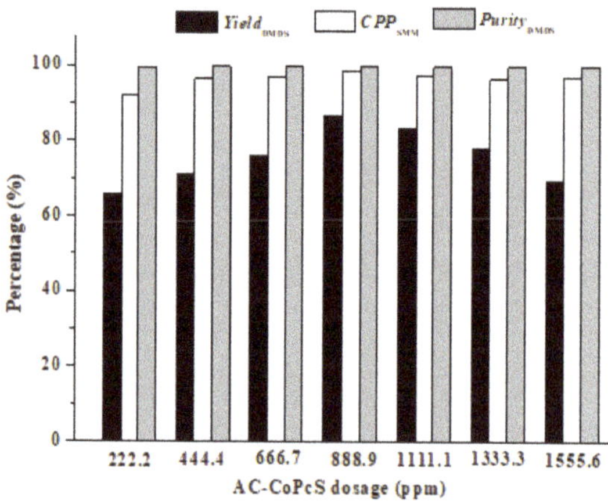

Figure 8. The effect of AC-CoPcS catalyst dosage on CPP_{SMM}, $Yield_{DMDS}$, and $Purity_{DMDS}$. Experimental conditions: SMM (18.0 wt%) 450 g, RR = 600 r/min, t = 60 min, T_{re} = 65 °C, $P(O_2)$ = 0.9 MPa.

3.2.4. The Effect of Reaction Temperature (T_{re})

Commonly, the reaction temperature has a great influence on the reaction rate. The effect of T_{re} on the CPP_{SMM}, $Yield_{DMDS}$, and $Purity_{DMDS}$ was investigated when other operation parameters kept constant, and the result is shown in Figure 9. The CPP_{SMM} and $Yield_{DMDS}$ increase gradually when T_{re} increases from 35 °C to 65 °C, but it decreases slightly when T_{re} continues to increase from 65 °C to 75 °C, so the optimum T_{re} is 65 °C where the best CPP_{SMM} and $Yield_{DMDS}$ are achieved. The T_{re} seems to have little effect on $Purity_{DMDS}$ when it is lower than 65 °C, however, when T_{re} increases from 65 °C to 75 °C the $Purity_{DMDS}$ decreases sharply from 99.8% to 98.0%. The reason for these results is that the reaction of the SMM and oxygen is endothermal, and the reagents are easier to activate, which accelerates the reaction rate when T_{re} increases, leading to the increase of the CPP_{SMM} and $Yield_{DMDS}$. However, extremely higher T_{re} of 75 °C brings not only the byproduct formation problems, which will seriously affect the purity of DMDS product, but also the decrease of $Yield_{DMDS}$ because of the volatilization of DMDS into the gaseous state [6]. When T_{re} is at 75 °C, a white cotton-like byproduct in the organic layer was observed in the experiment.

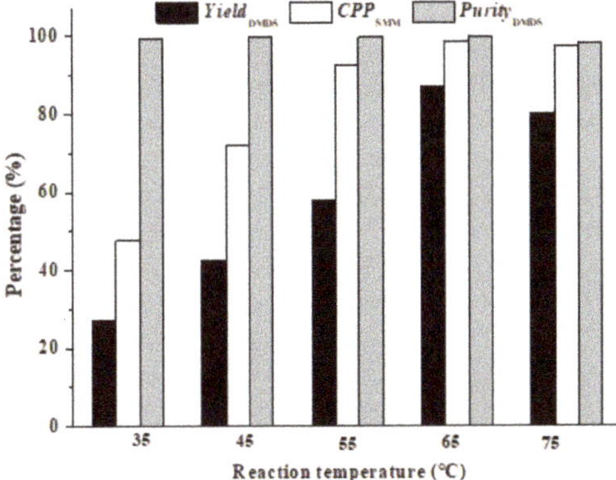

Figure 9. The effect of reaction temperature on CPP_{SMM}, $Yield_{DMDS}$, and $Purity_{DMDS}$ using the AC-CoPcS catalyst. Experimental conditions: SMM (18.0 wt%) 450 g, RR = 600 r/min, C_{ca} = 888.9 ppm, t = 60 min, $P(O_2)$ = 0.9 MPa.

3.2.5. The Effect of Oxygen Pressure ($P(O_2)$)

As is shown in Equation (6), the synthesis reaction of DMDS is a volume reduction reaction, so the increase of $P(O_2)$ will benefit the process. The effect of $P(O_2)$ on the CPP_{SMM}, $Yield_{DMDS}$, and $Purity_{DMDS}$ was investigated, and the result is shown in Figure 10. As is shown in the figure, the $Yield_{DMDS}$ increases sharply from 42.5% to 86.8% with the $P(O_2)$ increasing from 0.5 MPa to 0.9 MPa, and it decreases slightly at 1.0 MPa. However, at the same time, the CPP_{SMM} only increases from 90.5% to 98.7% when the $P(O_2)$ increases from 0.5 MPa to 0.9 MPa, and then also decreases slightly to 96.8% when $P(O_2)$ continues to increase to 1.0 MPa. These results indicate the fact that SMM is much easier to react with oxygen, but the reaction condition is much harsher to obtain a high yield of the DMDS product. Nevertheless, the purity of the DMDS was not affected by the $P(O_2)$, and the $Purity_{DMDS}$ is still as high, above 99.5%.

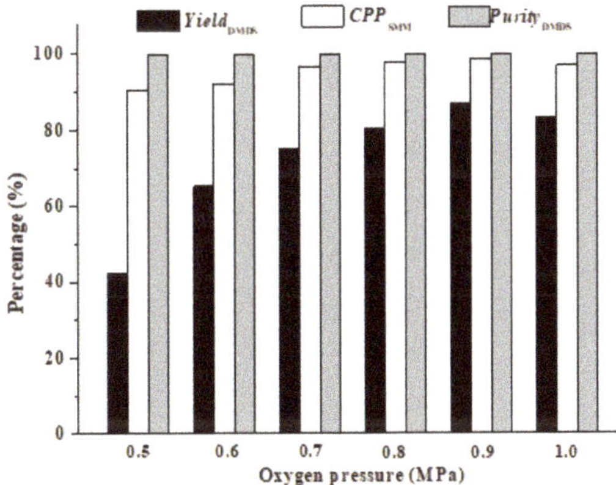

Figure 10. The effect of oxygen pressure on CPP_{SMM}, $Yield_{DMDS}$, and $Purity_{DMDS}$ using the AC-CoPcS catalyst. Experimental conditions: SMM (18.0 wt%) 450 g, RR = 600 r/min, C_{ca} = 888.9 ppm, t = 60 min, T_{re} = 65 °C.

3.3. The Performance Comparison between Commercial and Newly Prepared AC-CoPcS Catalysts

The same optimization method of the operation parameters was applied to the commercial AC-CoPcS catalyst, and the optimum parameters are listed in Table 3. The best catalyst dosage of commercial AC-CoPcS of 1333.3 ppm is much larger than the newly prepared one of 888.9 ppm, which will cost more for industrial application. The performance comparison experiment between the commercial and new AC-CoPcS catalysts was conducted under their optimum parameters, and the result is summarized in Table 3. When compared with commercial AC-CoPcS, the newly prepared AC-CoPcS catalyst shows higher CPP_{SMM} and $Yield_{DMDS}$ of 98.7% and 86.8% than the commercial one of 97.4% and 85.6%. However, the purity of the DMDS product using both the commercial and new AC-CoPcS catalysts are particularly high, 99.7% and 99.8%, respectively.

The most important advantage of the supported catalyst is its reusability. The reuse performance of the commercial and newly prepared AC-CoPcS catalyst is shown in Figures 11 and 12. As can be seen in the figures, CPP_{SMM} and $Yield_{DMDS}$ decrease with the increase of reuse times of the two catalysts. For the new AC-CoPcS catalyst, it can be reused seven times, and for the first four times of reuse the CPP_{SMM} and $Yield_{DMDS}$ only reduce from 98.7% and 86.8% to 92.0% and 72.7%, respectively. However, the significant decrease of the CPP_{SMM} and $Yield_{DMDS}$ occurs at the fifth reuse of the new catalyst of 78.2% and 46.6%, and the CPP_{SMM} and $Yield_{DMDS}$ is only 26.2% and 8% for the seventh reuse. The reuse performance of the commercial AC-CoPcS catalyst is slightly worse than the former one. It can only be reused five times. The CPP_{SMM} and $Yield_{DMDS}$ decrease from 97.4% and 85.6% to 90.2% and 71.3% for the second reuse, but decrease to 70.8% and 33.9% for the third reuse. As a result, the new and commercial AC-CoPcS catalysts can be properly reused four and two times, respectively, which still can achieve relatively high $Yield_{DMDS}$ and CPP_{SMM} of more than 70% and 90%. It is worth mentioning that the reuse times have little effect on the $Purity_{DMDS}$. The $Purity_{DMDS}$ of all the DMDS product remains above 99.3%. The decrease of catalytic performance of AC-CoPcS catalyst with the increase of reuse times can be attributed to the following reasons: Firstly, the more times AC-CoPcS catalyst is reused, the greater loss of its active components during the reaction process of high temperature and high pressure; secondly, the SMM reagent, DMDS product, or even the byproduct produced in the reaction may be absorbed by the AC carrier, which can block up the micropores of the AC, preventing the reagents and catalyst from contacting with each other; and, thirdly, the mechanical strength of the AC carrier will also drop, resulting in its hole collapse, which also reduces the connection of active spots of the catalyst with the reagents [30].

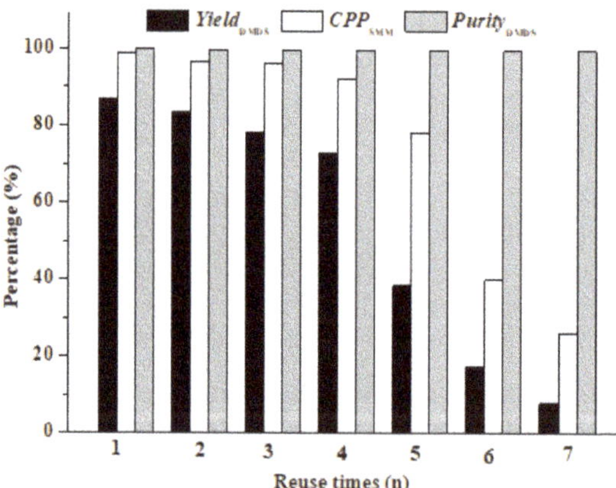

Figure 11. The effect of reuse times of new AC-CoPcS catalyst on CPP_{SMM}, $Yield_{DMDS}$, and $Purity_{DMDS}$. Experimental conditions: SMM (18.0 wt%) 450 g, RR = 600 r/min, C_{ca} = 888.9 ppm, t = 60 min, T_{re} = 65 °C, $P(O_2)$ = 0.9 MPa.

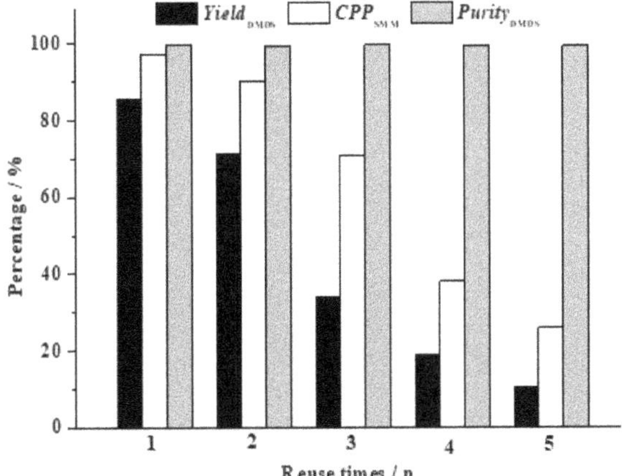

Figure 12. The effect of reuse times of commercial AC-CoPcS catalyst on CPP_{SMM}, $Yield_{DMDS}$, and $Purity_{DMDS}$. Experimental conditions: SMM (18.0 wt%) 450 g, RR = 600 r/min, C_{ca} = 1333.3 ppm, t = 60 min, T_{re} = 65 °C, $P(O_2)$ = 0.9 MPa.

Table 3. The performance comparison between commercial and newly-prepared AC-CoPcS catalysts.

Catalyst	Synthetic Method	CPP_{SMM}	$Yield_{DMDS}$	$Purity_{DMDS}$	Optimum Operation Parameters				Proper Reuse Times
					t	C_{ca}	T_{re}	$P(O_2)$	
AC-CoPcS (new)	chemical grafting	98.70%	86.8%	99.8%	60 min	888.9 ppm	65 °C	0.9 MPa	4
AC-CoPcS (commercial)	physical dipping	97.4%	85.6%	99.7%	60 min	1333.3 ppm	65 °C	0.9 Mpa	2

4. Conclusions

A new activated carbon (AC)-supported sulfonated cobalt phthalocyanine (AC-CoPcS) catalyst for a refined Merox process of SMM oxidation for the preparation of DMDS was successfully synthesized by the chemical grafting method. The new AC-CoPcS catalyst does not get into the DMDS product and can be easily separated for reuse at least four times to keep CPP_{SMM} and $Yield_{DMDS}$ higher than 90% and 70%. The operation parameters of the catalytic reaction were optimized, and the optimum conditions for the new reaction system are a reaction time of 60 min, AC-CoPcS dosage of 888.9 ppm, reaction temperature of 65 °C, and oxygen pressure of 0.9 MPa. Under the optimum conditions, the maximum CPP_{SMM} and $Yield_{DMDS}$ are 98.7% and 86.8%. The purity of the DMDS product can reach as high as 99.8%. Compared with commercial AC-CoPcS prepared by the physical dipping method, the newly prepared AC-CoPcS catalyst shows a better catalytic performance with higher CPP_{SMM} and $Yield_{DMDS}$. Additionally, it can be properly reused twice as often as the commercial one but only two-thirds of its catalyst dosage is needed.

5. Patents

The refined Merox process of Equations (5) and (6) for the preparation of DMDS is our patented technology (CN 105924372 B), and the synthetic method of the supported AC-CoPcS catalyst has already applied for a Chinese patent (CN 106984361 A). No person or organization is allowed to use these technologies for commercial purposes unless admitted by Chongqing Unis Chemical Company Ltd. (CUC).

Author Contributions: X.Q. and S.L. designed the work and revised the initial draft of the paper; M.D. and Z.C. conducted the experiment; Z.C. wrote the initial draft of the paper; and D.Z., Y.L., and R.Y. discussed data analysis and gave many constructive suggestions.

Funding: This research was funded by the Postdoctoral Research Project of Chongqing (Xm2016063), the Research Project of Chongqing Unis Chemical Company Ltd. (2016Q28), the Foundation and Frontier Research Project of Chongqing (cstc2015jcyjA20005), and the Scientific and Technological Research Program of Chongqing Municipal Education Commission (KJ1600927).

Conflicts of Interest: The authors declare no conflict of interest.

References

1. Ajwa, H.; Ntow, W.J.; Qin, R.; Gao, S. Chapter 9—Properties of Soil Fumigants and Their Fate in the Environment. In *Hayes' Handbook of Pesticide Toxicology*, 3rd ed.; Krieger, R., Ed.; Academic Press: New York, NY, USA, 2010; pp. 315–330. 4p.

2. Cabrera, J.A.; Wang, D.; Gerik, J.S.; Gan, J. Spot drip application of dimethyl disulfide as a post-plant treatment for the control of plant parasitic nematodes and soilborne pathogens in grape production. *Pest Manag. Sci.* **2014**, *70*, 1151–1157. [CrossRef] [PubMed]

3. Pecchia, S.; Franceschini, A.; Santori, A.; Vannacci, G.; Myrta, A. Efficacy of dimethyl disulfide (DMDS) for the control of chrysanthemum Verticillium wilt in Italy. *Crop Prot.* **2017**, *93*, 28–32. [CrossRef]

4. Stevens, M.; Freeman, J. Efficacy of dimethyl disulfide and metam sodium combinations for the control of nutsedge species. *Crop Prot.* **2018**, *110*, 131–134. [CrossRef]

5. Gereben, O.; Kohara, S.; Pusztai, L. The liquid structure of some food aromas: Joint X-ray diffraction, all-atom molecular dynamics and reverse Monte Carlo investigations of dimethyl sulfide, dimethyl disulfide and dimethyl trisulfide. *J. Mol. Liq.* **2012**, *169*, 63–73. [CrossRef]

6. Reid, E.E. Organic Chemistry of Bivalent Sulfur. Chemical Publishing, Co.: New York, NY, USA, 1960; Volume 3.

7. Wang, P. Manufacturing Process of Dimethyl Disulfide. CN 1031838A, 22 March 1989.

8. Custer, R.S.; Haines, P.G. Preparation of Dialkyl Disulfides. Eur. Patent 0171092A2, 12 February 1986.

9. Buchholz, B.; Dzierza, E.J.; Baltrus, J.R. Process for the Manufacture of Dialkyl Disulfides. Eur. Patent 0202420A1, 26 November 1986.

10. Pau, E.A. Process for the Manufacture of Dimethyl Disulphide. U.S. Patent US 5,312,993, 17 May 1994.

11. Fremy, G.; Jean-Michel, R. Process for Preparing Dialkyl Disulphides. U.S. Patent US 8,987,519,B2, 24 March 2015.

12. Roberts, J.S.; Hunt, H.R.; Drake, C.A. Continuous Process for the Production of Disulfudes. U.S. Patent US 5,202,494, 13 April 1993.

13. Liu, Y.; Chen, X.; Yao, Y.; Zheng, D. A New Method for Preparation of Dimethyl Disulfide. CN 1,059,243,72B, 22 December 2017.

14. Wang, J.; Fang, X.; Tian, Y.; Yang, J.; Ao, H.; Li, X.; Yi, X. A Method and Device of Methanthiol Solution Oxidation for Preparation of Dimethyl Disulfide. CN 105175296A, 23 December 2015.

15. Guo, K.; Guo, M.; Huang, Y.; Huang, F. A Method and Device for the Preparation of Dimethyl Disulfide. CN 105085338A, 25 November 2015.

16. Quan, X.; Dai, M.; Zheng, D.; Li, S.; Liu, Y.; Cheng, Z.; Yao, Y. Activated Carbon-Loaded Tetraamino Cobalt Phthalocyanine and Application Thereof in Preparing Dimethyl Disulfide as Catalyst. CN 106984361A, 28 July 2017.

17. Gleim, W.; Urban, P. Treatment of Hydrocarbon Distillates. U.S. Patent US 2988500, 13 June 1961.

18. Jiang, D.-e.; Zhao, B.; Huang, H.; Xie, Y.; Pan, G.; Ran, G.; Min, E. Dispersion of cobalt (II) phthalocyaninetetrasulfonate on active carbon. *Appl. Catal. A Gen.* **2000**, *192*, 1–8. [CrossRef]

19. Wang, Y.; Fang, Y.; Lu, W.; Li, N.; Chen, W. Oxidative removal of sulfa antibiotics by introduction of activated carbon fiber to enhance the catalytic activity of iron phthalocyanine. *Microporous Mesoporous Mater.* **2018**, *261*, 98–104. [CrossRef]

20. Huang, Z.; Bao, H.; Yao, Y.; Lu, W.; Chen, W. Novel green activation processes and mechanism of peroxymonosulfate based on supported cobalt phthalocyanine catalyst. *Appl. Catal. B Environ.* **2014**, *154–155*, 36–43. [CrossRef]

21. Chen, W.; Lu, W.; Yao, Y.; Xu, M. Highly Efficient Decomposition of Organic Dyes by Aqueous-Fiber Phase Transfer and in Situ Catalytic Oxidation Using Fiber-Supported Cobalt Phthalocyanine. *Environ. Sci. Technol.* **2007**, *41*, 6240–6245. [CrossRef]

22. Mirzaeian, M.; Rashidi, A.M.; Zare, M.; Ghabezi, R.; Lotfi, R. Mercaptan removal from natural gas using carbon nanotube supported cobalt phthalocyanine nanocatalyst. *J. Nat. Gas Sci. Eng.* **2014**, *18*, 439–445. [CrossRef]

23. Zhang, Y.; Li, H.; Huang, H.; Zhang, Q.; Guo, Q. Graphene Oxide–Supported Cobalt Phthalocyanine as Heterogeneous Catalyst to Activate Peroxymonosulfate for Efficient Degradation of Norfloxacin Antibiotics. *J. Environ. Eng.* **2018**, *144*, 04018052. [CrossRef]

24. Sorokin, A.B.; Tuel, A. Metallophthalocyanine functionalized silicas: catalysts for the selective oxidation of aromatic compounds. *Catal. Today* **2000**, *57*, 45–59. [CrossRef]

25. Wöhrle, D.; Hündorf, U.; Schulz-Ekloff, G.; Ignatzek, E. Phthalocyanines on Mineral Carriers, 2 Synthesis of Cobalt(II) and Copper(II)-phthalocyanines on γ-Al_2O_3 and SiO_2. *Z. Naturforsch. B* **1986**, *41*, 179. [CrossRef]

26. Wöhrle, D.; Buck, T.; Hündorf, U.; Schulz-Ekloff, G.; Andreev, A. Phthalocyanines on mineral carriers, 4. Low-molecular-weight and polymeric phthalocyanines on SiO_2, γ-Al_2O_3 and active charcoal as catalysts for the oxidation of 2-mercaptoethanol. *Die Makromol. Chem.* **1989**, *190*, 961–974. [CrossRef]

27. Schubert, U.; Lornez, A.; Kundo, N.; Stuchinskaya, T.; Gogina, L.; Salanov, A.; Zalkovskii, V.; Maizlish, V.; Shaposhmikov, G.P. Cobalt Phthalocyanine Derivatives Supported on TiO_2 by Sol-Gel Processing. Part 1: Preparation and Microstructure. *Chem. Ber.* **1997**, *130*, 1585–1589. [CrossRef]

28. Stuchinskaya, T.; Kundo, N.; Gogina, L.; Schubert, U.; Lorenz, A.; Maizlish, V. Cobalt phthalocyanine derivatives supported on TiO_2 by sol–gel processing: Part 2. Activity in sulfide and ethanethiol oxidation. *J. Mol. Catal. A Chem.* **1999**, *140*, 235–240. [CrossRef]

29. Zanjanchi, M.A.; Ebrahimian, A.; Arvand, M. Sulphonated cobalt phthalocyanine–MCM-41: An active photocatalyst for degradation of 2,4-dichlorophenol. *J. Hazard. Mater.* **2010**, *175*, 992–1000. [CrossRef]

30. Quan, X.; Dai, M.; Zheng, D.; Li, S.; Liu, Y.; Cheng, Z.; Yao, Y. Activated Carbon-Loaded Tetrasulfo Cobalt Phthalocyanine and Application Thereof in Preparing Dimethyl Disulfide as Catalyst. CN 106984362A, 28 July 2017.

31. Mugadza, T.; Nyokong, T. Synthesis and Characterization of Electrocatalytic Conjugates of Tetraamino Cobalt (II) Phthalocyanine and Single Wall Carbon Nanotubes. *Electrochim. Acta* **2009**, *54*, 6347–6353. [CrossRef]

32. Su, Y.; Wang, J.; Chen, G. Study on the enhancement of electrochemiluminescence of luminol–H_2O_2 system by sulphonated cobalt(II) phthalocyanine. *Anal. Chim. Acta* **2005**, *551*, 79–84. [CrossRef]

33. Seoudi, R.; El-Bahy, G.S.; El Sayed, Z.A. FTIR, TGA and DC electrical conductivity studies of phthalocyanine and its complexes. *J. Mol. Struct.* **2005**, *753*, 119–126. [CrossRef]

34. Zhang, L.; Li, H.; Fu, Q.; Xu, Z.; Li, K.; Wei, J. One-Step Solvothermal Synthesis of Tetranitro-Cobalt Phthalocyanine/Reduced Graphene Oxide Composite. *Nano* **2015**, *10*, 1550045. [CrossRef]

35. Leitão, A.; Rodrigues, A. Studies on the Merox process: kinetics of N-butyl mercaptan oxidation. *Chem. Eng. Sci.* **1989**, *44*, 1245–1253. [CrossRef]

Review

Magneto-Rheological Elastomer Composites.
A Review

Sneha Samal [1,*], Marcela Škodová [2], Lorenzo Abate [3] and Ignazio Blanco [4]

[1] FZU-Institute of Physics of Czech Academy of Sciences, Na Slovance 1999/2, 18221 Prague, Czech Republic
[2] Institute for Nanomaterials, Advanced Technologies and Innovation, Technical University of Liberec, Studentska 2, 461 17 Liberec, Czech Republic; marcela.skodova@tul.cz
[3] Department of Physical and Chemical Methodologies for Engineering, University of Catania, Viale A. Doria 6, 95125 Catania, Italy (Retired Professor); labate@dmfci.unict.it
[4] Department of Civil Engineering and Architecture and UdR-Catania Consorzio INSTM, University of Catania, Viale Andrea Doria 6, 95125 Catania, Italy; iblanco@unict.it
* Correspondence: samal@fzu.cz

Received: 1 July 2020; Accepted: 15 July 2020; Published: 17 July 2020

Abstract: Magneto-rheological elastomer (MRE) composites belong to the category of smart materials whose mechanical properties can be governed by an external magnetic field. This behavior makes MRE composites largely used in the areas of vibration dampers and absorbers in mechanical systems. MRE composites are conventionally constituted by an elastomeric matrix with embedded filler particles. The aim of this review is to present the most outstanding advances on the rheological performances of MRE composites. Their distribution, arrangement, wettability within an elastomer matrix, and their contribution towards the performance of mechanical response when subjected to a magnetic field are evaluated. Particular attention is devoted to the understanding of their internal micro-structures, filler–filler adhesion, filler–matrix adhesion, and viscoelastic behavior of the MRE composite under static (valve), compressive (squeeze), and dynamic (shear) mode.

Keywords: magneto-rheological (MR); composite; filler; elastomer; micro-structure; mechanical properties

---

1. Introduction

Magneto-rheological elastomer (MRE) composites belong to the category of smart materials, showing a fast (milliseconds) and reversible transformation, including a controllable stiffness and frequency-dependent viscoelastic behavior, when subjected to a magnetic field [1]. Such composites generally show variations in magneto-rheological (MR) effect for applying in magnetic flux densities of the order of magnitude of 0.8 T [2]. The conventional use of MR fluids creates the settling of particles within the matrix as the function of time. To solve the issue of sedimentation of particles, MR elastomer emerges as an alternative polymer for the MRE composite [3]. The properties could be influenced by the function of temperature. Figure 1 is a schematic representation about the role of temperature for the transformation of MR fluids to elastomer [4]. The MR effect in the composite opens a considerable property making them suitable for the application in mechanical systems where the active control of vibrations or transmission of torque are required. Such as in the areas of tunable vibration absorbers, shear type dampers, composite dampers, isolators, seismic vibration dampers, brakes, control valves, clutches, and artificial joints [5,6]. Furthermore, application fields using the magnetic control are those related with the thermal energy transfer, sound propagation, and isothermal magnetic advection, until you get to the biomedical applications [7,8]. Similarly, magnetic nanocomposites find important applications in the areas of remediation, catalysts, and microwave absorbers in response to external stimulus [9]. In addition, other than elastomeric polymers, current research also focuses on biocompatible and thermoresponsive polymers in the fabrication of magnetic field responsive composites [10].

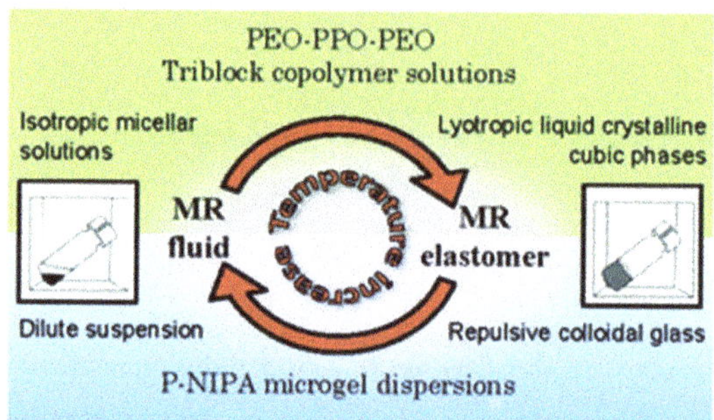

Figure 1. Magneto-rheological (MR) fluid to magneto-rheological elastomer (MRE) transition as a function of temperature. Reprinted from [3], © 2005 with permission from RSC.

Conventional MRE composites consist of two-phase matters obtained by dispersing magnetizable micron-sized solid particles in a liquid matrix, up to 30 vol %. Due to their considerable saturation magnetization ($\mu_0 M_s$ = 2.1 T), iron and carbonyl iron particles are commonly used to this purpose, whilst silicone oils are employed as carrier liquids. Additives are necessary to inhibit sedimentation and aggregation of filler particles within the matrix, in addition, they provide lubricating properties [11]. Silicone rubber, one of the most common elastomers, is used a as matrix consisting of two-phase liquids with base/binder in the 1:1 ratio. Generally, in the composites fabrication after the adding of the magnetic particles into the matrix, a curing is carried out either in the absence (isotropic MREs) or in the presence (anisotropic MREs) of an external magnetic field. The efficiency of the MRE composite, oxidative and chemical stability, and durability, strictly depends on various factors, among those, particles aggregation plays a key role, considering that the interface particles-matrix is determinant in the durability of the composite [12,13]. Thus, in designing MRE composites the main purpose is to obtain the highest MR effect, by inducing magnetization of the particles used as filler [14]. In fact, when the magnetic field is not present, it is possible to observe a random distribution of the particles within the matrix; thus, leading to an isotropic composite. Differently, the presence of a magnetic field, during the preparation, leads to the particles magnetization and attracting one another along the field lines, with the formation of an anisotropic aggregate that spans the system. The so obtained composite shows high yield stress and a large shear rate-dependent apparent viscosity, with increase in viscoelasticity due to the applied magnetic field. The magnetically induced shape change, because of the MRE based magnetostrictive sensors, is very effective with relatively soft materials. This effect, the so-called magnetostriction effect, is due to the decrease of the deformation (ε) as a function of the inverse square root of the elasticity modulus (E), ($\varepsilon \sim E^{-1/2}$). Classic elastomer composites, with reinforcing fillers, are prepared by exploiting the long chains cross-linking during the curing process. Generally, the above reported effect is very low for stiff elastomer composites, but the application of an external magnetic field can contribute to a considerable increase of the MRE's elastic modulus, up to 30%. As reported in literature by Mehnert at al., this affects the resonance frequency and the damping behavior of the adaptive components [15]. The mechanism of materials magnetization is described in Figure 2.

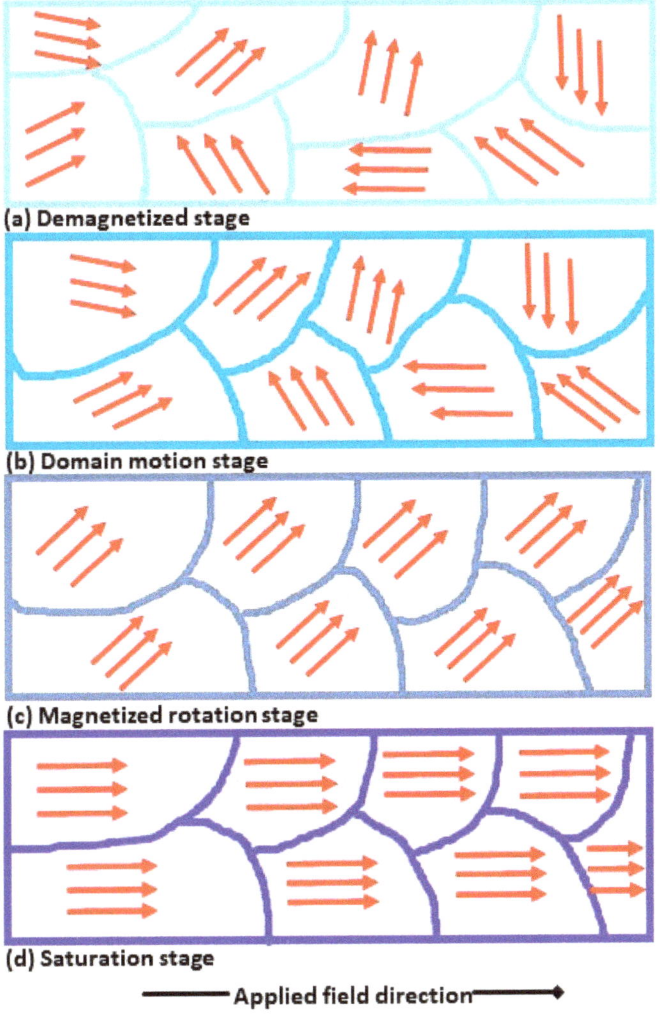

Figure 2. Schematic representation the describing the mechanism for the magnetization: (**a**) demagnetized stage, (**b**) low strain region—domain wall motion stage, (**c**) rotation (closest to the applied field) stage, and (**d**) saturation (alignment with applied field) stage.

The metal particles used as fillers for the MRE composites greatly affect their properties; thus, saturation magnetization and magnetic coercivity result key parameters for the adoptive components. Considering the easy reversibility of the magnetization direction, the energy dissipation in an alternating magnetic field is very low. At the same time, when an external magnetic field is applied, the particles magnetization and the magnetic flux density of the composite increase, and as a result interaction between neighboring particles increases. The MRE's triggering occurs at the application of the external magnetic field because the magnetization of the fillers lead to a rapid and stepwise change in magnetic flux. Among the adjacent magnetic particles dipole–dipole interactions are generated which stiffens the matrix; thus, forcing the fillers to arrange themselves in a columnar structure. This particular particle distribution is easily reached when a magnetic field is applied during the curing process. This distribution effect determines the MREs composite's switching ability and, since stiffness and hysteresis behaviors depend on the magnetic orientation, an anisotropic composite can be easily prepared by exploiting this behavior. The energy interaction of the magnetic particles (E_{MI}) increases as a function of the magnetic dipole moment square, and decreases with the third power of the average distance among the adjacent fillers $E_{MI} = V_p^2/\langle r \rangle^3$.

The particle volume fraction (ϕ) and the average distance (r) varies linearly with their diameter according to the following equation:

$$\langle r \rangle = d\left(\frac{\pi}{6\varnothing}\right)^{1/3}$$

(1)

Accordingly, EMI increases linearly as a function of the particle volume; thus, indicating a strict correlation among the magneto-rheological response of the whole sample and the filler's volume, so that the switching effect increases by increasing the particles size.

1.1. Physical Static Mechanism of MR Effect

Coupling the MR's magnetization with the anisotropic interaction we obtain the "particle magnetization model", which attributes the MR effect to the magnetic permeability mismatch among the constituent phases, both the continuous and dispersed one. Considering the solid phase as the magnetic one, it could be possible to observe that the larger size particles behave as magnetic multi-domains. Considering the particle's magnetic moment is field-induced, their Brownian motion is neglectable; furthermore, the solvent thermal forces ($\propto$ KT) are generally much smaller than magnetic and hydrodynamic ones.

A simplified model could have been achieved by neglecting multibody and multipole magneto static interactions among filler molecules. The magnetic moment of a particle with single domain and magnetizable sphere of radius, in a linear magnetization regime, is

$$m = 4\pi\mu_0\mu_{cr}\beta a^3 H_0$$

(2)

where μ_0 is the permeability of the vacuum and is equal to $\mu_0 = 4\pi \times 10^{-7} T_m A^{-1}$, μ_{cr} is the relative permeability of the continuous phase, and β is the contrast factor or coupling parameter $\beta = \mu_{pr} - \mu_{cr}/\mu_{pr} + 2\mu_{cr}$, μ_{pr} is the relative permeability of the particles, and H_0 is the magnetic field strength. Contrariwise, at higher magnitude, the particle magnetization saturates and the magnetic moment becomes independent of the field strength:

$$m = \frac{4}{3}\pi\mu_0\mu_{cr}\alpha^3 M_s$$

(3)

In a linear regime, the interaction energy's magnitude between two magnetic moments is

$$\lambda = \frac{\pi\mu_0\mu_{cr}\beta^2\alpha^3 H_0^2}{2k_B T}$$

(4)

where $0 < \beta < 1$ for strong MR fields (conventional conformation), and $-0.5 < \beta < 0$ for weak MR fields (not conventional conformation). The magnetostatic particles interaction dominates overall thermal motion in case of chain-like particle aggregates. The particles movement within the elastomer during fabrication of MRE composite results in kinetic aggregations into two different regimes. Firstly, the aggregates average length increases as a power law, then single width chain-like structure laterally aggregate forming columnar structures. Thermal fluctuations of particles in position correlate to their mechanism of lateral aggregation. Iron particles based MRE composite show coarsening defect of tip-to-tip stage of aggregation, leading to a local variation in the magnetic field surrounding to the chain-like structure. The lateral interactions between dipolar chains play an important role in the isotropic structure of MRE samples. In the MREs, the elastomer shows steady shear flow, as a result, the Mason number (M_n), is basically dimensionless; thus, shear rate can be defined as the ratio between the hydrodynamic drag and the magnetostatic forces (estimated as the dipole force magnitude) acting on the particles [16].

$$M_n = \frac{8\eta_c \dot{\gamma}}{\mu_0 \mu_{cr} B^2 H^2} \tag{5}$$

η_c is the continuous phase viscosity and $\dot{\gamma}$ is the magnitude of the shear rate tensor. MR vulcanized elastomer behave as Newtonian fluids in the absence of a magnetic field, whilst if a magnetic field is applied transverse to the direction of flow, it is possible to observe a yielding, shear, and viscoelastic behavior.

Yield stress is based on macroscopic and microscopic models. Macroscopic models employ minimum magnetic energy distribution that leads to homogenous structures of MRE composite reinforced with fillers such as spherical or layered particles in aggregates. This model contributes towards shear anisotropy of the strained aggregates under small deformation. Contrariwise, microscopic models treat inter-particle interactions, such as single particle chains sheared under an external magnetic field. The particle's magnetostatic interactions dominate the shear stress, and shear-induced deformation is assumed to be affine; thus, the magnetic field strengthens the network without affecting its shape significantly. Micro mechanical chain-like models are based on the balancing of magnetostatic and hydrodynamic forces. MRE system shows linear viscoelasticity when subjected to small strains. Instead, high order harmonics define the critical strain; thus, sanctioning the transition from the linear to the non-linear viscoelastic regime. Then, larger strain amplitudes define the second critical strain with a transition from the nonlinear viscoelasticity to viscoplasticity. In the presence of a magnetic field, and thus of a field-induced structure, it is possible to observe a storage modulus (G′), generally of an order of magnitude greater of the loss modulus (G″). At completely saturation of the particles magnetization, in the presence of a large field strength, G′ is independent by this field strength, G′ = 0.3 $\phi\mu_0 M_s^2$.

1.2. Dynamic of MRE

Dynamics of MRE are considered from the shear deformation for a linear viscoelastic material. The dynamic behavior is represented with a complex modulus related to the vibration frequency and controllable by external magnetic fields. The dynamic model or MRE's constitutive relationship [17] is given by

$$\sum_{i=0}^{n_1} a_i \frac{\partial^i \tau}{\partial t^i} = \sum_{k=0}^{n_2} b_k \frac{\partial^k \gamma}{\partial t^k} \tag{6}$$

where τ and γ represent the MRE's shear stress and strain, respectively; a_i and b_k are the constant coefficients dependent on external magnetic field strength; and n_1 and n_2 are integers. Considering the periodic harmonic strain and stress, the equation becomes

$$\bar{\bar{\tau}} = \left[e^{j\theta} \sum_{k=1}^{n_2} b_k (jw)^k \Big/ \sum_{i=1}^{n_1} a_i (jw)^i \right] \bar{\bar{\gamma}} = G(jw)\bar{\bar{\gamma}} \tag{7}$$

where $\bar{\bar{\tau}} = \bar{\gamma} e^{j\omega t}$, $\tau = \bar{\tau} e^{j(\omega t - \theta)}$, with $\bar{\gamma}$ and $\bar{\tau}$ being amplitudes of stress and strain, respectively. $j = \sqrt{-1}$, ω is the vibration frequency, θ is the delayed phase between stress and strain, and $G(j\omega)$ is a complex modulus that is a function of vibration frequency ω and MRE's external magnetic field strength. The real part of the complex modulus $G^R(\omega)$ is the storage modulus representing the viscoelastic stiffness. The imaginary part $G^I(\omega)$ is the loss modulus, with the ratio of loss to storage modulus, represented by the loss factor, describing the viscoelastic damping:

$$\Delta_w = \frac{G^I_{(\omega)}}{G^R_{(\omega)}} \tag{8}$$

The differential equation in Equation (6) is very difficult to solve for MRE systems. Therefore, the Equation (7) is commonly used to describe the MRE's dynamic stress–strain relationship. Experimental evidence shows mostly a linear increase of the storage modulus as a function of the vibration frequency, and both the storage and the loss modulus linearly increase with the external magnetic field strength in a certain range. Hence, the storage modulus $G^R(\omega)$ and the loss factor $\Delta(\omega)$ can be expressed approximately by:

$$G^R(\omega) = \alpha_0 + \alpha_1\omega \tag{9}$$

$$\Delta(\omega) = \beta_0 \tag{10}$$

with α_0, α_1 and β_0 depending only by the applied magnetic field strength.

2. Materials and Methods

We take into consideration a MRE composite prepared by mixing 30 vol % of iron and carbonyl iron particles in silicon elastomer matrix [18]. This latter was prepared from the combination of matrix/binder in the ratio of 50:50 in quantity. The iron particles incorporated into silicon elastomer had a size of 50–100 μm and purity > 95%. The used matrix was obtained by polyaddition, ZA 22, and by polycondensation, N 1522. The carbonyl iron was only considered in the matrix of ZA 22 (Figure 3).

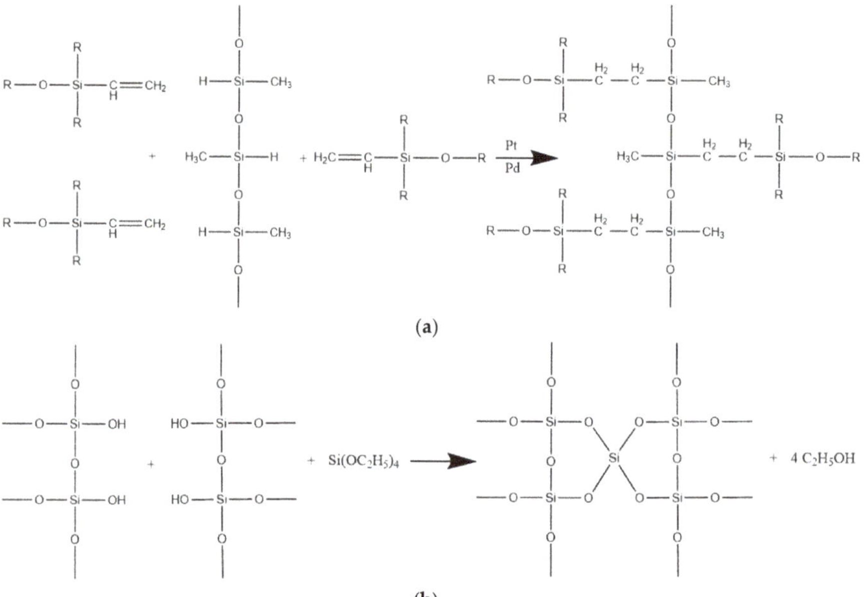

Figure 3. Chemical composition of elastomer matrix ZA 22 (**a**) and N1522 (**b**), poly addition and condensation, respectively.

In order to improve the particle's adhesion on the matrix surface, we used silicon oil. Therefore, particles treated with silicon oil, matrix, binder, and catalyst were slowly stirred, at room temperature, until homogenization (ca. 30 min). The obtained mixture was then treated under vacuum for removing air bubbles and then cured in the standard mold (ca. 24 h), without any influence of the magnetic field. Cylindrically samples were designed, with a length of 20 mm and a radius of 5 mm. The anisotropic composites were prepared using magnetic field along the perpendicular direction of the sample during curing time (Figure 4) [18].

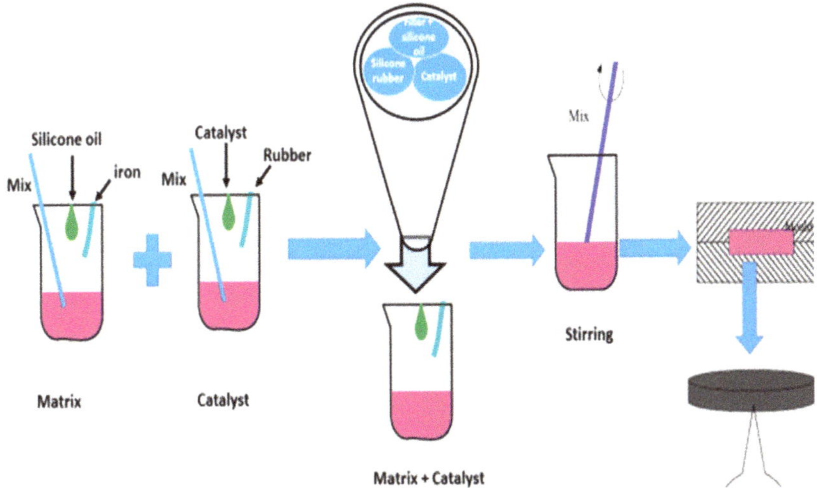

Figure 4. Fabrication of MRE composite with matrix and iron particles with presence of magnetic field. Reprinted from [19], © 2016 with permission from Elsevier.

The induced magnetic field was applied perpendicularly to the sample thickness during the static and compressive mode. However, the closed domain of the magnetic field was applied parallel to the sample position during shear or dynamic mode. Therefore, both the magnetic intensity and the sample orientation in the magnetic field have been set up. As shown in Figure 5, the MRE composite was put in contact with the copper slabs of the magnetic field circuit, both with its upper and lower surface. The simulation was performed by the means of MSC MARC, non-linear dynamics Software (Newport Beach, CA, USA). The simulation parameters were carried out at the boundary conditions of the magnetic potential and current of 1 A. After this test, SEM and μCT analyses were carried out on the MRE composites.

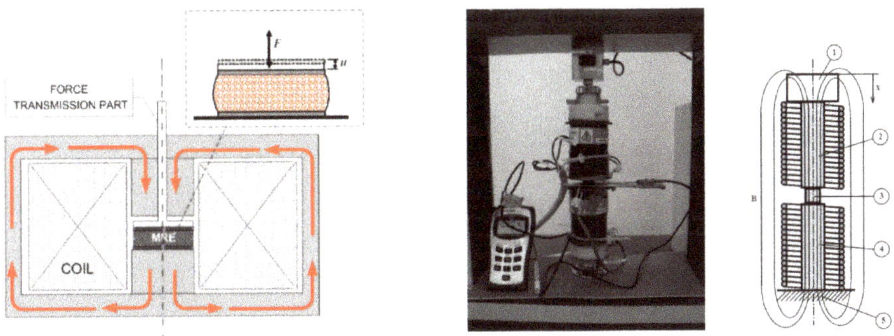

Figure 5. A static model of operation in squeeze form of deformation of MRE during compression with the magnetic domain. Reprinted from [19] © 2005 and [20], 2016 with permission from Elsevier.

A Hitachi TM-3000 (Hitachi High-Technologies Corporation, Tokyo, Japan) with a secondary electron detector and field emission source was used to perform scanning electron microscopy. Sample fragments were assembled on aluminum stubs, out-gassed in a desiccator (ca 48 h), and then coated with a 4 nm layer of platinum before SEM imaging. An acceleration voltage of 10 kV was used in SEM analyses. The μCT analysis was carried out, in sky scan mode with 3D reconstructed images from slices of the two-dimensional structure, on an open tube source with Tungsten (Bruker, R.M.I. s.r.o., Lázně

Bohdaneč, Czech Republic). The instrument setting and the measurement parameters are reported elsewhere [20–22]. Various modes of operation of magneto rheological elastomeric composite are tested in static, compressive (deformed), and dynamic (shear mode), to observe the effect of various magnetic flux density and the hysteresis behavior from force versus displacement response. Static and dynamic properties of MREs composite, of isotropic and anisotropic filler distribution, were investigated as a function of the magnetic intensity, frequency, strain amplitude, and damping properties in various modes [23]. The schematic diagram of the static and compressive (deformed mode) is shown in Figures 5 and 6.

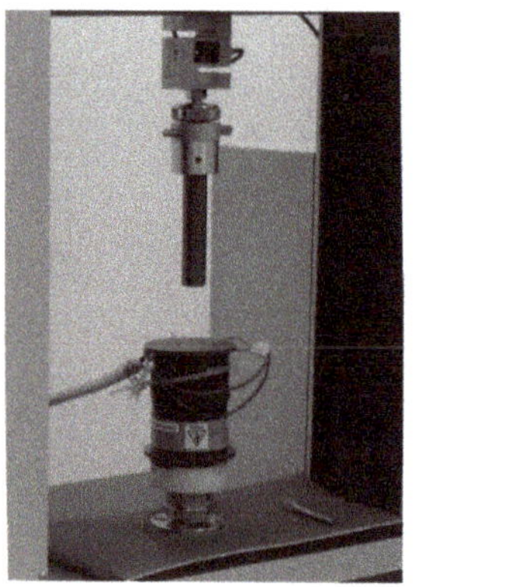
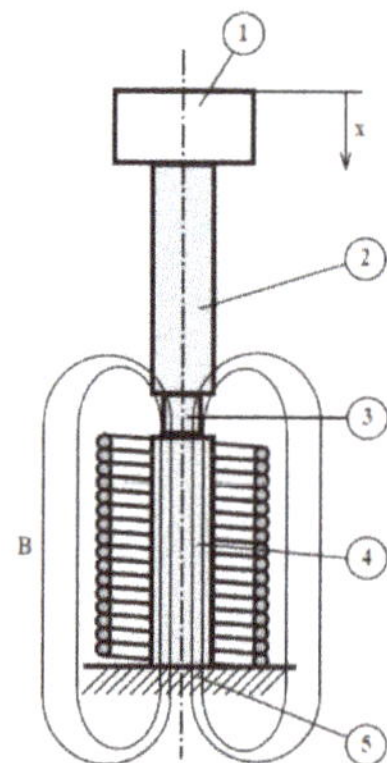

Figure 6. Static valve mode of operation using MRE on the top of the magnet and applying the loading force. Reprinted from [20], © 2016 with permission from Elsevier.

3. Results and Discussion

3.1. Simulation Mechanism of MRES Composite and Magnetic Flux Distribution

The magnetic flux distribution from the magnet towards MRE composite has been carried out using simulation method. Figure 7 shows the distribution of the magnetic induction in the system of the magnetic field with MRE included.

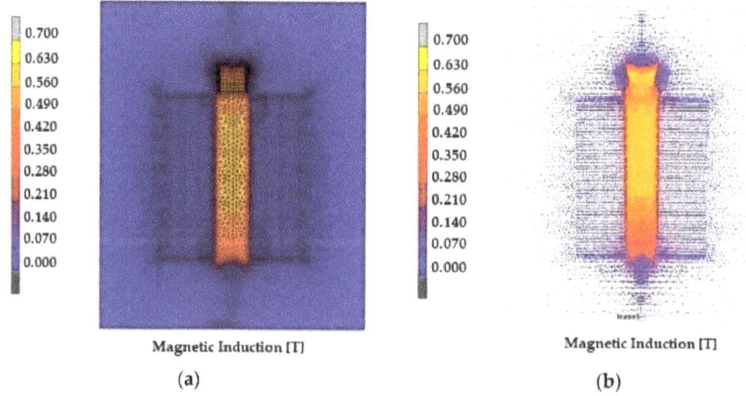

Figure 7. Distribution of magnetic induction on the above configuration of MRE in presence of magnetic field (a) with mesh and considering surrounding atmosphere; (b) without mesh and using lines of force as external parameter. Reprinted from [18], © 2018 with permission from MDPI.

The enclosed magnetic circuit records minimum energy losses due to the formation of a field path for the magnetic flux. In order to remove these energy losses, we need to maintain the gap between the two poles of the C-shape magnetic circuit as small as possible. This configuration allows for a uniform and strong magnetic field. As reported in Figure 8, the future MR elastomer devices require an increase in efficiency and compactness when designing magnetic circuits design; thus, launching a new challenge. For this reason, a different C-shape design in length and scale of components, such as coil and conductive path, is required [24].

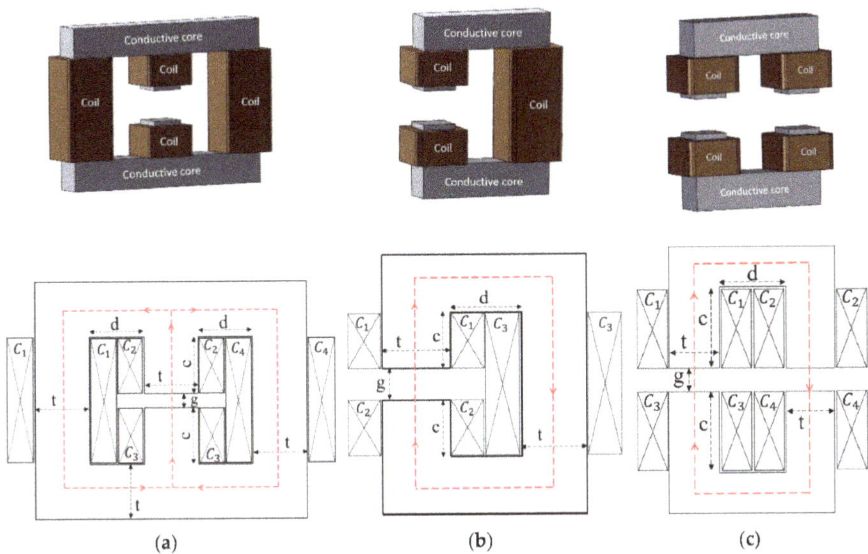

Figure 8. Three electromagnet designs considered for the proposed adaptive tuned vibration absorber (ATVA). (**a**) H-shaped electromagnet; (**b**) C-shaped electromagnet; and (**c**) U-shaped electromagnet. Reprinted from [25], © 2020 with permission from MDPI.

The first configuration is characterized by the presence of the field-sensitive material outside the magnetic coil with the material's motion direction parallel to the coil's axis. Thus, because of the field direction requirement, only small sections of MRE composite in the active area exhibit MR effect. In the second type of magnetic circuit configuration, to make the most of the magnetic field, the MR materials were attached on the surface of the coil, closely placed on the top or/and the bottom of the solenoid. This latter configuration was adopted in most of the MR elastomer-based vibration absorbers and isolators designs. The third configuration consisted in placing the MR materials inside the magnetic coil: thus, serving as the magnetic core of the circuit. In fact, the magnetic field inside the solenoid was strong and uniform, different to the one outside the coil, which was weak and divergent. The greatest advantage of the third configuration was represented by the strong magnetic field generated as well as a large active area. Despite the MR materials inside the solenoid being able to be fully energized by a uniform magnetic field, special care must be given due to the permeability of the magnetic core because of the low permeability of the elastomer component of the MR material. Liu et al., by using MR elastomer, reported about a squeeze mode configuration used in tunable spring element [26]. They applied the elastomer, with compressive loading, in the direction aligned with the magnetic flux. Thus, to enhance their stiffness, most of the MR elastomer devices use the magnetic field. To realize negative stiffness, a hybrid magnetic design for the MR elastomer devices has been used. In Figure 9 shows a tunable vibration absorber, designed by using both a permanent magnetic and an electromagnetic coil for field generation.

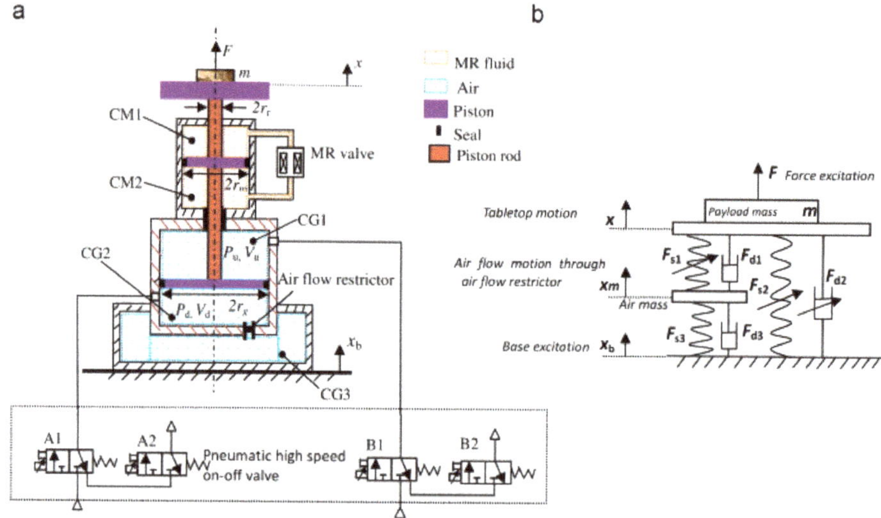

Figure 9. (**a**) The schematic structure of the magneto-rheological fluid embedded pneumatic vibration isolator (MrEPI) and (**b**) its equivalent mechanical system. Reprinted from [26], © 2015 with permission from Elsevier.

In such configuration, it is possible to enhance or weaken the magnetic field, by applying positive and negative current, towards the field generated by permanent magnetics. The resonant frequency of the device can be controlled by electrical currents from 55 Hz at 0 A to 81.25 Hz at 1.5 A, which is equivalent to an increase of 147%. Contrariwise, for an applied current of 1.0 A, the magnetic field decreases from 0.58 to 0.56 T (9% decrease).

An anisotropic MR elastomers design, with an iron particle mass ratio of 70%, is particularly valuable when certain stiffness is required for the device to maintain the serviceability and stability of the structures [27].

In Figure 10 an MRE composite in shear mode is shown, consisting of a particular design using a closed magnetic circuit, which was built with low-carbon steel parts, a wired coil for field generation and two anisotropic MR elastomers. In order to allow vertical motion, the reaction mass was placed among these two samples. Furthermore, to strengthen the magnetic field, the direction of the coils was arranged in an opportune way. Such design was realized by using anisotropic MR elastomers with 30 vol % of iron particles. This design focus on combining both shear and squeeze mode for the application of the MR elastomer device [28].

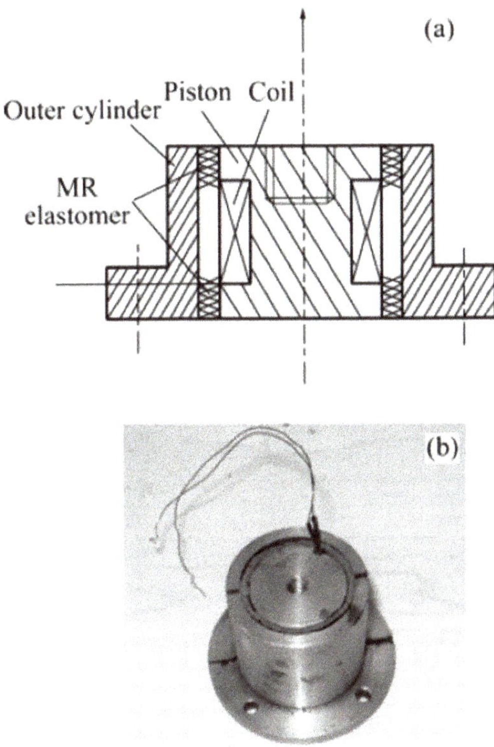

Figure 10. Schematic configuration (**a**) and photograph (**b**) of the MRE absorber. Reprinted from [28], © 2009 with permission from Elsevier.

3.2. Surface Characterization Using SEM and μCT Technique

Based on the filler particles distribution, isotropic and anisotropic MREs composite were characterized by surface morphology. Filler–matrix adhesion was investigated to verify the quality of the composite. The attachment of iron particles within the elastomer matrix was examined using both microscopy and internal volume of the composite [29].

The iron filler surface morphology displayed irregular surface features with filler distribution from 20–120 μm size (Figure 11a,b). The elastomer matrix and its respective MREs composites with fibril self-assembled structures are represented in the surface features (Figure 11c–f). The smaller particles show better adhesion in comparison to the larger ones. The adhesion of filler particles iron towards elastomer matrix is shown in Figure 11g,h. The distribution shows the random orientation of the filler particles within the elastomer matrix that leads to the isotropic distribution of the composite. As a result, the effect of the external magnetic field in isotropic MREs composite led to the magnetization of the small fragments generated during the self-assembled structure of the composite. The small fibrils generate magnetization force based on the particles spin orientation for the application of MRE composite [30,31]. However, the effect of magnetization was more prominent in the anisotropic MRE composites. During the preparation of the composite, the fillers showed a chain-like arrangement due to curing and locking pattern created in the presence of the external magnetic field. Similarly, anisotropic filler distribution in the MRE composite showed parallel chains of fillers along with the matrix of the composite during preparation in the presence of an external magnetic field. Anisotropic distribution of filler within the elastomer composite is shown in Figure 12.

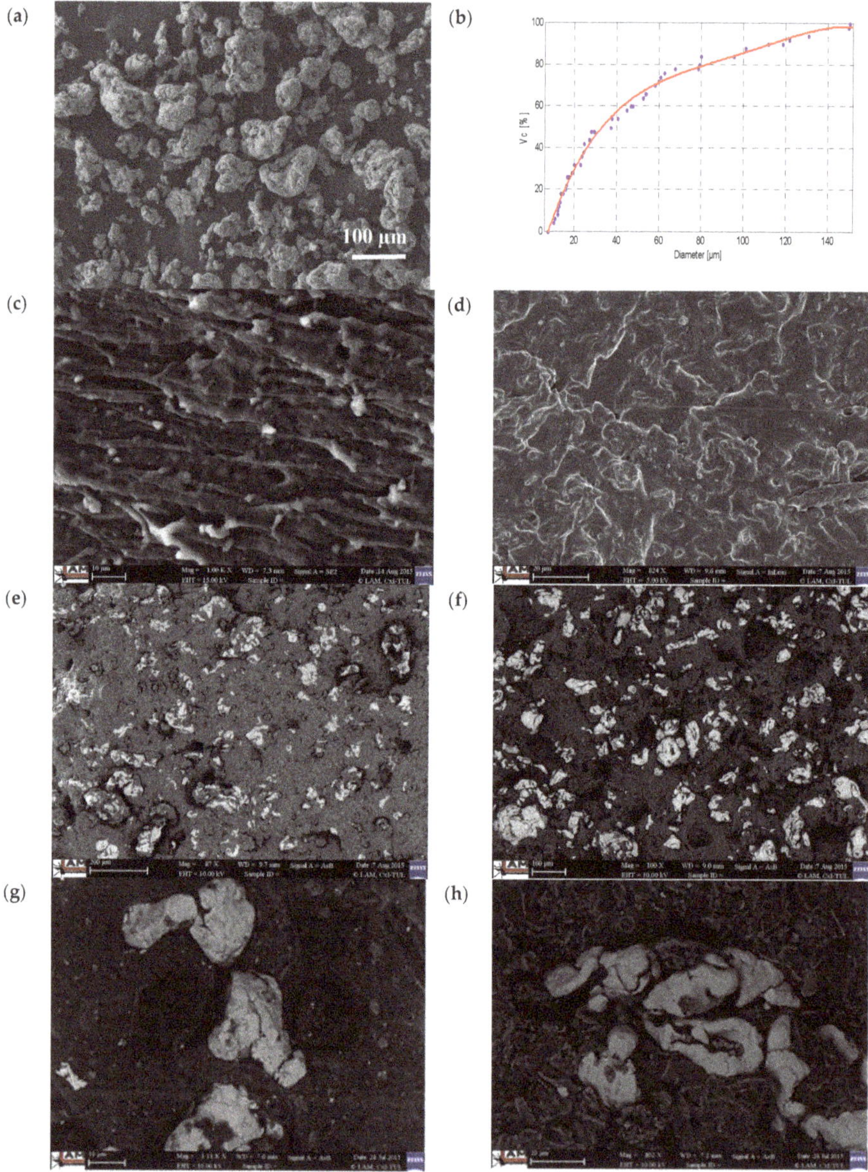

Figure 11. Filler, Elastomer, and MREs composite SEM images: (**a**) iron fillers, (**b**) particle size distribution in N1522 MRE composite, (**c**) Elastomer ZA22, (**d**) Elastomer N1522, (**e**) ZA22 Elastomer Composite, (**f**) N1522 Elastomer Composite, (**g**) Micro fibrils of ZA22 MRE composite, and (**h**) Micro fibrils for N1522 MRE composite. Reprinted from [18], © 2018 with permission from MDPI and from [21], © 2017 with permission from Wiley.

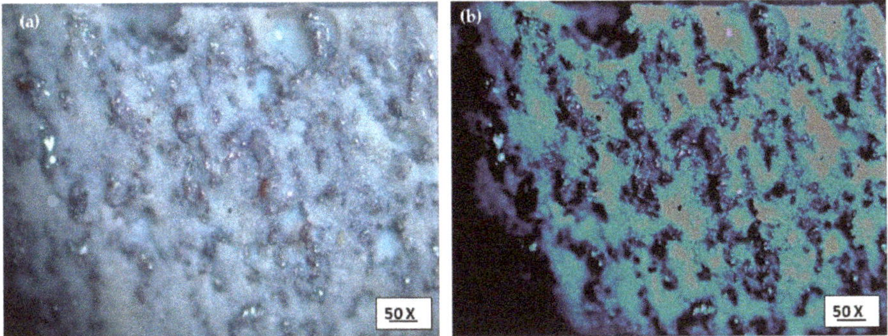

Figure 12. Anisotropic distribution filler within the elastomer composite in transverse (**a**) and longitudinal (**b**) direction from POM (polarizing optical microscope). Reprinted from [18], © 2018 with permission from MPDI.

Filler matrix adhesion induce a good wettability in the MRE composite that leads towards self-assembled particle–particle short fibrils network structure. The volumetric distribution of the particles within the elastomer matrix of the MREs composite was investigated by micro-computed tomography (μCT). This technique reveals a slice of MREs composite in the 2D frame as well as in 3D reconstructed images from all various slides of the composite (Figure 13).

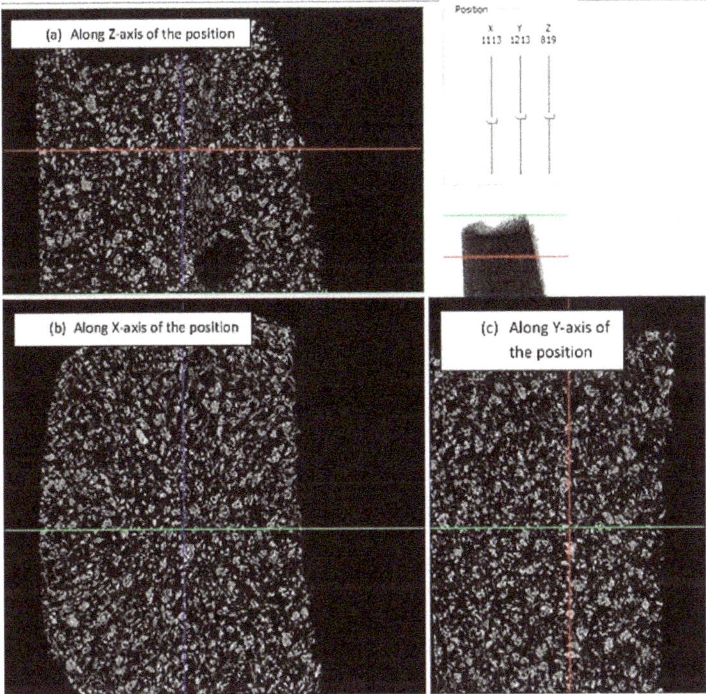

Figure 13. 2D frame of the filler distribution in MRE composite in Z (**a**), X (**b**), Y (**c**) direction. Reprinted from [18], © 2018 with permission from MPDI.

3D images were rebuilt from the various rotation angles of the composite and observed overall in volume fraction to examine their inner structure. This technique and image construction significantly reveal the closed and open pores within the volume and filler distribution in the various directions. The 3D images show one slice and overall images of the MRE composite with iron filler distribution, including the open and closed porosity in the whole structure.

Figure 14a,d displays the inner quality of the MRE composites in 3D planes. The whole volume of the MRE composites in shown in Figure 14a,c. The cut-up layer of the inner volume is shown in Figure 14b,d. This means a qualitative analysis of the composite with, and without, porosity in volume. Quality of the MRE composite is derived from the thermal and rheological behavior of the matrix during curing [32,33].

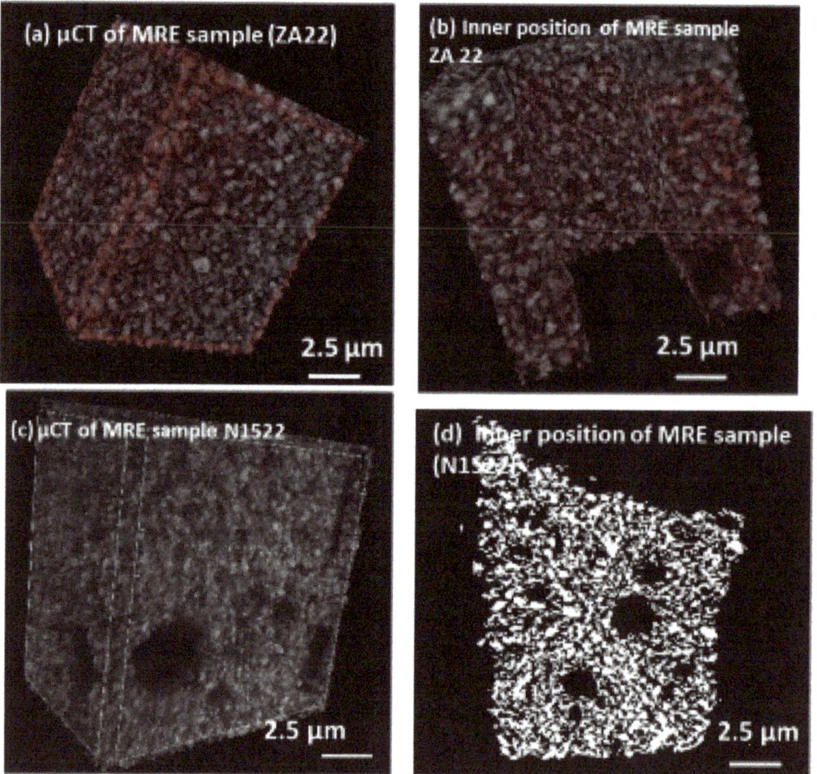

Figure 14. μCT surface features of MREs composite in 3D planes: (**a**) μCT of ZA22 MRE sample, (**b**) inner position of ZA22 MRE sample, (**c**) μCT of N1522 MRE sample, and (**d**) inner position of N1522 MRE sample. Reprinted from [18], © 2018 with permission from MPDI.

3.3. Static Characterization of MREs Composite, Single Field Active, Deformation (Squeeze) Mode

Elastomers exhibit Mullin effects from the results of evolution in the hard and soft domain microstructure induced by stress [34,35]. All MRE composites show a more or less pronounced Payne effect; i.e., the storage modulus decreases with increasing the strain due to the filler network breakdown. Thus, the in-rubber structure determines the strain–independent part of the modulus as a combination of polymer network, the contribution from the hydrodynamic effect, and the modulus [36,37].

MRE samples with isotropic filler distribution undergo mechanical testing in the presence of a magnetic field in a single field and squeeze mode of deformation using the domain of the magnetic field (Figures 4 and 5). The deformation mode of the MRE sample shows hysteresis at various magnetic flux intensities. It has been observed that on increasing the flux intensity of 0.5 T, the area under hysteresis curve increases showing the influence of magnetic domain in the MRE sample behavior (Figure 15).

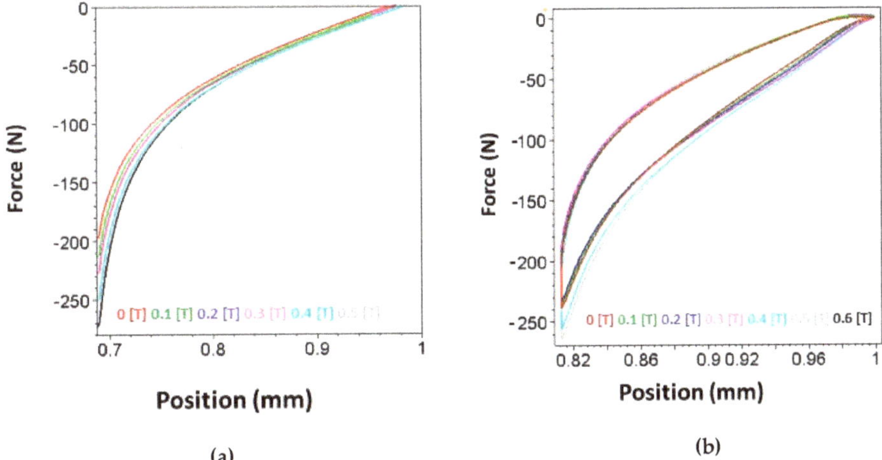

Figure 15. Force versus position for the MRE sample under static with single valve and deformation mode under various magnetic flux intensity (T). (**a**) Force versus displacement for compression; (**b**) relaxation stage at various magnetic intensity. Reprinted from [20], © 2016 with permission from Elsevier.

The maximum intensity of the magnetic domain 0.3 (T) shows the behavior of MRE composite and its response towards recovery to the initial position. The recovery of deformation is investigated by the single and double domains compression test and plotted as load versus position of the sample at various magnetic field intensities. The hysteresis response of the MRE composite and its deformation recovery in the presence of a magnetic field could open up the possibilities for their use as sensors and dampers.

3.4. Dynamic Characterization (Shear Test, Single, and Double Mode)

Dynamic characterization relates with damping behavior of MREs in a magnetic field. Effect of amplitude and frequency control the damping behavior of MREs in the presence of magnetic domain. Single and double lap shear test were implemented on single and double MRE samples with sandwich panel mode. Single lap shear test uses under the shear test for the response of dynamic properties of MRE sample under magnetic domain. The schematic presentation of single and double lap shear test is shown in Figure 16.

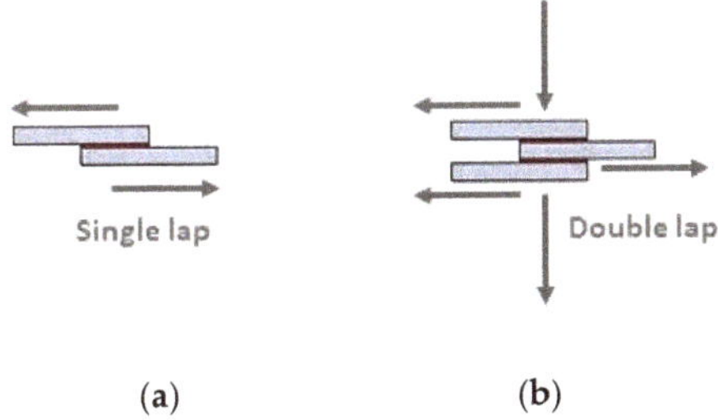

Figure 16. Single (**a**) and double (**b**) lap shear test scheme.

The effect of magnetic flux density on the measured stress–strain hysteresis loops for MRE, at different strain amplitude at 10 Hz (a) 2.5%, (b) 5%, (c) 10%, and (d) 20%, as the function of shear strain are reported in Figure 17 showing the strong correlation of the magnetic flux density effect and vol % of filler towards the shear behavior of the MRE composite.

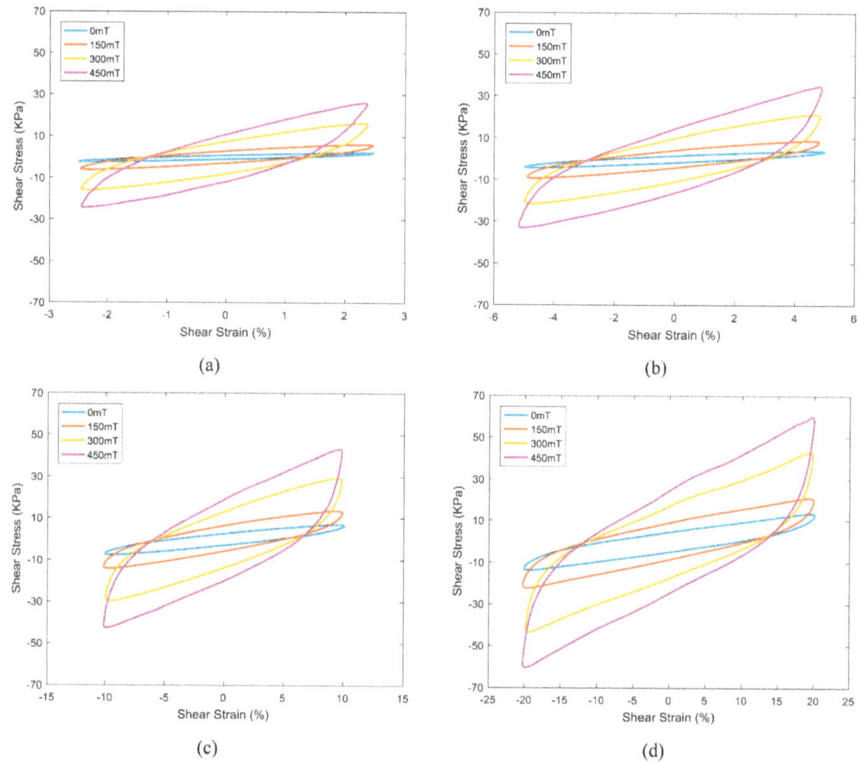

Figure 17. Effect of magnetic flux density on the measured stress-strain hysteresis loops of the MRE as a function of different strain amplitude at 10Hz: (**a**) 2.5%; (**b**) 5%; (**c**) 10%; and (**d**) 20%. Reprinted from [24], © 2019 with permission from Elsevier.

4. Future Prospectus on the Applications of MREs Composite

MRE composites have opportunities in the field of elastomer devices in practical way of engineering applications. MRE composites under tension, compression, and shear test shows possible uses in the application areas of vibration isolation and absorption capacities. Combined approach of shear and compression test may allow application in the area of sensing of elastomer devices. MR elastomer devices include large adaptive working frequency range, low power consumption, and compact configuration. Thus, given a sufficient magnetic field is provided, devices with large changes in their modulus and damping (i.e., MR effect), between field-on and field-off status, can offer a great adaptive range. Effect of filler size, shape, and nature of the matrix has a significant role in the quality of MREs composite. The volume fraction of 30% of filler particles are considered to be optimized for significant distribution of fillers within the matrix. The circular and smaller particles such as carbonyl iron with higher magnetization contribute maximum MR effect in the MREs composite. Soft elastomer matrix such as rubber matrix could be considered as a most suitable one; however, the vulcanization time is longer that may hinder the durability of the devices.

Several cross-linking systems such as sulfurs, organic peroxide, phenolic resins, amino acids, ultrasonic waves, and microwaves have been used for the rubber vulcanization. Among these sulfur and peroxides are the most widely used in the vulcanization process. The mechanism of the accelerated sulfur vulcanization is still controversial partly due to the complexity of the formulation. Therefore, the need to consider the trade-off among MR effects and other performance criteria should be carefully considered for these elastomers. Their application in civil engineering structures, especially for base isolation systems, has recently attracted increasing attention towards adaptive base isolators made of MREs composite.

A decrease in the storage modulus as a function of the strain increase, when the filler network is subject to a breakdown, the so-called Payne effect, is shown by all MRE composites, some by more and some by less. In particular, the isotropic MREs show, at low particles content, a systematic decrease in the small-strain storage modulus G' [38,39]. By contrast, an increase of the small-strain storage modulus G' is generally recorded for the anisotropic MREs. With 30% particle content, anisotropic MRE composite change was more pronounced for the storage modulus (20.4%) than for the loss modulus (8.8%), both changes were determined at 30 Hz with the maximum applied magnetic field being about 0.775 T. This behavior is explained with the magnetized particle arrangement in anisotropic strings during the curing process. The new arrangement that contributes to the stiffness of the compound. Just the micro-sized iron particles, with high saturation magnetization, show a considerable orientation effect within a magnetic field. In fact, due to their large magnetic moment, they generate strong attractive dipole–dipole interactions when the magnetic field is applied [40,41]. Furthermore, close contact is also decisive to get these dipole–dipole interactions in the magnetic field [42]. MRE's devices are generally classified into two categories, those having either fixed poles (pressure-driven mode) and those having relatively moveable poles (direct-shear mode and squeeze-film mode). The first category includes servo-valves, dampers, and shock absorbers, whilst the second one includes clutches, brakes, chucking, and locking devices. A third mode of operation, also known as biaxial elongation flow mode, appears in slow motion and/or high force applications.

5. Conclusions

MRE composites with improved mechanical properties were designed and prepared by using different magnetic fillers at different sizes with, varying composition and keeping constant (i.e., 30 vol %) the total fillers volume fraction. Vulcanization has been carried out, in the presence of an external magnetic field, to obtain anisotropic MREs, and thus improving the magneto-rheological behavior of the composites. Since the viscosity of the polymer matrix is strongly reduced during curing at high temperature, the mobility of the particles used as filler is increased. Therefore, the particles can move readily in the presence of a magnetic field, influencing the distribution of the magnetic filler in the polymer matrix. Non-crosslinked samples were investigated by magneto-rheological viewpoint, showing particles alignment into strings, which was attributed to the strong magnetic dipole–dipole interactions within the micro-sized filler. Various modes of application in the MRE field are discussed that may lead to the application in engineering fields. The surface morphology and inner structure of the MRE composite with filler-polymer, filler–filler adhesion is configuring from 3D structure. This special microstructure allows the preparation of samples with improved shear stiffness and a magnetically conductive filler network highly sensitivity for magneto-rheological applications.

Author Contributions: S.S. and M.Š. conceived and designed the study; discussion on article write up and necessary modifications were carried out by all the authors in a team effort. I.B. and L.A. has revised the manuscript briefly. L.A. corrected the manuscript and prepared the formatted version in text and figures. All authors have read and agreed to the published version of the manuscript.

Funding: This work was carried out within the Institute of Physics and Institute of Plasma Physics under the Solid-21 project (SOLID21: CZ.02.1.01/0.0/0.0/16_019/0000760, SOLID21-Fyzika pevných látek pro 21. Století, Fyzikální ústav AV ČR, v. v. i. (2018–2023). Ignazio Blanco is grateful to the University of Catania within the "Bando-CHANCE" n° 59722022250, for supporting the project HYPERJOIN-HYBRID HIGH PERFORMANCE INNOVATIVE JOINTS.

Conflicts of Interest: The authors declare no conflict of interest.

References

1. Li, Y.; Li, J.; Li, W.; Du, H. A state-of-the-art review on magnetorheological elastomer devices. *Smart Mater. Struct.* **2014**, *23*, 123001. [CrossRef]
2. Zhang, J.; Pang, H.; Wang, Y.; Gong, X. The magneto-mechanical properties of off-axis anisotropic magnetorheological elastomers. *Compos. Sci. Technol.* **2020**, *191*, 108079. [CrossRef]
3. Samal, S. Effect of shape and size of filler particle on the aggregation and sedimentation behavior of the polymer composite. *Powder Technol.* **2020**, *366*, 43–51. [CrossRef]
4. Shahrivar, K.; de Vicente, J. Thermoresponsive polymer-based magneto-rheological (MR) composites as a bridge between MR fluids and MR elastomers. *Soft Matter* **2013**, *48*, 11451–11456. [CrossRef]
5. Kim, Y.K.; Koo, J.H.; Kim, K.S.; Kim, S. Developing a real time controlled adaptive MRE-based tunable vibration absorber system for a linear cryogenic cooler. In Proceedings of the IEEE/ASME International Conference on Advanced Intelligent Mechatronics (AIM), Budapest, Hungary, 3–7 July 2011; pp. 287–290. [CrossRef]
6. Ginder, J.M.; Scholotter, W.F.; Nichols, M.E. Magnetorheological elastomers in tunable vibration absorbers. In Proceedings of the SPIE's 8th Annual International Symposium on Smart Structures and Materials, Newport Beach, CA, USA, 4–8 March 2001; Volume 4331. [CrossRef]
7. Jolly, M.R.; Carlson, J.D.; Munoz, B.C. A model of the behaviour of magnetorheological materials. *Smart Mater. Struct.* **1996**, *5*, 607–614. [CrossRef]
8. Balasoiu, M.; Bica, I. Composite magnetorheological elastomers as dielectrics for plane capacitors: Effects of magnetic field intensity. *Results Phys.* **2016**, *6*, 199–202. [CrossRef]
9. Flatau, A.B.; Dapino, M.J.; Calkins, F.T. High bandwidth tenability in a smart vibration absorber. *J. Intell. Mater. Syst. Struct.* **2000**, *11*, 923–929. [CrossRef]
10. Thevenot, J.; de Oliveira, H.; Sandre, O.; Lecommandoux, S. Magnetic responsive polymer composite materials. *Chem. Soc. Rev.* **2013**, *42*, 7099–7116. [CrossRef]
11. Chertovich, A.V.; Stepanov, G.V.; Kramarenko, E.Y.; Khokhlov, A.R. New Composite Elastomers with Giant Magnetic Response. *Macromol. Mater. Eng.* **2010**, *295*, 336–341. [CrossRef]
12. Krautz, M.; Werner, D.; Schrödner, M.; Funk, A.; Jantz, A.; Popp, J.; Eckert, J.; Waske, A. Hysteretic behavior of soft magnetic elastomer composites. *J. Magn. Magn. Mater.* **2017**, *426*, 60–63. [CrossRef]
13. Filipcsei, G.; Csetneki, I.; Szilágyi, A.; Zrínyi, M. Magnetic Field-Responsive Smart Polymer Composites. In *Oligomers-Polymer Composites-Molecular Imprinting*; Springer: Berlin/Heidelberg, Germany, 2007; Volume 206. [CrossRef]
14. Farshad, M.; Benine, A. Magnetoactive elastomer composites. *Polym. Test.* **2004**, *23*, 347–353. [CrossRef]
15. Mehnert, M.; Hossain, M.; Steinmann, P. Towards a thermo-magneto-mechanical framework for magneto-rheological elastomers. *Int. J. Solids Struct.* **2017**, *128*, 117–132. [CrossRef]
16. de Vicente, J.; Klingenberg, D.J.; Hidalgo-Alvarez, R. Magnetorheological fluids: A review. *Soft Matter* **2011**, *8*, 3701–3710. [CrossRef]
17. Chen, L.; Jerrams, S. A rheological model of the dynamic behaviour of magnetorheological elastomers. *Aip J. Appl. Phys.* **2011**, *110*, 013513. [CrossRef]
18. Samal, S.; Kolinova, M.; Blanco, I. The Magneto-Mechanical Behavior of Active Components in Iron-Elastomer Composite. *J. Compos. Sci.* **2018**, *2*, 54. [CrossRef]
19. Kallio, M. *The Elastic and Damping Properties of Magnetorheological Elastomers: Dissertation*; VTT Technical Research Centre of Finland; VTT: Espoo, Finland, 2005; p. 146. Available online: http://www.vtt.fi/inf/pdf/publications/2005/P565.pdf (accessed on 27 June 2020).
20. Samal, S.; Vlach, J.; Kavan, P. Improved mechanical properties of magneto rheological elastomeric composite with isotropic iron filler distribution. *Ciência Tecnol. Dos Mater.* **2016**, *28*, 155–161. [CrossRef]
21. Samal, S.; Vlach, J.; Kolinova, M.; Kavan, P. Micro-computed tomography characterization of isotropic filler distribution in magnetorheological elastomeric composites. *Adv. Process. Manuf. Technol. Nanostructured Multifunct. Mater.* **2017**, *7*, 57–69. [CrossRef]
22. Samal, S.; Kolinova, M.; Rahier, H.; Dal Poggetto, G.; Blanco, I. Investigation of the Internal Structure of Fiber Reinforced Geopolymer Composite under Mechanical Impact: A Micro Computed Tomography (μCT) Study. *Appl. Sci.* **2019**, *9*, 516. [CrossRef]

23. Samal, S.; Škodová, M.; Blanco, I. Effects of Filler Distribution on Magnetorheological Silicon-Based Composites. *Materials* **2019**, *12*, 3017. [CrossRef] [PubMed]

24. Dargahi, A.; Sedaghati, R.; Rakheja, S. On the properties of magnetorheological elastomers in shear mode: Design, fabrication and characterization. *Compos. Part B Eng.* **2019**, *159*, 269–283. [CrossRef]

25. Asadi Khanouki, M.; Sedaghati, R.; Hemmatian, M. Multidisciplinary Design Optimization of a Novel Sandwich Beam-Based Adaptive Tuned Vibration Absorber Featuring Magnetorheological Elastomer. *Materials* **2020**, *13*, 2261. [CrossRef] [PubMed]

26. Liu, C.; Jing, X.; Daley, S.; Li, F. Recent advances in micro-vibration isolation. *Mech. Syst. Signal Process.* **2015**, *56–57*, 55–80. [CrossRef]

27. Kallio, M.; Lindroos, T.; Aalto, S.; Järvinen, E.; Kärnä, T.; Meinander, T. Dynamic compression testing of a tunable spring element consisting of a magnetorheological elastomer. *Smart Mater. Struct.* **2007**, *16*, 506–514. [CrossRef]

28. Dong, X.-M.; Yu, M.; Liao, C.-R.; Chen, W.-M. A new variable stiffness absorber based on magneto-rheological elastomer. *Trans. Nonferrous Met. Soc. China* **2009**, *19*, s611–s615. [CrossRef]

29. Samal, S.; Kolinova, M.; Blanco, I.; Dal Poggetto, G.; Catauro, M. Magnetorheological Elastomer Composites: The Influence of Iron Particle Distribution on the Surface Morphology. *Macromol. Symp.* **2020**, *389*, 1900053. [CrossRef]

30. Seo, J.; Kim, S.; Samal, S.; Kim, H. Viscous behaviour of Bi2O3–B2O3–ZnO glass composites with ceramic fillers. *Adv. Appl. Ceram.* **2014**, *113*, 334–340. [CrossRef]

31. Fu, S.-Y.; Feng, X.-Q.; Lauke, B.; Mai, Y.-W. Effects of particle size, particle/matrix interface adhesion and particle loading on mechanical properties of particulate–polymer composites. *Compos. Part B* **2008**, *39*, 933–961. [CrossRef]

32. Nurul, A.Y.; Saiful, A.M.; Ubaidillah, S.A.; Salihah, T.S.; Nurul, A.A.W. Thermal Stability and Rheological Properties of Epoxidized Natural Rubber-Based Magnetorheological Elastomer. *Int. J. Mol. Sci.* **2019**, *20*, 746. [CrossRef]

33. Samal, S.; Stuchlík, M.; Petrikova, I. Thermal behavior of flax and jute reinforced in matrix acrylic composite. *J. Therm. Anal. Calorim.* **2018**, *131*, 1035–1040. [CrossRef]

34. Cantournet, S.; Desmorat, R.; Besson, J. Mullins effect and cyclic stress softening of filler elastomers by internal sliding and friction thermodynamics model. *Int. J. Solids Struct.* **2009**, *46*, 2255–2264. [CrossRef]

35. Clough, J.; Creton, C.; Craig, S.; Sijbesma, R. Covalent bond scission in the Mullins effect of a filler elastomer: Real-time Visualization with Mechanoluminescence. *Adv. Funct. Mater.* **2016**, *26*, 9063–9074. [CrossRef]

36. Chazeau, L.; Brown, J.D.; Yanyo, L.C.; Sternstein, S.S. Modulus recovery kinetics and other insights into the Payne effect for filled elastomers. *Polym. Compos.* **2000**, *21*, 202–223. [CrossRef]

37. Frohlich, J.; Niedermeier, W.; Luginsland, H.-D. The effect of filler-filler and filler-elastomer interaction on rubber reinforcement. *Compos. Part A Appl. Sci. Manuf.* **2005**, *36*, 449–460. [CrossRef]

38. Jolly, M.R.; Carlson, J.D.; Munoz, B.C.; Bullions, T.A. The magnetoviscoelastic effect of elastomer composites consisting of ferrous particles embedded in a polymer matrix. *J. Intell. Mater. Syst. Struct.* **1996**, *7*, 613–622. [CrossRef]

39. Stepanov, G.V.; Borin, D.Y.; Raikher, L.Y.; Melenev, P.V.; Perov, N.S. Motion of ferroparticles inside the polymeric matrix in magnetoactive elastomers. *J. Phys. Condens. Matter.* **2008**, *20*, 204121. [CrossRef]

40. Glavan, G.; Kettl, W.; Brunhuber, A.; Shamonin, M.; Drevenšek-Olenik, I. Effect of material composition on tunable surface roughness of magnetoactive elastomers. *Polymers* **2019**, *11*, 594. [CrossRef]

41. Li, W.H.; Zhang, X.Z.; Du, H. Magnetorheological elastomers and their applications. In *Advances in Elastomers I: Blends and Interpenetrating Networks*; Visakh, P.M., Thomas, S., Chandra, A.K., Mathew, A.P., Eds.; Springer: Berlin, Germany, 2013; pp. 357–374.

42. Boczkowska, A.; Awietjan, S. Microstructure and Properties of Magnetorheological Elastomers. *Adv. Elastomers–Technol. Prop. Appl.* **2012**, *595*. [CrossRef]

Review

A Short Review on the Effect of Surfactants on the Mechanico-Thermal Properties of Polymer Nanocomposites

Ahmad Adlie Shamsuri [1,*] and Siti Nurul Ain Md. Jamil [2,3,*]

[1] Laboratory of Biocomposite Technology, Institute of Tropical Forestry and Forest Products, Universiti Putra Malaysia, UPM Serdang, Selangor 43400, Malaysia
[2] Department of Chemistry, Faculty of Science, Universiti Putra Malaysia, UPM Serdang, Selangor 43400, Malaysia
[3] Centre of Foundation Studies for Agricultural Science, Universiti Putra Malaysia, UPM Serdang, Selangor 43400, Malaysia
* Correspondence: adlie@upm.edu.my (A.A.S.); ctnurulain@upm.edu.my (S.N.A.M.J.)

Received: 29 May 2020; Accepted: 16 June 2020; Published: 16 July 2020

Abstract: The recent growth of nanotechnology consciousness has enhanced the attention of researchers on the utilization of polymer nanocomposites. Nanocomposite have widely been made by using synthetic, natural, biosynthetic, and synthetic biodegradable polymers with nanofillers. Nanofillers are normally modified with surfactants for increasing the mechanico-thermal properties of the nanocomposites. In this short review, two types of polymer nanocomposites modified by surfactants are classified, specifically surfactant-modified inorganic nanofiller/polymer nanocomposites and surfactant-modified organic nanofiller/polymer nanocomposites. Moreover, three types of surfactants, specifically non-ionic, anionic, and cationic surfactants that are frequently used to modify the nanofillers of polymer nanocomposites are also described. The effect of surfactants on mechanico-thermal properties of the nanocomposites is shortly reviewed. This review will capture the interest of polymer composite researchers and encourage the further enhancement of new theories in this research field.

Keywords: surfactant; mechanico-thermal; polymer; nanocomposites

---

1. Introduction

Recently, the development of polymer nanocomposites in the composite industry has grown rapidly because they have superb mechanico-thermal properties and are very promising replacements for conventional polymer composites. Polymer nanocomposites are polymers incorporated with nanometer scale fillers, regardless of the origin of the materials, whether it is inorganic or organic. Synthetic polymers (examples are displayed in Table 1) are frequently utilized for the creation of polymer nanocomposites, including HDPE [1], PP [2], PS [3,4], DGEBA [5–7], PEVA [8,9], PBO [10], SR [11], PAN [12], PUA [13], etc. Synthetic biodegradable polymers (examples are also displayed in Table 1), such as PLA [14–18], PBAT [19], PBS [20], PVA [21], PCL [22], etc., are also utilized for nanocomposite preparation.

Table 1. Examples of synthetic polymers and synthetic biodegradable polymers used in the preparation of polymer nanocomposites.

Synthetic Polymer	Abbreviation	Synthetic Biodegradable Polymer	Abbreviation
High-density polyethylene	HDPE	Poly(lactic acid)	PLA
Polypropylene	PP	Poly(butylene adipate-co-terephthalate)	PBAT
Polystyrene	PS	Poly(butylene succinate)	PBS
Diglycidyl ether of bisphenol A	DGEBA	Polyvinyl alcohol	PVA
Poly(ethylene-co-vinyl acetate)	PEVA	Poly(ε-caprolactone)	PCL
Polybenzoxazine	PBO		
Silicone rubber	SR		
Polyacrylonitrile	PAN		
Polyurethane acrylate	PUA		
Thermoplastic polyurethane	TPU		
Epoxy-terminated dimethylsiloxane	EDTS		

On the other hand, nanofillers (examples are displayed in Table 2), such as MMT [2,9,12,20], Ag NPs [17,18], ZnO NPs [15], TiO$_2$ NRs [3], CNTs [10,23,24], CNFs [5,25,26], G NPs [11,13,23], GO NSs [6,27], LDHs [22], and HNTs [4], are used in the fabrication of polymer nanocomposites, as they possess a high aspect ratio, large surface area, high stiffness, and low density [17,20]. The application of bio-based nanofillers, for example, CNCs [15,18,28,29], CNPs [30], NCFs [8,19], and CNWs [7,16], also contribute to the production of nanocomposites because they are renewable and sustainable materials.

Table 2. Examples of nanofillers and surfactants used in the preparation of polymer nanocomposites.

Nanofiller	Abbreviation	Surfactant	Abbreviation
Montmorillonite	MMT	Hexadecyltrimethylammonium bromide	HTAB
Silver nanoparticles	Ag NPs	Didodecyldimethylammonium bromide	DDAB
Zinc oxide nanoparticles	ZnO NPs	Sodium dodecyl sulfate	SDS
Titanium dioxide nanorods	TiO$_2$ NRs	Polypropylene glycol ethoxylated and propoxylated	Ultraric PE 105
Carbon nanotubes	CNTs	Silicon-based surfactant	Niax L-595
Carbon nanofibers	CNFs	Bis-(2-hydroxyethyl)methyltallowalkylammonium chloride	HMAC
Graphene nanoplatelets	G NPs	Polyoxyethylene octyl phenyl ether	Triton X-100
Graphene oxide nanosheets	GO NSs	Acid phosphate ester of ethoxylated nonylphenol	Beycostat A B09
Layered double hydroxides	LDHs	Lauric arginate	LAE
Halloysite nanotubes	HNTs	Poly(ethylene glycol) monooleate	PEGMONO
Expanded graphite nanoplatelets	EG NPs	Triblock copolymers of poly(ethylene oxide) and poly(propylene oxide)	Pluronic
Cellulose nanocrystals	CNCs	Oleic acid	OA
Cellulose nanoparticles	CNPs	Polysiloxane-polyether copolymer	AK8805
Nanocellulose fibers	NCFs	Stearic acid	SA
Cellulose nanowhiskers	CNWs	Palmitic acid	PA
		Sorbitan monostearate	Span 60
		Benzimidazolium-N,N′-hexadecane-2-hydroxy-ethyl bromide	BHHB
		Poly(ethylene glycol)	PEG
		Dodecylbenzene sulfonic acid	DBSA
		Sodium cholate	SC

Surfactants (examples are also displayed in Table 2), for instance, HTAB [13,20,31], DDAB [12,32], SDS [4,21], Ultraric PE 105 [19], Niax L-595 [8], HMAC [9], Triton X-100 [6,10,11,24], Beycostat A B09 [15,17,18,28], LAE [29], PEGMONO [14,16], Pluronic [7,14], OA [3], AK8805 [5], SA [2], PA [22], Span 60 [30], and BHHB [1], are commonly employed for modifying nanofillers in the preparation of polymer nanocomposites.

Surfactants have an amphiphilic character, owing to hydrophilic and hydrophobic functional groups [33,34]. Previous studies have indicated that the surfactants could act as interaction links between hydrophilic and hydrophobic polymers [35–37]. Furthermore, the presence of surfactants in polymer nanocomposites not only increased the uniformity of the nanofillers dispersion [8], but also improved the compatibility and wettability between polymer and nanofiller [11], as well as enhanced the final properties of the nanocomposites [18].

In the past few decades, many modification methods have been suggested for the purpose of increasing the mechanico-thermal properties (e.g., tensile strength, flexural strength, impact strength, degradation temperature, glass transition temperature, melting temperature, etc.) of polymer nanocomposites. The non-covalent surface modification method, by utilizing surfactants, is an effective way to modify nanofillers for enhancing the mechanico-thermal properties of polymer nanocomposites. Nonetheless, to the authors' knowledge, no short review has been made covering the work on surfactant-modified inorganic and organic nanofillers for polymer nanocomposites. That is the aim of conducting a systematic review in this paper. This short review is broad, albeit not comprehensive, but is completed with other relevant literatures.

2. Types of Surfactant-Modified Polymer Nanocomposites

2.1. Surfactant-Modified Inorganic Nanofiller/Polymer Nanocomposites

Table 3 indicates the examples of inorganic nanofillers, surfactants, polymer matrices, and preparation processes of polymer nanocomposites. An inorganic filler, such as MMT, could be modified by a cationic surfactant, like HMAC, to produce organoclay (OMMT) [9]. OMMT was incorporated into PEVA via melt blending and compression molding processes for the preparation of OMMT/PEVA nanocomposites [9]. TiO_2 NRs have also been used for the preparation of polymer nanocomposites. TiO_2 NRs could be modified or capped by an anionic surfactant, such as OA, to prevent aggregation in solution [3]. The OA-capped TiO_2 NRs were mixed with PS solution via the solvent blending technique, followed by a drop casting process to prepare the TiO_2/PS nanocomposites [3]. Other than clay mineral and metal oxide nanomaterials, the CNTs, G NPs, and GO NSs are also categorized as inorganic nanofillers. Although carbon is an organic material, the allotropes of carbon, such as CNTs, G NPs, and GO NSs, are considered as inorganic nanomaterials [38]. CNTs could be modified by a non-ionic surfactant, like Triton X-100, to improve dispersion in a PBO matrix [10]. CNT/PBO nanocomposites were prepared through solvent-free blending and pour casting processes at an elevated temperature [10].

Table 3. Examples of inorganic nanofillers, surfactants, polymer matrices, and preparation processes of polymer nanocomposites.

Inorganic Nanofiller	Surfactant	Polymer Matrix	Mixing Process	Final Process	References
MMT	HMAC	PEVA	Melt blending	Compression molding	[9]
TiO_2 NRs	OA	PS	Solvent blending	Drop casting	[3]
CNTs	Triton X-100	PBO	Solvent-free blending	Pour casting	[10]
CNFs	AK8805	DGEBA	Solution blending	Compression molding	[5]
G NPs	Triton X-100	SR	Solution blending	Compression molding	[11]
GO NSs	Triton X-100	DGEBA	Solution blending	Pour casting	[6]

CNFs are also regarded as an inorganic nanofiller due to their preparation procedure and structure, which are the same as in CNTs, but CNFs have a larger diameter than CNTs [25]. CNFs could be modified by a silicone surfactant, like AK8805, to improve the dispersion of CNFs in an epoxy matrix, such as DGEBA [5]. CNF/DGEBA nanocomposites were prepared through solution blending and compression molding processes at an elevated temperature [5]. On the other hand, G NPs could also be modified by Triton X-100 via sonication in the solvent [11]. Modified G NPs were mixed with an SR solution through the solution blending process, followed by a compression molding process for vulcanization to obtain G NP/SR nanocomposites [11]. In addition, Triton X-100 could treat GO NSs to promote dispersion in a DGEBA matrix [6]. GO NS/DGEBA nanocomposites could also be prepared by means of solution blending and pour casting processes [6]. Table 3 demonstrates that the surfactant-modified inorganic nanofiller/polymer nanocomposites could be prepared by using typical polymer composite processing procedures without requiring complicated or complex preparation processes.

2.2. Surfactant-Modified Organic Nanofiller/Polymer Nanocomposites

Table 4 indicates the examples of organic nanofillers, surfactants, polymer matrices, and preparation processes of polymer nanocomposites. Organic fillers, such as CNCs, could be modified by an anionic surfactant, like Beycostat A B09, to produce surfactant-modified CNCs (s-CNCs) [15]. The s-CNCs were mixed with PLA via solution blending and solvent casting processes for the preparation of s-CNC/PLA nanocomposites [15]. CNPs have also been used for the preparation of polymer nanocomposites. CNPs could be modified by a non-ionic surfactant, such as Span 60, to improve the dispersion of hydrophilic CNPs in a hydrophobic PS matrix [30]. Span 60-modified CNPs were mixed with a PS solution via the solution blending process, followed by a solvent casting process to prepare CNP/PS nanocomposites [30]. On the other hand, NCFs could be suspended in a silicon-based surfactant, such as Niax L-595, with dispersant oils to facilitate the dispersion of NCFs in a PEVA matrix [8]. The NCF suspension was added to the PEVA through the melt blending process, followed by a compression molding process to obtain NCF/PEVA nanocomposites [8].

Table 4. Examples of organic nanofillers, surfactants, polymer matrices, and preparation processes of polymer nanocomposites.

Organic Nanofiller	Surfactant	Polymer Matrix	Mixing Process	Final Process	References
CNCs	Beycostat A B09	PLA	Solution blending	Solvent casting	[15]
CNPs	Span 60	PS	Solution blending	Solvent casting	[30]
NCFs	Niax L-595	PEVA	Melt blending	Compression molding	[8]
CNWs	Pluronic	DGEBA	Solvent-free blending	Pour casting	[7]
NCFs	Ultraric PE 105	PBAT	Solution blending	Solvent casting	[19]
CNWs	PEGMONO	PLA	Solution blending	Solvent casting	[16]

In addition, Pluronic could treat CNWs to improve the interactions between the CNWs and the DGEBA epoxy [7]. CNW/DGEBA nanocomposites could be prepared by means of solvent-free blending and pour casting processes [7]. NCFs could also be modified by a non-ionic surfactant, like Ultraric PE 105, for promoting the interfacial interaction between the NCF and PBAT phases [19]. NCFs were mixed with the surfactant and PBAT via solution blending and solvent casting processes for the preparation of NCF/PBAT nanocomposites [19]. CNWs could also be modified by a non-ionic surfactant, such as PEGMONO, to improve CNW dispersion in a non-polar PLA matrix [16]. PEGMONO-modified CNWs were mixed with PLA solution via the solution blending process, followed by a solvent casting process to prepare CNW/PLA nanocomposites [16]. Table 4 demonstrates that the surfactant-modified organic nanofiller/polymer nanocomposites could mostly be prepared through a solvent casting process by using organic solvents.

3. Effect of Surfactants on Mechanico-Thermal Properties

3.1. Types of Surfactants

Surfactants have frequently been categorized into four types, namely non-ionic, anionic, cationic, and amphoteric. The categories are based on the polarity of the surfactant head group, for example, non-ionic, anionic, cationic, and amphoteric or zwitterionic. There is no charge on a head group of non-ionic surfactants, while anionic and cationic surfactants have negative and positive charges on their head groups, respectively. On the other hand, there are both negative and positive charges for the amphoteric surfactants [33]. Table 5 displays the types of surfactants, types of nanofillers, and modification methods for the preparation of polymer nanocomposites. Non-ionic surfactants, such as Triton X-100 (chemical structure showed in Figure 1a), could modify inorganic nanofillers like CNTs via an ultrasonication method to serve as a bridge between CNTs and DGEBA [24]. The NCF organic nanofiller could be modified with the non-ionic surfactant Ultraric PE 105 by mechanical stirring to improve the dispersion of NCFs in the nanocomposite [19].

Table 5. Types of surfactants, types of nanofillers, and modification methods for the preparation of polymer nanocomposites.

Surfactant	Type of Surfactant	Nanofiller	Type of Nanofiller	Modification Method	References
Triton X-100	Non-ionic	CNTs	Inorganic	Ultrasonication	[24]
Ultraric PE 105	Non-ionic	NCFs	Organic	Stirring	[19]
SDS	Anionic	G NPs	Inorganic	Ultrasonication	[21]
Beycostat A B09	Anionic	CNCs	Organic	Stirring	[18]
DDAB	Cationic	MMT	Inorganic	Agitation	[12]
HTAB	Cationic	CNCs	Organic	Stirring	[31]

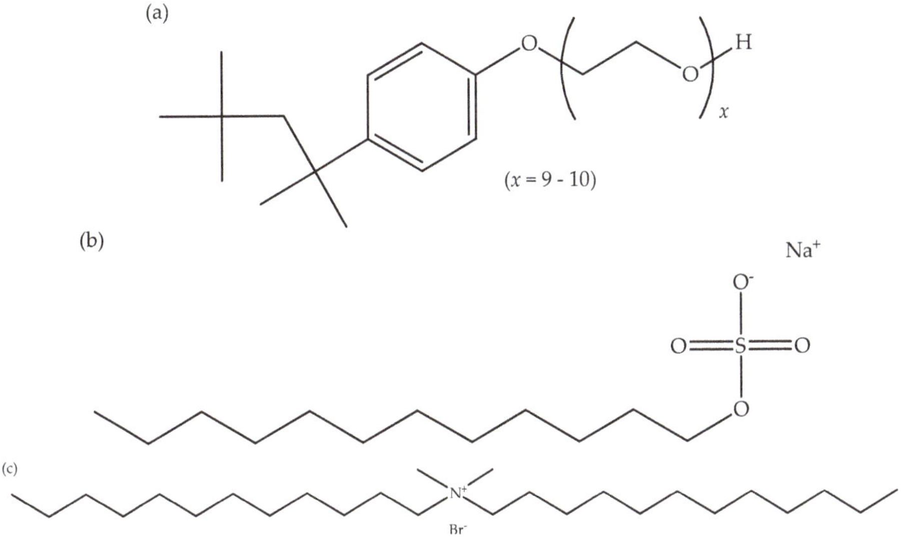

Figure 1. Chemical structures of (**a**) Triton X-100, (**b**) SDS, and (**c**) DDAB.

An anionic surfactant, for example, SDS (chemical structure showed in Figure 1b), is commonly employed in the modification of inorganic nanofillers, such as G NPs, through an ultrasonication method to assist G NP dispersion in a PVA matrix [21]. The CNC organic nanofiller could be modified with the anionic surfactant Beycostat A B09 by mechanical stirring to obtain stable dispersions of CNCs in nanocomposites [18]. On the other hand, cationic surfactants, for instance, DDAB (chemical structure showed in Figure 1c), could be utilized for the modification of inorganic nanofillers, such as MMT, via an agitation process [12]. CNCs could be modified with the cationic surfactant HTAB, also by mechanical stirring, for the preparation of such nanocomposites [31]. Table 5 also exhibits that the non-ionic, anionic, and cationic surfactants could modify both inorganic and organic nanofillers without any limitations. However, the modification of nanofillers by amphoteric surfactants is not only complicated [39], but also limited to inorganic nanofillers, such as MMT [40,41].

3.2. Effect of Non-Ionic Surfactants

Table 6 shows the mechanico-thermal properties of polymer nanocomposites modified by non-ionic surfactants. Pluronic (chemical structure showed in Figure 2a) could modify organic nanofillers like CNWs and modified CNWs have been used for the preparation of CNW/DGEBA nanocomposites [7]. The modification of CNWs has improved the mechanico-thermal properties of the nanocomposites. The mechanical properties, such as tensile strength, tensile modulus, and elongation at break, of the nanocomposites have increased by up to 79%, 27%, and 77%, respectively, compared to the unmodified CNW/DGEBA nanocomposite. This is attributed to the improved dispersion and enhanced interfacial interaction between the Pluronic-modified CNWs and the DGEBA matrix [7]. However, the degradation temperature of the nanocomposites decreased because of the presence of small aggregates of Pluronic-modified CNWs in the nanocomposites, which degraded easily in comparison with the large agglomerates. Furthermore, the glass transition temperature of the nanocomposites also decreased due to possible Pluronic micelle formation inside the DGEBA matrix [7].

Table 6. Mechanico-thermal properties of polymer nanocomposites modified by non-ionic surfactants.

Non-ionic Surfactant	Nanofiller	Polymer Matrix	Mechanico-Thermal Properties *										References
			TS	TM	EB	FS	FM	IS	T_d	T_g	T_m	T_c	
Pluronic	CNWs	DGEBA	↑	↑	↑	-	-	-	↓	↓	-	-	[7]
PEGMONO	CNCs	PLA	↓	-	↓	-	-	-	↑	-	-	-	[14,16]
Triton X-100	CNTs	PBO	-	-	-	↑	↑	↑	-	↑	↑	-	[10]
AK8805	CNFs	DGEBA	↑	↑	-	-	-	↑	↑	↑	-	-	[5]
Triton X-100	G NPs	SR	↑	↑	↑	-	-	-	↑	-	↓	↓	[11]
Ultraric PE 105	NCFs	PBAT	↑	↑	-	-	-	-	↓	-	-	-	[19]

TS = tensile strength, **TM** = tensile modulus, **EB** = elongation at break, **FS** = flexural strength, **FM** = flexural modulus, **IS** = impact strength, T_d = degradation temperature, T_g = glass transition temperature, T_m = melting temperature, and T_c = crystallization temperature. * The symbol ↑ corresponds to an increase in the properties and ↓ a decrease in the properties while "-" means "not available".

(a)

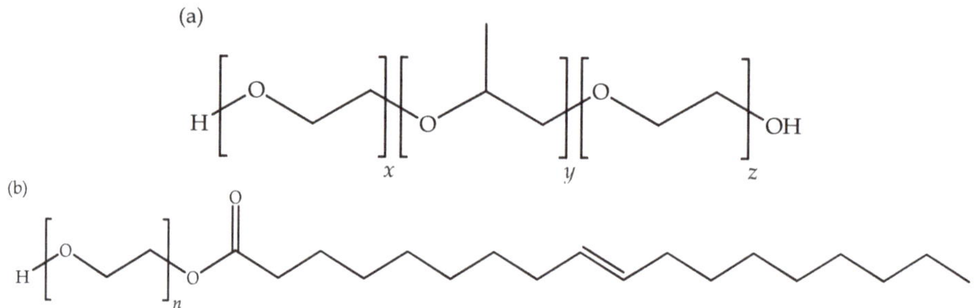

(b)

Figure 2. Chemical structures of (**a**) Pluronic and (**b**) PEGMONO.

PEGMONO (chemical structure showed in Figure 2b) could modify organic nanofillers such as CNCs and modified CNCs have been utilized for the preparation of CNC/PLA nanocomposites [14]. The modification improved the maximum degradation temperature of the nanocomposites up to 0.6% compared to neat PLA [16]. This is attributed to the higher thermal stability of PEGMONO, which covered the surface of the CNCs and prevented the modified CNCs from decomposing quickly [14]. Nevertheless, the tensile strength of the nanocomposites decreased in comparison with the nanocomposite without PEGMONO; this is because of the presence of CNC/PEGMONO aggregates. Moreover, the elongation at break of the nanocomposites slightly decreased, as PEGMONO improved CNC interaction with the PLA matrix [14].

On the other hand, Triton X-100 could modify inorganic nanofillers like CNTs and modified CNTs have been applied for the preparation of CNT/PBO nanocomposites [10]. The modification of CNTs improved the mechanico-thermal properties of the nanocomposites, such as flexural strength, flexural modulus, impact strength, glass transition temperature, and melting temperature. The flexural strength, flexural modulus, and impact strength increased by up to 31%, 10%, and 13%, respectively, compared to the unmodified CNT/PBO nanocomposite, which was due to strong hydrophobic attraction between the hydrophobic segment of Triton X-100 and the surface of the CNTs, whereas the hydrophilic segment of Triton X-100 interacted with the PBO matrix through hydrogen bonding [10]; both interactions improved the wettability and reduced surface tension of the CNTs in the nanocomposites. Moreover, the glass transition temperature and melting temperature increased by up to 1.0% and 0.6%, respectively, which was also because of the improved interfacial interaction between the Triton X-100-modified CNTs and the PBO matrix, as well as the improved degree of dispersion of the modified CNTs in the nanocomposites [10].

AK8805 could modify inorganic nanofillers like CNFs and the modified CNFs have been used for the preparation of CNF/DGEBA nanocomposites [5]. The modification of CNFs improved the mechanico-thermal properties of the nanocomposites. The mechanical properties, such as tensile strength, tensile modulus, and impact strength, of the nanocomposites increased by up to 20%, 165%, and 15%, respectively, compared to the neat DGEBA, which was due to enhancement of the interaction between the modified CNFs and the DGEBA matrix [5]. Furthermore, the degradation temperature and glass transition temperature of the nanocomposites also increased by up to 4.5% and 2.7%, respectively, because of the better dispersion of CNFs caused by the enhanced interaction [5].

Triton X-100 could also modify inorganic nanofillers such as G NPs and modified G NPs have been utilized for the preparation of G NP/SR nanocomposites [11]. The modification of G NPs improved the tensile strength, tensile modulus, and elongation at break by up to 20%, 5.4%, and 22%, respectively. This is attributed to the capability of the surfactant to act as a bridge between the G NPs and the SR matrix, which subsequently improved the compatibility and wettability of the graphene platelets and this provided a good adhesion to the SR [11]. The maximum degradation temperature of the nanocomposites also improved by up to 3.0% due to the good Triton X-100-G NP/SR interface interaction that offered a better barrier effect [11]. Nonetheless, the melting temperature and crystallization temperature of the nanocomposites insignificantly decreased compared to the nanocomposite without Triton X-100 because of Triton X-100 does not link with the crystal structure.

On the other hand, Ultraric PE 105 could modify organic nanofillers like NCFs and modified NCFs have been applied for the preparation of NCF/PBAT nanocomposites [19]. The modification of NCFs slightly improved the mechanical properties of the nanocomposites, such as tensile strength and Young's modulus, by up to 8.9% and 4.4%, respectively, compared to the unmodified NCF/PBAT nanocomposite. The minor increase was probably related to the excess of Ultraric PE 105 in the nanocomposites, which influenced the exerted force on the nanocomposites [19]. Moreover, the degradation temperature of the nanocomposites decreased because Ultraric PE 105 has a lower molecular weight than PBAT, which consequently affected its thermal stability [19]. Table 6 clearly displays that the modification of inorganic and organic nanofillers by non-ionic surfactants improved the mechanico-thermal properties of the prepared polymer nanocomposites, regardless of the type of the non-ionic surfactant.

3.3. Effect of Anionic Surfactants

Table 7 shows the mechanico-thermal properties of polymer nanocomposites modified by anionic surfactants. SA (chemical structure showed in Figure 3a) could modify inorganic nanofillers like MMT and modified MMTs have been used for the preparation of MMT/PP nanocomposites [2]. The modification of MMT improved the mechanico-thermal properties of the nanocomposites. The mechanical properties, such as tensile strength, elongation at break, and impact strength, of the nanocomposites increased by up to 5.3%, 125%, and 50%, respectively. This is due to the dispersion state of SA-MMT being very good [2]. In contrast, the Young's modulus of the nanocomposites decreased because the tougher the composites, the less stiff their character. The melting temperature of the nanocomposites slightly increased by up to 1.0% due to the strengthening of mechanical properties of the nanocomposites. However, the crystallization temperature of the nanocomposites decreased compared to the nanocomposite without SA because of the very good dispersion of SA-MMT in the nanocomposites [2].

Table 7. Mechanico-thermal properties of polymer nanocomposites modified by anionic surfactants.

Anionic Surfactant	Nanofiller	Polymer Matrix	Mechanico-Thermal Properties *										References
			TS	TM	EB	IS	SM	LM	T_d	T_g	T_m	T_c	
SA	MMT	PP	↑	↓	↑	↑	-	-	-	-	↑	↓	[2]
OA	TiO$_2$ NRs	PS	-	-	-	-	↓	↑	↑	↓	-	-	[3]
PA	LDHs	PCL	↓	↑	-	↑	↑	↑	-	↓	↕	↑	[22]
SDS	HNTs	PS	-	-	-	↑	↑	-	↑	↓	-	-	[4]
Beycostat A B09	CNCs	PLA	↑	↑	↓	-	-	-	↕	↓	↓	↓	[18]
SDS	G NPs	PVA	↑	↑	↑	-	-	-	↑	-	↓	↑	[21]

TS = tensile strength, **TM** = tensile modulus, **EB** = elongation at break, **IS** = impact strength, **SM** = storage modulus, **LM** = loss modulus, T_d = degradation temperature, T_g = glass transition temperature, T_m = melting temperature, and T_c = crystallization temperature. * The symbol ↑ corresponds to an increase in the properties and ↓ a decrease in the properties while "-" and ↕ mean "not available" and "unchanged", respectively.

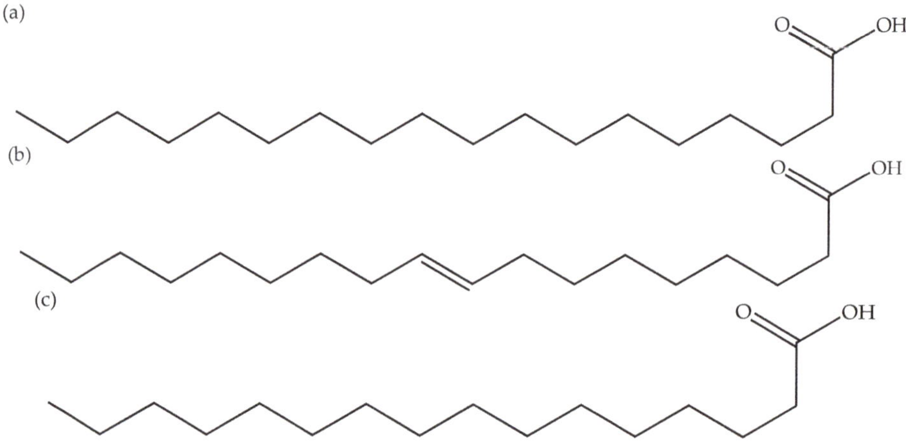

Figure 3. Chemical structures of (**a**) SA, (**b**) OA, and (**c**) PA.

OA (chemical structure showed in Figure 3b) could modify inorganic nanofillers, such as TiO$_2$ NRs, to prevent the agglomeration of TiO$_2$ NRs and modified TiO$_2$ NRs have been utilized for the preparation of TiO$_2$ NR/PS nanocomposites [3]. The modification of TiO$_2$ NRs improved the loss modulus of the nanocomposites by up to 22% compared to neat PS. This is attributed to the PS chains becoming softer due to the presence of OA-modified TiO$_2$. [3]. Moreover, the glass transition temperature of the nanocomposite decreased because of the soft PS chains caused by the OA molecules. In addition, the storage modulus of the nanocomposite decreased due to the plasticization effect of OA, which is present at the TiO$_2$ NR surface [3]. Nevertheless, the maximum degradation temperature of the nanocomposites increased by up to 1.3% because the thermal stability of the TiO$_2$ NRs was more than 400 °C.

On the other hand, PA (chemical structure showed in Figure 3c) could modify inorganic nanofillers like LDHs and modified LDHs have been applied for the preparation of LDH/PCL nanocomposites [22]. The modification of LDHs improved the mechanico-thermal properties of the nanocomposites, such as tensile modulus, impact strength, storage modulus, and crystallization temperature. The tensile modulus and impact strength properties increased by up to 12% and 65%, respectively, compared to the neat PCL, which was due to the reinforcement property of the PA-modified LDHs (PA-LDHs), and they could also act as an impact strength modifier in the nanocomposite system. Moreover, the storage modulus of the nanocomposites increased by up to 9.1%, which was because of a favorable dispersion and interaction between the PA-LDH nanofiller and the PCL matrix [22]. The loss modulus of the nanocomposites also increased by up to 10% due to the plasticizing effect of the PA-LDHs. Additionally, the crystallization temperature of the nanocomposites increased by up to 3.0%, which could possibly be attributed to the PA-LDHs, which acted as heterogeneous nucleation sites in the PCL matrix. However, the tensile strength of the nanocomposites decreased because of the presence of tactoids at a high content of PA-LDHs, which may act as stress concentrations [22]. In addition, the glass transition temperature of the nanocomposites also decreased due to improved PCL chain mobility, and it is probable that it would enhance in free volume.

SDS could modify inorganic nanofillers such as HNTs and modified HNTs have been used for the preparation of HNT/PS nanocomposites [4]. The modification of HNTs improved the mechanico-thermal properties of the nanocomposites. The mechanical properties, for example, impact strength and storage modulus, increased by up to 203% and 39%, respectively, with the addition of HNTs compared to neat PS. This is because the HNTs could improve of the stiffness of the nanocomposites [4]. Furthermore, the degradation temperature of the nanocomposites also increased by up to 14% due to the entrapment of decomposition products by the HNT lumen. Nonetheless, the glass transition temperature of the nanocomposites decreased because of the presence of interactions between the HNTs and PS [4].

Beycostat A B09 could modify organic nanofillers like CNCs and modified CNCs have been utilized for the preparation of CNC/PLA nanocomposites [18]. The modification of CNCs improved the mechanical properties of the nanocomposites, such as tensile strength and tensile modulus. The tensile strength and tensile modulus increased by up to 63% and 50%, respectively, which was attributed to the Beycostat A B09 efficiently dispersing CNCs in the PLA matrix [18]. However, the decrease in the elongation at break of the nanocomposites was due to the CNCs initiating considerable local stress concentrations and then failure at lowered strain values. Moreover, the decrease in crystallization temperature was because of the presence of Beycostat A B09 on the CNC surface, which caused a better dispersion of the CNCs in the PLA matrix, which certainly increased the nucleation effect on the nanocomposites [18]. Nonetheless, the other thermal properties, for instance, the glass transition temperature and melting temperature, decreased insignificantly compared to the nanocomposite without Beycostat A B09.

On the other hand, SDS could also modify inorganic nanofillers like G NPs and modified G NPs have been applied for the preparation of G NP/PVA nanocomposites [21]. The modification of G NPs improved the mechanical properties of the nanocomposites, such as tensile strength and tensile modulus, by up to 75% and 154%, respectively, compared to neat PVA. The substantial increases were due to the SDS aiding the dispersion of G NPs, which maximized the load transfer from PVA to the G NPs. Moreover, the elongation at break of the nanocomposites also increased by up to 53% because of the slippage of the intercalated state of G NP dispersion in the PVA matrix during tensile testing [21]. Additionally, the degradation temperature and crystallization temperature of the nanocomposites also increased by up to 0.8% and 1.6%, respectively, even at a low content of SDS-modified G NPs. This due to a heterogenous nucleating effect, which was caused by the adsorption of SDS into the G NPs [21]. Nevertheless, the melting temperature of the nanocomposites decreased, which was related to the decrease in their degree of crystallinity. Table 7 clearly displays that the modification of inorganic and organic nanofillers by anionic surfactants improved the mechanico-thermal properties of the prepared polymer nanocomposites, regardless of the type of anionic surfactant.

3.4. Effect of Cationic Surfactant

Table 8 shows the mechanico-thermal properties of polymer nanocomposites modified by cationic surfactants. HTAB (chemical structure showed in Figure 4a) could modify inorganic nanofillers like MMT and modified MMTs have been used for the preparation of MMT/PBS nanocomposites [20]. The modification of MMT improved the mechanico-thermal properties of the nanocomposites. The mechanical properties, such as tensile strength, tensile modulus, elongation at break, flexural strength, flexural modulus, and impact strength, of the nanocomposites increased by up to 34%, 7.3%, 210%, 34%, 3.6%, and 66%, respectively, compared to the unmodified MMT/PBS nanocomposite. This is due to the presence of trimethyl groups in HTAB that eased the MMT dispersion and offered a better reinforcement effect. Additionally, the high aspect ratio of HTAB-modified MMTs provided a higher surface area to interact with the PBS matrix [20]. Furthermore, the melting temperature also improved by up to 2.3% because of the higher degree of crystallinity of the nanocomposites. In addition, the crystallization temperature also increased by up to 6.5% due to the improved modified MMTs and the PBS matrix interactions and better filler dispersion [20].

Table 8. Mechanico-thermal properties of polymer nanocomposites modified by cationic surfactants.

Cationic Surfactant	Nanofiller	Polymer Matrix	Mechanico-Thermal Properties *											References
			TS	TM	EB	FS	FM	IS	SM	T_d	T_g	T_m	T_c	
HTAB	MMT	PBS	↑	↑	↑	↑	↑	↑	-	-	-	↑	↑	[20]
LAE	CNCs	PLA	↑	↑	↑	-	-	-	↑	-	↑	↑	↓	[29]
HMAC	MMT	PEVA	-	-	-	-	-	-	↑	-	-	↓	↓	[9]
HTAB	G NPs	PUA	-	-	-	-	-	-	↑	↑	↑	-	-	[13]
BHHB	MMT	HDPE	↑	↑	↓	-	-	-	↑	↑	-	-	-	[1]
DDAB	MMT	PS	↑	↑	↑	-	-	-	-	↑	-	-	-	[32]

TS = tensile strength, **TM** = tensile modulus, **EB** = elongation at break, **FS** = flexural strength, **FM** = flexural modulus, **IS** = impact strength, **SM** = storage modulus, T_d = degradation temperature, T_g = glass transition temperature, T_m = melting temperature, and T_c = crystallization temperature. * The symbol ↑ corresponds to an increase in the properties and ↓ a decrease in the properties while "-" means "not available".

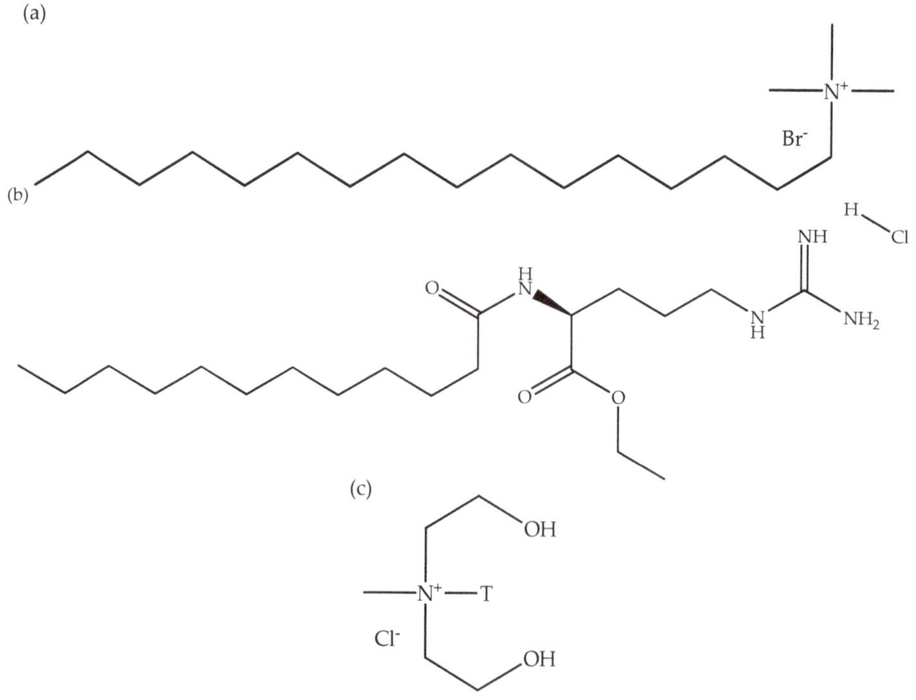

(a)

(b)

(c)

Figure 4. Chemical structures of (**a**) HTAB, (**b**) LAE, and (**c**) HMAC.

LAE (chemical structure showed in Figure 4b) could modify organic nanofillers such as CNCs and modified CNCs have been utilized for the preparation of CNC/PLA nanocomposites [29]. The modification of CNCs increased the tensile strength, Young's modulus, and strain at break by up to 91%, 78%, and 58%, respectively. This could probably be attributed to the result of improved crystallinity and the enhanced dispersion/interface compatibility of LAE-modified CNCs within the PLA matrix. Additionally, the storage modulus of the nanocomposites increased by up to 69%; this could be associated with the creation of a stiff continuous modified CNC percolation network and the modified CNCs, which prompted the limitation of the PLA chain movement [29]. Furthermore, the glass transition temperature and melting temperature also improved by up to 15%, and 1.4%, respectively, due to the increased modified CNC dispersion and interface compatibility, which efficiently restricted the movement of PLA chains. In contrast, the crystallization temperature of the nanocomposites decreased compared to the nanocomposite without LAE because of the strong nucleating effect of modified CNCs, which promoted crystallization [29].

On the other hand, HMAC (chemical structure showed in Figure 4c) could modify inorganic nanofillers like MMT and modified MMTs have been applied for the preparation of MMT/PEVA nanocomposites [9]. The modification of MMT improved the mechanical properties of the nanocomposites, such as storage modulus. The storage modulus of the nanocomposites increased by up to 340%, due to strong HMAC-modified MMT and PEVA interactions compared to the nanocomposite without HMAC [9]. However, the melting temperature of the nanocomposites decreased due to the presence of HMAC, which offered interactions between the modified MMTs and the PEVA matrix [9]. Moreover, the crystallization temperature of the nanocomposites also decreased because the crystallization kinetics were altered by modified MMTs, which diminished the crystal perfection of the PEVA matrix.

HTAB could also modify inorganic nanofillers like G NPs and modified G NPs have been used for the preparation of G NPs/PUA nanocomposites [13]. The modification of G NPs improved the mechanico-thermal properties of the nanocomposites. The storage modulus of the nanocomposites increased by up to 104% compared to the neat PUA; this could be attributed to a high aspect ratio of HTAB-modified G NPs with high surface areas dispersed evenly in the PUA matrix and forming strong interfacial interactions with the matrix via hydrogen bonding [13]. Furthermore, the degradation temperature of the nanocomposites improved by up to 23% due to the high thermal stability of modified G NPs that acted as an efficient physical barrier, which hindered thermal decomposition by slowing down the vaporization of volatile molecules. Moreover, the glass transition temperature of the nanocomposites also increased by up to 554% because of the physical barrier effect of the crumpled modified G NPs, which induced mechanical interlocking with PUA chains, and subsequently confined segmental movements at the modified G NP–PUA interface [13].

BHHB could modify inorganic nanofillers such as MMT and modified MMTs have been utilized for the preparation of MMT/HDPE nanocomposites [1]. The modification of MMT improved the tensile strength and Young's modulus by up to 31% and 66%, respectively, compared to the neat HDPE; this is attributed to the homogeneous distribution and a large aspect ratio of BHHB-modified MMTs that induced the strong interfacial interaction between modified MMTs and the HDPE matrix for better interfacial stress transfer efficiency. Nonetheless, the elongation at break of the nanocomposites significantly decreased because of the cross-section of HDPE resisted deformation, which was efficiently reduced by the modified MMTs [1]. The storage modulus of the nanocomposites increased by up to 99% because an interconnected network of modified MMTs within the HDPE matrix formed, which restricted the long-range movement of HDPE chains. Moreover, the degradation temperature of the nanocomposite improved by up to 5.3% because the modified MMTs acted as a mass transport barrier to the volatile products generated during thermal decomposition [1].

On the other hand, DDAB could also modify inorganic nanofillers like MMT and modified MMTs have been applied for the preparation of MMT/PS nanocomposites [32]. The modification of MMT improved the mechanical properties of the nanocomposites, such as tensile strength, Young's modulus, and elongation at break, by up to 32%, 30%, and 18%, respectively, compared to the unmodified MMT/PS nanocomposite. The increases were related to the exceptional dispersion of DDAB-modified MMTs in the PS matrix and the remarkable adhesion between the modified MMTs and the PS [32]. Moreover, the degradation temperature of the nanocomposites also increased by up to 7.1%, possibly because DDAB, with two long alkyl chains, could provide outstanding thermal stability [32]. Table 8 clearly displays that the modification of inorganic and organic nanofillers by cationic surfactants has improved the mechanico-thermal properties of the prepared polymer nanocomposites, regardless of the type of the cationic surfactant.

3.5. Effect of Surfactants on Thermal Conductivity

Table 9 displays the effect of different types of surfactants on the thermal conductivity of polymer nanocomposites. The thermal conductivity of G NP/SR nanocomposites increased by up to 14% when the G NPs were modified with Triton X-100. The increase is due to the modified G NPs that have a good dispersion and good interface in the nanocomposites [11]. The thermal conductivity of CNT/PS nanocomposites increased by up to 202% when the CNTs were modified with a non-ionic surfactant, such as PEG (chemical structure showed in Figure 5a). The increase is because the modified CNTs have a more even dispersion character, which was acquired through ultrasonic dispersion with PEG [42]. The thermal conductivity of CNT/SR nanocomposites increased by up to 38% when the CNTs were modified with an anionic surfactant like DBSA (chemical structure showed in Figure 5b). The increase is because the modified CNTs have a good homogeneity, which creates a large interface area with SR [43].

On the other hand, the thermal conductivity of CNT/TPU nanocomposites increased by up to 2971% when the CNTs were modified with an anionic surfactant, such as SC (chemical structure showed in Figure 5c). The increase is because the modified CNTs were smaller and had a fine distribution that generated effective interactions with the TPU matrix [44]. The thermal conductivity of EG NP/PEVA nanocomposites increased by up to 137% when the EG NPs were modified with SDS. The increase is because the modified EG NPs have a better dispersion, which formed the constant EG NP pathways [45]. The thermal conductivity of GO NS/EDTS nanocomposites increased by up to 404% when the GO NSs were modified with HTAB. The increase is because the modified GO NSs formed a thin, finely dispersed layer, which created an uninterrupted heat flux pathway and interacted with the EDTS matrix [46]. Table 9 clearly shows that the modification of inorganic nanofillers by non-ionic, anionic, and cationic surfactants has increased the thermal conductivity of the prepared polymer nanocomposites.

Table 9. Effect of different types of surfactants on the thermal conductivity of polymer nanocomposites.

Surfactant	Nanofiller	Polymer Matrix	Thermal Conductivity	References
Triton X-100	G NPs	SR	↑ (14%)	[11]
PEG	CNTs	PS	↑ (202%)	[42]
DBSA	CNTs	SR	↑ (38%)	[43]
SC	CNTs	TPU	↑ (2971%)	[44]
SDS	EG NPs	PEVA	↑ (137%)	[45]
HTAB	GO NSs	EDTS	↑ (404%)	[46]

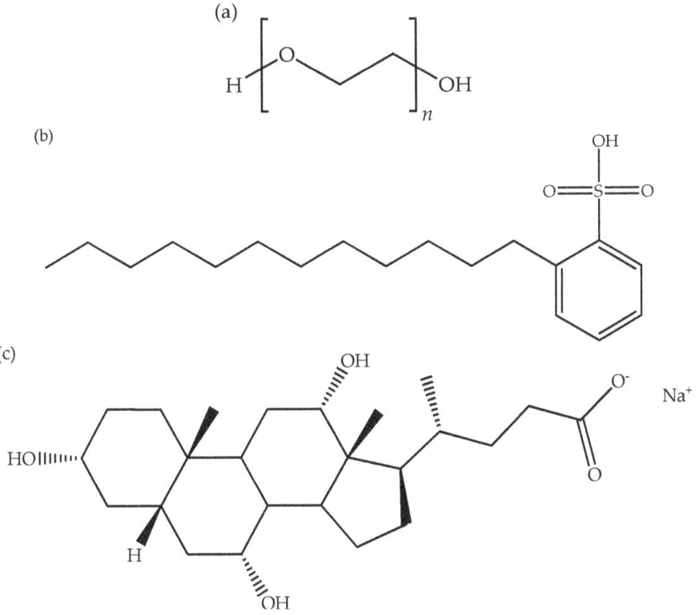

Figure 5. Chemical structures of (**a**) PEG, (**b**) DBSA, and (**c**) SC.

4. Conclusions

Nanofillers, polymer matrices, preparation procedures, and the mechanico-thermal properties of polymer nanocomposites modified with surfactants have been briefly reviewed in this paper. The primary mechanico-thermal properties, for instance, tensile strength, flexural strength, impact strength, degradation temperature, glass transition temperature, and the melting temperature of the nanocomposites, have also been described in this short review. Surfactants have regularly been applied for the modification of nanofillers because they possess an amphiphilic character. Surfactants utilized for different types of polymer nanocomposites are mostly based on their chemical structures. In addition, non-ionic, anionic, and cationic surfactants have been the three most significant surfactants for the improvement of the mechanico-thermal properties of polymer nanocomposites. Non-ionic surfactants are typically used in the modification of organic nanofillers, such as nanocelluloses. Anionic and cationic surfactants are frequently employed in the modification of inorganic nanofillers, like metal oxides, MMT, etc. The proper modification of nanofillers by surfactants could improve the dispersion of the nanofillers in the polymer matrices. Moreover, the surfactant-modified inorganic and organic nanofillers could effectively form strong interactions between the nanofillers and the polymer matrices. This short review might be beneficial, not only for polymer composite researchers, but also be useful for the commercialization of polymer nanocomposites for various applications.

Author Contributions: Conceptualization, A.A.S. and S.N.A.M.J.; methodology, A.A.S.; validation, A.A.S. and S.N.A.M.J.; formal analysis, S.N.A.M.J.; investigation, A.A.S.; resources, S.N.A.M.J.; data curation, A.A.S.; writing—original draft preparation, A.A.S.; writing—review and editing, S.N.A.M.J.; project administration, A.A.S.; funding acquisition, S.N.A.M.J. All authors have read and agreed to the published version of the manuscript.

Funding: This short review was funded by the Universiti Putra Malaysia under the Grant Putra IPM Scheme (project number: GP-IPM/2017/9579900).

Acknowledgments: The authors would like to thank the Laboratory of Biocomposite Technology, Institute of Tropical Forestry and Forest Products, Universiti Putra Malaysia.

Conflicts of Interest: The authors declare no conflict of interest. The funder had no role in the design of the review; in the collection, analyses, or interpretation of data; in the writing of the manuscript, or in the decision to publish the results.

References

1. El Achaby, M.; Ennajih, H.; Arrakhiz, F.Z.; El Kadib, A.; Bouhfid, R.; Essassi, E.; Qaiss, A. Modification of montmorillonite by novel geminal benzimidazolium surfactant and its use for the preparation of polymer organoclay nanocomposites. *Compos. Part B* **2013**, *51*, 310–317. [CrossRef]

2. Gonzalez, L.; Lafleur, P.; Lozano, T.; Morales, A.B.; Garcia, R.; Angeles, M.; Rodriguez, F.; Sanchez, S. Mechanical and thermal properties of polypropylene/montmorillonite nanocomposites using stearic acid as both an interface and a clay surface modifier. *Polym. Compos.* **2014**, *35*, 1–9. [CrossRef]

3. Patra, N.; Salerno, M.; Cozzoli, P.D.; Athanassiou, A. Surfactant-induced thermomechanical and morphological changes in TiO2-polystyrene nanocomposites. *J. Colloid Interface Sci.* **2013**, *405*, 103–108. [CrossRef] [PubMed]

4. Lin, Y.; Ng, K.M.; Chan, C.M.; Sun, G.; Wu, J. High-impact polystyrene/halloysite nanocomposites prepared by emulsion polymerization using sodium dodecyl sulfate as surfactant. *J. Colloid Interface Sci.* **2011**, *358*, 423–429. [CrossRef]

5. Li, J.; Zhang, G.; Shang, Z.; Fan, X.; Zhang, H.; Zhou, L.; Shi, X. Enhanced electromagnetic interference shielding and mechanical properties of foamed epoxy nanocomposites containing carbon nanofiber treated with silicone surfactant. *J. Appl. Polym. Sci.* **2018**, *135*, 46833–46843. [CrossRef]

6. Zang, J.; Wan, Y.J.; Zhao, L.; Tang, L.C. Fracture behaviors of TRGO-filled epoxy nanocomposites with different dispersion/interface levels. *Macromol. Mater. Eng.* **2015**, *300*, 737–749. [CrossRef]

7. Emami, Z.; Meng, Q.; Pircheraghi, G.; Manas-Zloczower, I. Use of surfactants in cellulose nanowhisker/epoxy nanocomposites: Effect on filler dispersion and system properties. *Cellulose* **2015**, *22*, 3161–3176. [CrossRef]

8. Zimmermann, M.V.G.; da Silva, M.P.; Zattera, A.J.; Campomanes Santana, R.M. Effect of nanocellulose fibers and acetylated nanocellulose fibers on properties of poly (ethylene-co-vinyl acetate) foams. *J. Appl. Polym. Sci.* **2017**, *134*, 44760–44771. [CrossRef]

9. Marini, J.; Branciforti, M.C.; Lotti, C. Effect of matrix viscosity on the extent of exfoliation in EVA/organoclay nanocomposites. *Polym. Adv. Technol.* **2010**, *21*, 408–417. [CrossRef]

10. Kaleemullah, M.; Khan, S.U.; Kim, J.K. Effect of surfactant treatment on thermal stability and mechanical properties of CNT/polybenzoxazine nanocomposites. *Compos. Sci. Technol.* **2012**, *72*, 1968–1976. [CrossRef]

11. Zhang, G.; Wang, F.; Dai, J.; Huang, Z. Effect of functionalization of graphene nanoplatelets on the mechanical and thermal properties of silicone rubber composites. *Materials* **2016**, *9*, 92. [CrossRef] [PubMed]

12. Hwang, J.J.; Ma, T.W. Preparation, morphology, and antibacterial properties of polyacrylonitrile/ montmorillonite/silver nanocomposites. *Mater. Chem. Phys.* **2012**, *136*, 613–623. [CrossRef]

13. Xu, J.; Cai, X.; Shen, F. Preparation and property of UV-curable polyurethane acrylate film filled with cationic surfactant treated graphene. *Appl. Surf. Sci.* **2016**, *379*, 433–439. [CrossRef]

14. Góis, G.S.; Nepomuceno, N.C.; França, C.H.; Almeida, Y.M.; Hernandéz, E.P.; Oliveira, J.E.; Oliveira, M.P.; Medeiros, E.S.; Santos, A.S. Influence of morphology and dispersion stability of CNC modified with ethylene oxide derivatives on mechanical properties of PLA-based nanocomposites. *Polym. Compos.* **2019**, *40*, E399–E408. [CrossRef]

15. Luzi, F.; Fortunati, E.; Jiménez, A.; Puglia, D.; Chiralt, A.; Torre, L. PLA nanocomposites reinforced with cellulose nanocrystals from Posidonia oceanica and ZnO nanoparticles for packaging application. *J. Renew. Mater.* **2017**, *5*, 103–115. [CrossRef]

16. Gois, G.D.S.; Andrade, M.F.D.; Garcia, S.M.S.; Vinhas, G.M.; Santos, A.S.; Medeiros, E.S.; Oliveira, J.E.; Almeida, Y.M.B.D. Soil biodegradation of PLA/CNW nanocomposites modified with ethylene oxide derivatives. *Mater. Res.* **2017**, *20*, 899–904. [CrossRef]

17. Fortunati, E.; Armentano, I.; Zhou, Q.; Puglia, D.; Terenzi, A.; Berglund, L.A.; Kenny, J.M. Microstructure and nonisothermal cold crystallization of PLA composites based on silver nanoparticles and nanocrystalline cellulose. *Polym. Degrad. Stab.* **2012**, *97*, 2027–2036. [CrossRef]

18. Fortunati, E.; Armentano, I.; Zhou, Q.; Iannoni, A.; Saino, E.; Visai, L.; Berglund, L.A.; Kenny, J.M. Multifunctional bionanocomposite films of poly (lactic acid), cellulose nanocrystals and silver nanoparticles. *Carbohydr. Polym.* **2012**, *87*, 1596–1605. [CrossRef]

19. Bauli, C.R.; Rocha, D.B.; dos Santos Rosa, D. Composite films of ecofriendly lignocellulosic nanostructures in biodegradable polymeric matrix. *SN Appl. Sci.* **2019**, *1*, 774–785. [CrossRef]

20. Phua, Y.J.; Chow, W.S.; Mohd Ishak, Z.A. Organomodification of montmorillonite and its effects on the properties of poly (butylene succinate) nanocomposites. *Polym. Eng. Sci.* **2013**, *53*, 1947–1957. [CrossRef]

21. Thayumanavan, N.; Tambe, P.B.; Joshi, G. Effect of surfactant and sodium alginate modification of graphene on the mechanical and thermal properties of polyvinyl alcohol (PVA) nanocomposites. *Cellul. Chem. Technol.* **2015**, *49*, 69–80.

22. Moyo, L.; Makhado, E.; Sinha Ray, S. Anomalous impact strength for layered double hydroxide-palmitate/poly (ε-caprolactone) nanocomposites. *J. Appl. Polym. Sci.* **2014**, *131*, 41109–41118. [CrossRef]

23. Tkalya, E.E.; Ghislandi, M.; de With, G.; Koning, C.E. The use of surfactants for dispersing carbon nanotubes and graphene to make conductive nanocomposites. *Curr. Opin. Colloid Interface Sci.* **2012**, *17*, 225–232. [CrossRef]

24. Geng, Y.; Liu, M.Y.; Li, J.; Shi, X.M.; Kim, J.K. Effects of surfactant treatment on mechanical and electrical properties of CNT/epoxy nanocomposites. *Compos. Part A* **2008**, *39*, 1876–1883. [CrossRef]

25. Rana, S.; Alagirusamy, R.; Joshi, M. Mechanical behavior of carbon nanofibre-reinforced epoxy composites. *J. Appl. Polym. Sci.* **2010**, *118*, 2276–2283. [CrossRef]

26. Rana, S.; Alagirusamy, R.; Joshi, M. Development of carbon nanofibre incorporated three phase carbon/epoxy composites with enhanced mechanical, electrical and thermal properties. *Compos. Part A* **2011**, *42*, 439–445. [CrossRef]

27. Wan, Y.J.; Tang, L.C.; Yan, D.; Zhao, L.; Li, Y.B.; Wu, L.B.; Jiang, J.X.; Lai, G.Q. Improved dispersion and interface in the graphene/epoxy composites via a facile surfactant-assisted process. *Compos. Sci. Technol.* **2013**, *82*, 60–68. [CrossRef]

28. Fortunati, E.; Peltzer, M.; Armentano, I.; Torre, L.; Jiménez, A.; Kenny, J.M. Effects of modified cellulose nanocrystals on the barrier and migration properties of PLA nano-biocomposites. *Carbohydr. Polym.* **2012**, *90*, 948–956. [CrossRef]

29. Chi, K.; Catchmark, J.M. Enhanced dispersion and interface compatibilization of crystalline nanocellulose in polylactide by surfactant adsorption. *Cellulose* **2017**, *24*, 4845–4860. [CrossRef]

30. Kim, J.; Montero, G.; Habibi, Y.; Hinestroza, J.P.; Genzer, J.; Argyropoulos, D.S.; Rojas, O.J. Dispersion of cellulose crystallites by nonionic surfactants in a hydrophobic polymer matrix. *Polym. Eng. Sci.* **2009**, *49*, 2054–2061. [CrossRef]

31. Kaboorani, A.; Auclair, N.; Riedl, B.; Landry, V. Physical and morphological properties of UV-cured cellulose nanocrystal (CNC) based nanocomposite coatings for wood furniture. *Prog. Org. Coat.* **2016**, *93*, 17–22. [CrossRef]

32. Lu, F.L.; Shen, M.X.; Xue, Y.J.; Zeng, S.H.; Chen, S.N.; Hao, L.Y.; Yang, L. Application of calcium montmorillonite on flame resistance, thermal stability and interfacial adhesion in polystyrene nanocomposites. *e-Polymers* **2019**, *19*, 92–102. [CrossRef]

33. Dzulkefly, K.; Khoh, H.F.; Ahmad, F.B.H.; Adlie Ahmad, S.; Lim, W.H. Solvent-free esterification process for the synthesis of glucose bolaform surfactants. *Orient. J. Chem.* **2010**, *26*, 747–752.

34. Shamsuri, A.A.; Daik, R.; Zainudin, E.S.; Tahir, P.M. Compatibilization of HDPE/agar biocomposites with eutectic-based ionic liquid containing surfactant. *J. Reinf. Plast. Compos.* **2014**, *33*, 440–453. [CrossRef]

35. Shamsuri, A.A.; Azid, M.K.A.; Ariff, A.H.M.; Sudari, A.K. Influence of surface treatment on tensile properties of low-density polyethylene/cellulose woven biocomposites: A preliminary study. *Polymers* **2014**, *6*, 2345–2356. [CrossRef]

36. Sudari, A.K.; Shamsuri, A.A.; Zainudin, E.S.; Tahir, P.M. Exploration on compatibilizing effect of nonionic, anionic, and cationic surfactants on mechanical, morphological, and chemical properties of high-density polyethylene/low-density polyethylene/cellulose biocomposites. *J. Thermoplast. Compos. Mater.* **2017**, *30*, 855–884. [CrossRef]

37. Shamsuri, A.A.; Md. Jamil, S.N.A. Compatibilization effect of ionic liquid-based surfactants on physicochemical properties of PBS/rice starch blends: An initial study. *Materials* **2020**, *13*, 1885. [CrossRef]

38. Urie, R.; Ghosh, D.; Ridha, I.; Rege, K. Inorganic nanomaterials for soft tissue repair and regeneration. *Annu. Rev. Biomed. Eng.* **2018**, *20*, 353–374. [CrossRef]

39. McLauchlin, A.R.; Thomas, N.L. Preparation and thermal characterisation of poly (lactic acid) nanocomposites prepared from organoclays based on an amphoteric surfactant. *Polym. Degrad. Stab.* **2009**, *94*, 868–872. [CrossRef]

40. Wang, X.; Wang, Y.; Zhao, Z.; Zhu, X.; Nie, C.; Du, S. Preparation and characterization of exfoliated poly (ethyleneterephthalate)/montmorillonite nanocomposites using modified MMTs with variable content of antimony acetate. *Chin. J. Chem.* **2011**, *29*, 1278–1284. [CrossRef]

41. McLauchlin, A.R.; Thomas, N.L. Preparation and characterization of organoclays based on an amphoteric surfactant. *J. Colloid Interface Sci.* **2008**, *321*, 39–43. [CrossRef] [PubMed]

42. Yang, Y.; Gupta, M.C.; Zalameda, J.N.; Winfree, W.P. Dispersion behaviour, thermal and electrical conductivities of carbon nanotube-polystyrene nanocomposites. *IET Micro Nano Lett.* **2008**, *3*, 35–40. [CrossRef]

43. Vilčáková, J.; Moučka, R.; Svoboda, P.; Ilčíková, M.; Kazantseva, N.; Hřibová, M.; Mičušík, M.; Omastová, M. Effect of surfactants and manufacturing methods on the electrical and thermal conductivity of carbon nanotube/silicone composites. *Molecules* **2012**, *17*, 13157–13174. [CrossRef]

44. Yuan, S.; Bai, J.; Chua, C.K.; Wei, J.; Zhou, K. Highly enhanced thermal conductivity of thermoplastic nanocomposites with a low mass fraction of MWCNTs by a facilitated latex approach. *Compos. Part A* **2016**, *90*, 699–710. [CrossRef]

45. Sefadi, J.S.; Luyt, A.S.; Pionteck, J.; Piana, F.; Gohs, U. Effect of surfactant and electron treatment on the electrical and thermal conductivity as well as thermal and mechanical properties of ethylene vinyl acetate/expanded graphite composites. *J. Appl. Polym. Sci.* **2015**, *132*, 42396–42405. [CrossRef]

46. Oh, H.; Kim, K.; Ryu, S.; Kim, J. Enhancement of thermal conductivity of polymethyl methacrylate-coated graphene/epoxy composites using admicellar polymerization with different ionic surfactants. *Compos. Part A* **2019**, *116*, 206–215. [CrossRef]

Review

Recent Progress in Hybrid Solar Cells Based on Solution-Processed Organic and Semiconductor Nanocrystal: Perspectives on Device Design

Sihang Xie [1], Xueqi Li [1], Yasi Jiang [1], Rourou Yang [1], Muyi Fu [1], Wanwan Li [1], Yiyang Pan [1], Donghuan Qin [1,2,*], Wei Xu [1,2,*] and Lintao Hou [3,*]

[1] School of Materials Science and Engineering, South China University of Technology, Guangzhou 510640, China; 201730321308@mail.scut.edu.cn (S.X.); 201730321155@mail.scut.edu.cn (X.L.); 201730321056@mail.scut.edu.cn (Y.J.); 201766321334@mail.scut.edu.cn (R.Y.); mstranswoodfmy@mail.scut.edu.cn (M.F.); mswanwanli@mail.scut.edu.cn (W.L.); 201730321247@mail.scut.edu.cn (Y.P.)

[2] Institute of Polymer Optoelectronic Materials & Devices, State Key Laboratory of Luminescent Materials & Devices, South China University of Technology, Guangzhou 510640, China

[3] Guangdong Provincial Key Laboratory of Optical Fiber Sensing and Communications, Guangzhou Key Laboratory of Vacuum Coating Technologies and New Energy Materials, Siyuan Laboratory, Department of Physics, Jinan University, Guangzhou 510632, China

* Correspondence: qindh@scut.edu.cn (D.Q.); xuwei@scut.edu.cn (W.X.); thlt@jnu.edu.cn (L.H.)

Received: 31 May 2020; Accepted: 16 June 2020; Published: 22 June 2020

Abstract: Solution-processed hybrid solar cells have been well developed in the last twenty years due to the advantages of low cost, low material-consuming and simple fabricating technology. However, the performance, stability and film quality of hybrid solar cells need to be further improved for future commercial application (with a lifetime up to 20 years and power conversion efficiency higher than 15%). By combining the merits of organic polymers and nanocrystals (NC), the reasonable design of interface engineering and device architecture, the performance coupled with stability of hybrid solar cells can be significantly improved. This review gives a brief conclusive introduction to the progress on solution-processed organic/inorganic semiconductor hybrid solar cells, including a summary of the development of hybrid solar cells in recent years, the strategy of hybrid solar cells with different structures and the incorporation of new organic hole transport materials with new insight into device processing for high efficiency. This paper also puts forward some suggestions and guidance for the future development of high-performance NC-based photovoltaics.

Keywords: semiconductor nanocrystal; hybrid; solar cells; organic semiconductor; solution processed

1. Introduction

The usage of semiconductor nanocrystals (NCs) as the light harvesting materials in thin film solar cells has attracted intense research over a period of nearly two decades [1–6]. The solution-processed hybrid solar cells (HSCs) are recently developed and offer the potential advantages of low cost, low materials consumption, simple fabrication process, etc. Among all kinds of the NCs in solution-processed HSCs, CdSe [7], CdTe [8], ZnO [9], TiO$_2$ [10] and PbS [11] have been well used due to their appropriate bandgap, stability and high optical absorption coefficient. In general, solution-processed HSCs consist of a transparent conducting metallic oxide layer (such as ITO, FTO or AZO), an electron transfer (ETL), a photoactive layer, a hole transfer layer (HTL) and a back contact electrode layer. To prepare HSCs, a polymer and NCs are firstly dispersed into a solvent such as octane, toluene, alcohol, chloroform or water with typical concentrations around 10 mg/mL [12,13]. The hybrid solution is then deposited on a substrate by spin-casting or printing. The photovoltaic performance of

HSCs can be easily adjusted at the atomic or molecular level. Compared to organic solar cells (OSCs), the HSC emerged very late and the progress has been relatively slow, which is mainly due to the compatibility issue between organic polymers and NCs. For example, the solubility of the two materials in some solvents is quite distinctive due to the large difference of polarity, and the operation mechanism or physical essence for organic/NC junctions is complex and still controversial [14,15]. There are many strategies to improve the HSC properties, in which the device architecture of the HSC shows a significant effect on the efficiency of the HSC. In this review, we report the advances in hybrid solar cells based on the solution-processed semiconductor NC/polymer and focus particularly on the optimized device design for improving HSC performance. Here, we begin with introducing the fabrication of HSCs by using a single bulk heterojunction (BHJ) active layer. Then, the fundamentals of hybrid bulk heterojunction active layers are depicted. The HSCs with multi-photoactive layers and complex structure are then presented. The hole/electron transfer equilibrium is also a key factor for improving NC solar cell performance. Here, we emphasize to address the numerous strategies using organic materials as hole/electron transfer layers for the efficient collection of holes and electrons. To extend the range of the spectrum response for HSCs, a tandem structure is preferred to further improve the device performance. In the end, we will emphasize the challenges and future prospects in the architecture design of HSCs towards ~15% power conversion efficiency (PCE) aiming for commercial application.

2. Hybrid BHJ Solar Cells with a Large Bandgap Semiconductor NC as an Acceptor

The active layer of HSCs is an interpenetrating network structure in the early time, which is similar to OSCs [16]. It is well known that the OSCs consist of a BHJ active layer which includes a polymer as a donor and (6,6)-phenyl-C61-butyric acid methyl ester (PCBM) as an acceptor [17]. The drawback in this structure is the relative low mobility of electron acceptor PCBM that may affect electron transfer and collection in the OSCs. By replacing PCBM with semiconductor NCs as the acceptor, the polymer NC hybrid thin film may combine the merits of high light absorbance of organic polymers and high electron mobility of the stable semiconductor NC, which is promising for achieving inexpensive, stable and high-performance solar cells. In the early work of HSCs, Alivisatos et al., for the first time, reported a novel HSC with a device configuration of ITO/poly (3-hexylthiophene) (P3HT): CdSe NC/Al by solution processing [18]. Right here, electrons transferred along CdSe nanorods while holes were transported in the polymer. Figure 1a–c shows the structure of regioregular P3HT, energy diagram of the polymer and NCs and device structure. Photogenerated electrons are transferred from the conduction band of the polymer into the conduction band of the NC while holes from NC to polymer. It was found that the CdSe NC and P3HT in the blend film have a complementary absorption property in the visible spectrum. In addition, by altering the diameter of CdSe NCs, the onset of the absorption spectrum can be tuned in a large range. By carefully controlling both the diameter and length of CdSe NCs, a PCE of 1.7% was obtained in a champion device. The experimental results indicate that the length of nanorods has a bigger effect on the PCE of HSCs than the diameter of NCs. The onset of the spectrum response was extended up to ~720 nm by improving the diameter of NCs.

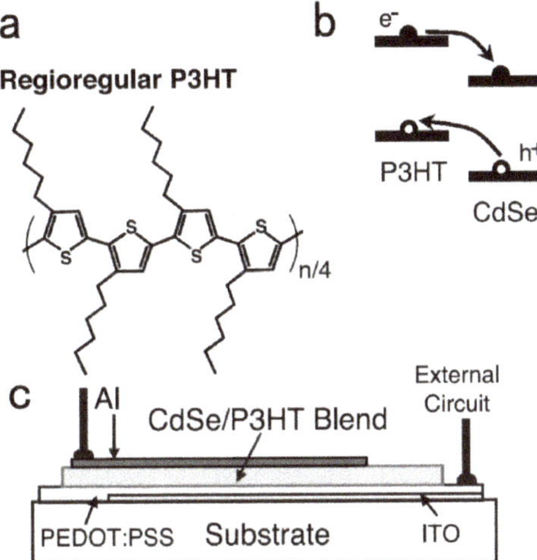

Figure 1. (**a**) The molecular structure of regioregular P3HT. (**b**) The energy level diagram of CdSe nanorods and P3HT with a schematic drawing of electron transfer to CdSe and hole transfer to P3HT. (**c**) The device structure consists of a 200 nm film sandwiched between an aluminum electrode and a transparent hole transfer layer (HTL) of PEDOT:PSS (Bayer AG, Pittsburgh, PA), which was deposited on the indium tin oxide electrode. The active area of the device is 3 mm^2. This film was spin-cast from a pyridine chloroform solution of 90 wt% CdSe nanorods in P3HT. Reproduced with permission from [18], Copyright © 2002, The American Association for the Advancement of Science.

Couderc et al. [19] adopted the ultrafast transient spectroscopy to investigate the charge transfer (CT) dynamics between the polymer and NCs of the hybrid film. Through systematically analyzing the bleaching signal of CdSe NCs, the electron transfer was found to occur in a very fast way (less than 65 fs). The morphology of CdSe NCs has a significant effect on HSC performance. Sun et al. [20] found that the HSC processed from the mixture of multi-armed CdSe NCs and a traditional polymer P3HT showed enhanced properties than the device from nanorod/polymer blends, which was attributed to the improved electron extraction since CdSe tetrapods with four arms can facilitate the electron transfer vertical to the plane of the active layer. When the polymer P3HT was replaced by poly(p-phenylenvinylene) derivative (OC1C10-PPV), [21] and a solvent of 1,2,4-trichlorobenzene with a high boiling point (up to 200 °C), instead of chloroform used to prepare the HSC, a high PCE of 2.8% was obtained. Subsequently, the crystalline structure and the solubility of CdSe NCs were found to greatly affect the HSC performance [22,23]. Han et al. [22] developed a new receipt for the fabrication of spherical CdS NCs using cadmium carboxyl as the Cd precursor. From Figure 2a,b, the photoluminescence (PL) intensity of hybrid blends decays linearly with the increase of CdSe NC content, which is irrespective to the CdSe NC synthesis method. The HSC based on MEH-PPV: CdSe NC blend film showed improved efficiency when post-annealing at 180 °C was carried out (Figure 2c,d).

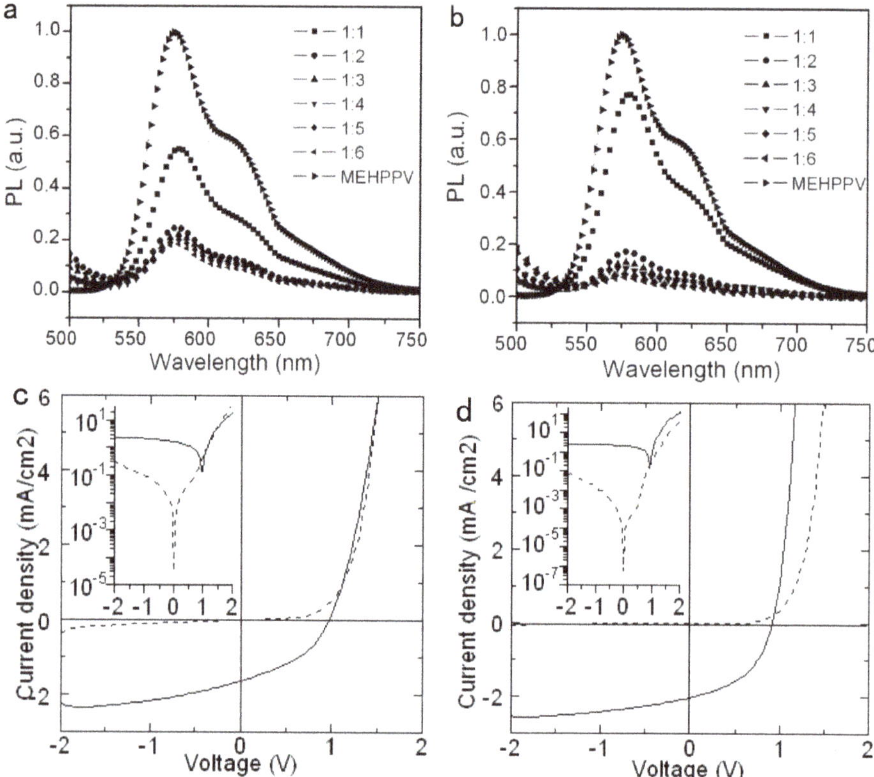

Figure 2. Photoluminescence intensity of MEH-PPV: CdSe nanocrystal (NC) blends as a function of the weight fraction of CdSe NCs under excitation at 500 nm: (**a**) CdSe NCs prepared by Se powder, (**b**) CdSe NCs prepared by trioctylphosphine/Se injection. Current density–voltage (*J–V*) characteristics of a MEH-PPV: CdSe NC device with the weight ratio of 1:8 (**c**) without and (**d**) with annealing at 180 °C for 20 min. The dotted line was measured in the dark and the solid line under 1.5 AM solar illumination. The insets are the *J–V* curves with current density (*J*) expressed in logarithmic coordinates. Reproduced with permission from [22], Copyright © 2006, IOP Publishing.

In most cases, CdSe NCs are prepared by the solvothermal method. Long-chain organic ligands such as alky acid, alkyl phosphoric acid or trioctylphosphine oxide (TOPO) are used to stabilize the NCs and prevent their aggregation [24–28]. However, the side chains of polymers (usually alkyl chains) cannot cap on the surface of NCs, which makes the separation of the organic/inorganic phase uncontrollable in the hybrid active layer and hinders the transfer of charges between the NC and polymer and thus reduces device performance. To address this problem, Liu et al. [29] fabricated P3HT with different functional groups that can easily control the morphology of polymer/CdSe NCs. Figure 3a shows the synthesis process of P3HT with a –NH$_2$ functional group. When blended with an NC acceptor, the amino end group of the P3HT donor caps on the surface of the NC and forms intimate nanocomposites with a favorable morphology. The HSC shows a significantly higher PCE than the control device (Figure 3b). More importantly, the highest efficiency occurred at lower concentrations of CdSe NCs, which is mainly determined by the high degree of homogeneity of hybrid blend layers.

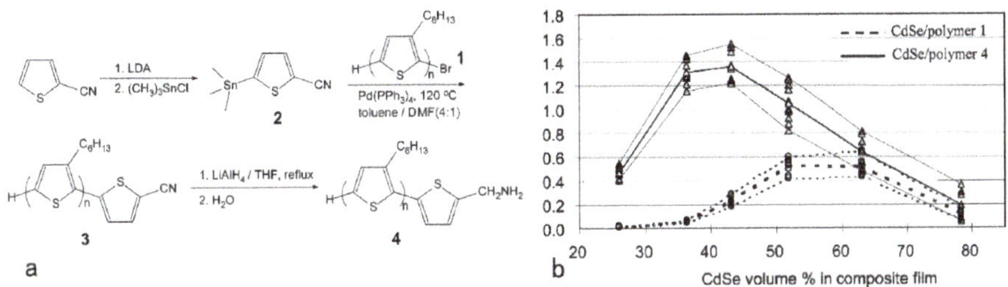

Figure 3. (**a**) Synthesis route for P3HT with amino end-functionality. (**b**) Plots of power conversion efficiency (PCE) versus the volume ratio of CdSe with polymer 4 (solid lines) and polymer 1 (dashed lines) as the donor, respectively. Note the very high reproducibility of the lowest, highest and average performance for numerous duplicating devices. Reproduced with permission from [29], Copyright © 2004, American Chemical Society.

An alternative way to improve the quality of the HSC active layer is to eliminate the insulated ligands (such as oleic acid) on the CdSe NC, which will reduce the interface recombination between the NC and polymer, and improve the carriers' collecting ratio. In recent years, solution-processed PbS colloidal quantum dot (CQD) solar cells have attracted much attention and excellent achievement has been obtained (with PCEs exceeding 12%), which is mainly attributed to the ligand chemistry engineering. The insulated ligands oleic acid (OA) can be replaced by 1,2-ethanedithiol (EDT), mercaptopropionic acid (MPA), NH_4I, etc. [30–33]. Inspired by these fruitful works, Ren et al. [34] demonstrated a facile method to fabricate nanoscale morphology of P3HT nanowires (NWs) and CdS NC blends adopting the ligand exchange method and solvent-assisted chemical grafting. It was found that an interpenetrating BHJ network was formed in the blends, which increased the carriers' separating and collecting ratios, leading to efficient PCE. The HSC with a configuration of ITO/poly(3,4-ethylenedioxythiophene):poly(styrenesulfonate) (PEDOT:PSS)/P3HT NW:CdS/bathocuproine (BCP)/Mg:Ag delivered an energy conversion efficiency exceeding 4%. It should be noted that P3HT or PPV was widely used as the donor material for HSCs in early works [19,22,29]. Unfortunately, the onset of the spectrum response for these materials (with bandgaps exceeding 1.9 eV) was below 650 nm, which significantly reduced photon absorption in the longer wavelengths and thus rendered the short-circuit current density (J_{sc}) of HSCs. Poly [2,6-(4,4-bis-(2-ethylhexyl)-4H-cyclopenta [2,1-b;3,4-b0]-dithiophene)-alt-4,7-(2,1,3-benzothiadiazole)](PCPDTBT), a conjugated polymer with a low bandgap, has been widely used in OSCs [35]. PCPDTBT used as a donor in HSCs should expand the spectrum responsibility into the near-infrared field and thus high J_{sc} is expected. Moreover, in order to adjust the interface morphology among the blend of PCPDTBT/CdSe NCs, different substituents ($-NH_2$, $-OCH_3$, $-CH_3$, $-F$, $-(CF3)_2$, $-NO_2$) have been developed to replace the terminal group of the benzenethiols (BTs) [36]. It is important to select the ligand unit with an appropriate molecular dipole as it has an intense effect on the molecular and electronic framework at the polymer/NC boundary and subsequently on HSC properties. A maximum efficiency of 4.0% for HSCs was obtained when 4-uorobenzenethiol was selected to make the ligand exchange. Zhou et al. [37] developed a post EDT treatment strategy for hybrid blend film using a device architecture of ITO/PEDOT:PSS/P3HT:CdSe NCs/Al (Figure 4a,b). After EDT treatment, from the Fourier transform infrared (FTIR) measurement results, one could see that the absorption peaks for the -CH$_3$, P-O and P=O groups were significantly decreased, which implies the TOPO or TDPA can be effectively removed by ligand exchange.

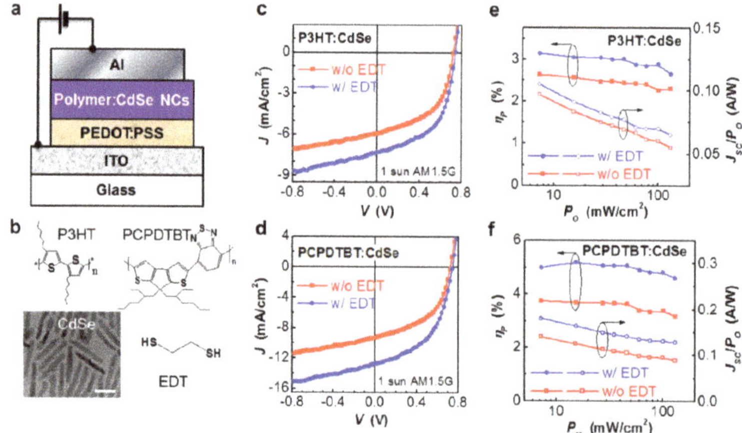

Figure 4. Performance enhancement in polymer: NC hybrid solar cells (HSCs) upon EDT treatment. (**a**) Schematic structure of polymer: NC HSCs. (**b**) TEM image of CdSe nanorods (scale bar: 20 nm) and chemical structures of conjugated polymers (P3HT and PCPDTBT) and EDT. The nanorods have an aspect ratio of 7. *J-V* characteristics of (**c**) P3HT:CdSe and (**d**) PCPDTBT:CdSe HSCs with/without EDT treatment. (**e,f**) Dependence of illumination power on PCE and J_{sc}/P_0 for P3HT:CdSe and PCPDTBT:CdSe HSCs with/without EDT treatment, respectively. Reproduced with permission from [37], Copyright © 2013, American Chemical Society.

3. Hybrid Bulk Heterojunction Solar Cells Using Low Bandgap Nanocrystals

HSCs based on polymer: CdSe (or other wide bandgap semiconductors such as ZnO, CdS and TiO_2 [38,39]) NC BHJ have been well explored, and PCEs up to 4% have been obtained by optimizing hybrid layer morphology and ligand engineering. However, HSCs can only absorb visible light and the onset of the spectrum response is mainly determined by donor materials, which restrict HSCs for largely usage of long-wavelength infrared photons. Recently, PbS [40], PbSe [41] and PbS_xSe_{1-x} [42,43] alloy NCs have demonstrated potential application in low cost, highly efficient photovoltaic products as they contain special properties, for example, a size-tunable bandgap, high charge carrier mobility, multiple charge generation and solution processable ability. A PCE of 12% for PbS NC bulk homojunction photovoltaics has been certificated recently by surface ligand engineering and an optimized design of device structure [44]. By incorporating PbS or PbSe into HSCs, the efficient use of both short and long (>700 nm) wavelength photons was achieved [45]. In 2010, Noone et al. [46] described new HSCs containing a mixture of colloidal PbS NCs and donor polymers including poly(2,3-didecyl-quinoxaline-5,8-diyl-alt-N-octyldithieno[3,2-b:2′,3′-d] pyrrole) (PDTPQx), PDTPBT and P3HT. It was found that when the polymer PDTPQx was employed, the PCE was 10–100 times higher than that based on traditional polymers of P3HT or PDTPBT. It was confirmed that the PDTPQx:PbS NC HSC behaved as the real bulk heterojunction rather than as other junctions such as Schottky diodes.

Although the early ligand exchange method (for example, the alky acid ligands of PbS NCs were substituted by amine ligands before mixing) can improve HSC performance, the butyl amine ligand still affects the polymer matrix during the film deposition. Seo et al. [47] demonstrated a simple method to improve the performance of HSCs using PbS NCs and PDTPBT as the blends. In this case, the blend film was treated by direct post-deposition ligand exchange using 1,2-ethanedithiol (EDT). The OA ligand was selectively replaced by EDT without affecting the polymer matrix. A PCE as high as 3.78% was achieved after further optimizing the device structure by inserting an ETL TiO_2 (device structure: ITO/PEDOT:PSS/PDTPBT:PbS EDT/TiO_2/LiF/Al). In truth, the observed interparticle spacing is reduced after the alky acid ligand is substituted by EDT or MPA ligands (~4.37 nm for OA ligands and ~3.36 nm for EDT or MPA ligands), which has been confirmed in the previous work [48]. The –SH can be tightly bonded to the surface of the PbS NC and replace the OA ligand, which can reduce the distance between neighboring NCs and decrease interface recombination. Moreover, Piliego et al. [49] demonstrated that the charge separation efficiency was nearly the same as the organic-organic blends for the PbS-polymer hybrid blends. The low performance of HSCs is greatly affected by worse morphology and more traps in the hybrid blends. Most recently, Lu et al. [50] used iodidum such as lead iodide (PbI_2) and ammonium iodide (NH_4I) coupled with n-butylamine to make ligands exchange for PbS NCs. HSCs based on Si-PCPDTBT and NH_4I- or PbI-capped PbS NCs were fabricated without further post ligand exchange. From the results presented by the time-resolved PL spectroscopy, one could see that the PbI_2-exchanged HSC showed higher energy or a charge transfer process than the NH_4I-exchanged HSC. A PCE of 4.8% was achieved for the HSC with the PbI_2-exchanged PbS acceptor, ranking among the best value ever presented for polymer-PbS NC BHJ solar cells.

4. HSC with More Active Layers

In the case of the BHJ photovoltaic device, the blend mixture includes a conjugated polymer and a semiconductor NC. The carrier recombination between the hybrid layer and the contact electrode is serious, which renders the performance of the HSC. It was confirmed that the self-assembly property of the active solution in the coating process significantly affects the phase separation of the dried hybrid layer, which is important for the carrier departure and charge migrate [51,52]. As for the polymer/fullerene blend solar cells, it is well known that the donor/acceptor phase separation in the vertical distribution in the polymer/fullerene blend film is also critical to device performance [53,54]. Therefore, an optimized device architecture for HSCs may consist of multi-film such as donor: acceptor (BHJ)-donor-acceptor structure (D:A-D-A) for optimizing interface engineering. With this strategy, the charge recombination is reduced, which has been demonstrated in small molecule OSCs [55,56]. For HSCs, Liu et al. [57] designed a device with a configuration of $ITO/TiO_2/CdTe/MEH-PPV:CdTe/MoO_3/Au$ based on aqueous-processed CdTe NCs and MEH-PPV. Figures 5a and 5b show the schematic device structure and band alignment of semiconductor NCs and the polymer.

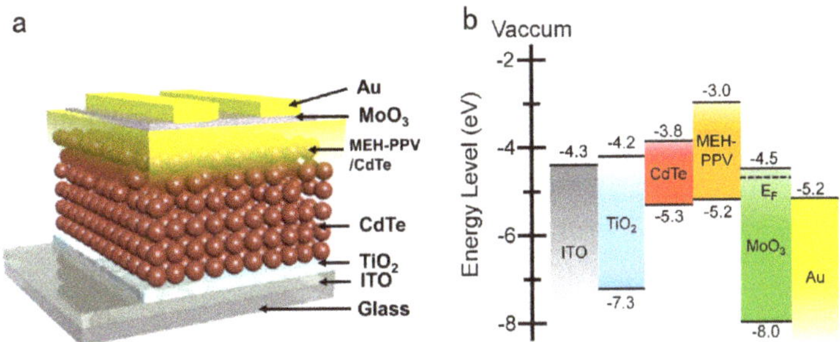

Figure 5. (a) HSC structure and (b) band alignment of different layers. EF represents the Fermi level. Reproduced with permission from [57], Copyright © 2016, Royal Society of Chemistry.

The ratio of MEH-PPV to CdTe in the mixture layer was found to affect the nanoscale morphology of hybrid blend film, which then impacts CT and transport. A PCE as high as 4.20% was attained by controlling the ratio of MEH-PPV: NCs in the blend film (the optimized weight ratio for MEH-PPV: CdTe is 1:12) and annealing temperature (the optimized temperature is 315 °C). Later on, the PCE was further improved to 4.32% when the MEH-PPV was replaced by another polymer P3HT [58]. To enhance the internal quantum efficiency and extend the range of the depletion field, an HSC with both sides bulk heterojunction (BHJ) (device architecture $ITO/TiO_2/CdTe:TiO_2/CdTe/PPV:CdTe/MoO_3/Au$) was recently reported by Jin et al. [59]. They found that employing different acceptors with more active layers can simultaneously suppress the interface recombination and promote the carrier extraction in the HSC. Benefiting from the advantages of increased carrier collection and lifetime, a PCE up to 6% was obtained in a champion device. Following this, this research group selected ZnO NCs and sol-gel ZnO as an acceptor/ETL to build an HSC with a device architecture of ITO/ZnO/CdTe:ZnO NC/CdTe/PPV:CdTe/MoO$_3$/Au [60]. A high PCE of 6.51% was achieved, which was among the highest PCEs already reported for aqueous-processed CdTe NC HSCs. In the field of PbS NC HSCs, the main reason for high device properties in these cases is mainly attributed to the exchange ligands, which reduce the defect density by passivating the bulk and interface states. It is noted that the PbS NCs become insoluble in most organic solvent after ligand-exchange, which prevents the dissolvement of the polymer: PbS NC layer when the PbS NC layer is deposited on the hybrid blend. Therefore, an optimized structure with the configuration of D-D:A-A can be realized for PbS NC HSCs. Inspired by efficient small molecule OSC design [55], Liu et al. [61] developed a D-D:A-A HSC structure of ITO/PEDOT/DDTPBT: NC/NC/LiF/Al. A PCE as high as 5.50% coupled with improved FF was achieved in this HSC, which is mainly owed to the outstanding properties of the PbS$_x$Se$_{1-x}$ alloy NCs and the optimized device structure for efficient carrier separation and collecting. The band alignment of the polymer and PbS NCs was also found to be critical for charge separation. In the case of HSCs with the configuration of ITO/PEDOT–PSS/polymer: PbS/PbS/LiF/Al, a maximum PCE was obtained when using PDBT as the donor materials in the blend mixture [62]. Although various strategies for HSCs have been studied to improve the PCE by exploiting the merits of conjugated polymer and semiconductor materials, the PCE of these HSCs is still far below 10%; the value is remarkably lower than pure PbS NCs or OSCs. This is mainly attributed to the poor charge transfer and carrier extraction efficiencies in the hybrid blend film. Most recently, Baek et al. [63] developed a new HSC with the device structure of ITO/ZnO/PbS NC/PBDTTT-E-T: 2,2'-((2Z,2'Z)-((5,5'-(4,4,9,9-tetrakis(4-hexylphenyl)-4,9-dihydros-indaceno[1,2-b:5,6-b']dithiophene-2, 7-diyl)bis(4-((2-ethylhexyl)-oxy)thiophene-5,2-diyl))bis(methanylylidene))bis(3-oxo-2,3-dihydro-1H-indene-2,1-diylidene))dimalononitrile (IEICO)/MoO$_3$/Ag. There are two active layers in this device configuration, viz. PbS NCs and polymer/SM blend film. The usage of small-molecule acceptors instead of NCs increases the efficiency reported for all solution-processed organic/inorganic NC HSCs.

5. HSCs Using Organic Materials as the HTL

For NC solar cells, the carrier recombination between the active layer and the contact electrode is serious, which renders the weak collection of carriers. For example, the CdTe thin film has a high work function (up to 5.3 eV) due to its self-compensation effects [64], which makes it difficult to attain ohmic contact between CdTe NC thin film and a metal electrode. There are several ways to obtain ohmic contact for CdTe-based solar cells with inverted structure. Among which, introducing a low cost and stable HTL for reducing carriers recombining is preferred. It is well known that MoO$_x$ [65], V$_2$O$_5$ [66], NiO [67], WO$_3$ [68] and other metal-organic dielectric materials [69,70] are widely used as HTLs in CdTe thin film solar cells and improved performance is usually obtained. However, researchers have found that these metal oxide materials are not stable when exposed in ambient conditions and the energy levels are difficult to be changed, which is not preferred for commercial application. On the contrary, there are many advantages for organic hole transport materials with a high HOMO level, such as being stable, low energy-consuming and they can be fabricated by a solution process. Recently, Wang [71] et al. developed PEDOT: PSS as an HTL between CdTe thin film and a contact electrode, and the device showed improved efficiency compared to the control devices. Spiro-OMeTAD(2,2,7,7-tetrakis(N,N-di-4-methoxyphenylamino)-9,9-spirobitluorene), as an important hole transfer material, has been applied successfully in perovskite solar cells, which is also preferred for NC solar cells as the high work function (~5.2 eV) [72]. Based on potential change results presented by Kelvin probe microscopy between Spiro and the CdTe NC film, it was found that a dipole field is formed between CdTe NC and HTL, which strengthens the build-in electric field and increases the NC device's performance. However, like many metallic oxides, PEDOT: PSS or Spiro-OMeTAD is not stable under wet environments. Guo [73] and coworkers, for the first time, incorporated a crosslinkable conjugated polymer as an HTL for decreasing carrier recombination in the interface of the NC thin film and contact metal. Compared to other organic HTLs, crosslinkable conjugated polymers are very stable after the crosslinking reaction, and can be adhered to the NC surface by forming N-Cd bonds. Moreover, this material has high carrier mobility and easy tunability of energy levels by molecular design. In this case, a poly(diphenylsilane-co-4-vinyltriphenylamine) HTL is deposited on the CdTe film and forms a Cd-N covalent bond, which reduces the interface trap state and carrier recombination. More importantly, a dipole layer is formed between CdTe NC film and the HTL, which facilitates holes' transport with a small barrier and improves carriers' collecting efficiency. As high as 8.34% PCE has been obtained in NC solar cells with a Si-TPA interlayer, which is significantly higher than the controlled device or device with PEDOT:PSS or PVK as an HTL. In addition, it was demonstrated that CdTe solar cells with CdSe as the n type partner may be subjected to low V_{oc} owing to a large electron injection barrier and interface recombination in metal oxide/CdSe$_x$Te$_{1-x}$ alloys [74]. By incorporating a CdS/CdSe NC double ETL and TPA HTL, the interface recombination was decreased and high V_{oc} expected [75]. The device structure and preparing process are presented in Figure 6. These two aspects' optimization led to reduced interface trap-assisted recombination both in the ZnO/CdS and CdTe/contact electrode. As shown in Figure 7, a very high PCE of 9.2% was attained by simultaneously optimizing the ETL and HTL of the HSC. Besides, it is noted that the EQE spectrum of the optimized device showed a high EQE value from 400 to 900 nm, which is significantly higher than other control devices. The high performance obtained in this case was recorded for all CdTe NC HSCs with inverted structure.

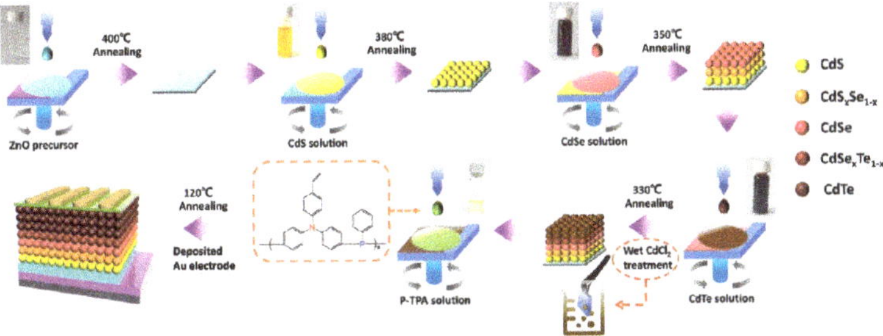

Figure 6. A schematic outline of the layer-by-layer process for making CdTe NC solar cells with a crosslinkable HTL. Reproduced with permission from [75], Copyright © 2019, WILEY-VCH Verlag GmbH & Co. KGaA, Weinheim.

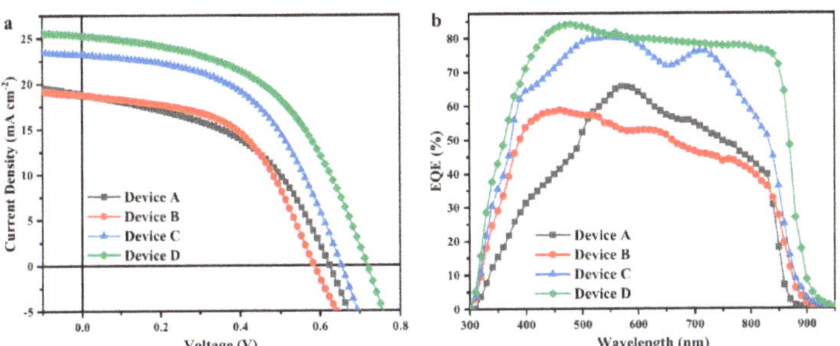

Figure 7. (**a**) *J–V* curves of CdTe NC solar cells with different structures (Device A: ITO/ZnO/CdS/CdTe/Au; Device B: ITO/ZnO/CdSe/CdTe/Au; Device C: ITO/ZnO/CdS/CdSe/CdTe/Au; Device D: ITO/ZnO/CdS/CdSe/CdTe/HTL/Au; (**b**) corresponding EQE spectra. Reproduced with permission from [75], Copyright © 2019, WILEY-VCH Verlag GmbH & Co. KGaA, Weinheim.

6. HSCs with Tandem Structure

With the rapid development of molecular and device design, up to 18% efficiency has been realized in single junction OSCs [76]. However, the further improvement of single junction solar cell performance is limited, and thus double, triple or more junction solar cells should be carried out to improve the HSC performance [77–79]. For solar cells with tandem structures, broader spectrum photons can be harvested after using different bandgapsvactive materials. Organic and some NCs such as PbS have similar fabricating technology, and low-cost, light-weight solar modules can be realized using the same coating method. Compared to organic thin film, PbS NC layers benefit from a tunable bandgap by adjusting the size of PbS quantum dots, which is appropriate for tandem cell design. In 2014, Speirs and coworkers [80], for the first time, reported a new tandem HSC employing PbS NCs as the front sub-cell while a P3HT: PCBM blend was the bottom cell (device structure: ITO/PbS/WO$_3$/Al/P3HT:PCBM/Al), which allowed the usage of sunlight from ultraviolet to near-infrared. However, only a 1.8% PCE was attained for this device, which shows also lower performance than the PbS NC or P3HT:PCBM solar cells. The main reason for low efficiency originates from the interface recombination and inadequate fabricating parameter optimization for each single junction. Later on, Kim et al. [81] developed a different device design consisting of PbS NCs as the bottom cell and a polymer (with large bandgap): fullerene blend as the top cell (device structure: FTO/TiO$_2$/PbS NC/MoO$_x$/ZnO/PFN/polymer:fullerene/MoO$_x$/Ag). The HSC-based tandem structure delivered a high V_{oc} (1.3 V) and PCE greater than 5% after optimizing the tunnel junction layer and individual organic/inorganic cells. The effective connection of two single sub-cells in hybrid tandem solar cells is very important for decreasing interface recombination and internal losses. It was found that when MoO$_x$/Au/ZnO was selected as the interface recombination layer in the hybrid tandem cell with a device architecture of ITO/AZO/PbS/MoO$_3$/Au/PFN/PCBM:PDPP3T/MoO$_3$/Ag, an 8.3% PCE was obtained after optimizing the optical and electrical properties of each sub-cell [82]. However, in the previous reports, one noteworthy limitation was the inaccurate arrangement of the PbS NC (with low bandgap) layer as the front cell, which limited the light harvested and prevented the tandem cells to create more photocurrent. The main reason for this structure design is the incompatibility of the PbS NC solution to organic bulk heterojunction layer [83,84]. Secondly, in order to improve the PbS NC film quality, ligand (such as MPA or EDT) exchange must be taken out to eliminate trap states, which cause additional damage of underlayers [85]. To overcome these challenges, PbS NC ink utilizes hexane and the deposition of PbS NCs is simplified by eliminating the many steps required for PbS NC layer deposition. Figure 8a presents the device structure of an HSC with a tandem configuration. The organic PTB7:PCBM blend device acts as the front cell while PbS NCs act as the bottom cell. Figure 8b shows the band alignment of the hybrid tandem solar cell. Here, an ultrathin (0.5 nm) Au film was inserted between the AZO and MoO$_x$ layer, as the effective recombination centers to decrease the collection of photo-generated carriers in the active layers. After optimizing device fabricating parameters, the tandem solar cell reached a high PCE of 9.4%.

Most recently, Aqoma et al. [86] presented a hybrid tandem cell in which organic BHJ worked as the back-cell collected the NIR photons while the CQD layer absorbed the visible light photons. This device with a configuration of ITO/ZnO/PbS/EDT-PbS/Au/ZnO/PTBT-Th:IEICO-4F/MoO$_3$/Ag reinforced the photon-to-current conversion from visible to NIR wavelengths (from 350 to 1000 nm). A champion device achieved a PCE of 12.8% after optimizing the J_{sc} balance in conjunction with excellent series connection, which ranks among the highest PCE ever reported in PbS hybrid tandem solar cells.

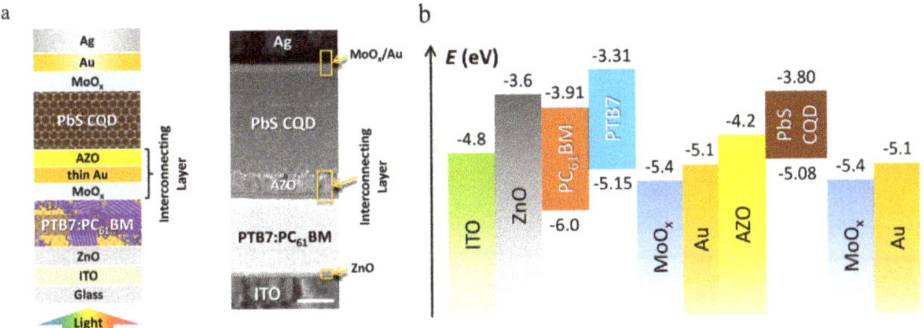

Figure 8. (**a**) Hybrid tandem solar cell configuration and TEM cross-sectional image of the corresponding device. The scale bar is 100 nm. (**b**) Energy level diagram of the hybrid tandem solar cell. Reproduced with permission from [85], Copyright © 2018, American Chemical Society.

7. Conclusions

The hybrid organic/inorganic NC solar cell is considered to be a promising candidate as a next generation photovoltaic product. Recent developments of organics and NCs for photovoltaic applications have been promoted including device design, interface engineering, new molecular design and synthesis, film processing techniques, ligands exchange technics, etc. These have enabled the solution-processed HSC to achieve a PCE of over 12%. The efficient HSC with all kinds of the device configurations reported in recent years are summarized in Table 1.

Compared to pure NCs or OSCs, HSCs exhibit several advantages such as combining the merits of high electron mobility of semiconductor NCs and high absorption coefficients of organic polymers, complementary light absorption spectra, the use of stable inorganic semiconductor NCs and high practicability of film fabrication based on solution processing. HSCs are expected to be applied in flexible and low-cost photovoltaic products in the future. There are several urgent issues to be handled to further improve the PCE and stability of HSCs. Firstly, it is critical to design or select an appropriate polymer to permit well band alignment with NCs for BHJ structure. Secondly, the interfacial junctions between the polymer HTL and NC layer should be studied deeply as the trap states in the interface could behave as non-radiative recombination centers. Thirdly, more junctions should be introduced into the HSC design to expand the spectrum response to 1,200–1,300 nm, for example, by using PbS quantum dots with a small bandgap (~1.0 eV) as bottom sub-cells for IR harvesting in the optimized device structure. Finally, the lifetime of HSCs has not been studied extensively. Compared to the commercial inorganic thin film photovoltaic module (~12% for amorphous silicon and ~19% for CdTe and CuInGaSe2 [87]), the PCE of HSCs (~13% for the best cells with a small area) needs to be further improved. Moreover, the stability of HSCs should be improved by materials purify, device design, interface passivation and device packing technology. Anyway, much attention should be paid to practical applications. With the development of HSCs, solution-processed HSCs must have a promising future for solar cell applications.

Table 1. HSC solar cell devices with all kinds of device configurations. Device type: (**A**) Hybrid bulk heterojunction solar cells with semiconductor NCs as acceptors. (**B**) Hybrid solar cells with more active layers. (**C**) Hybrid solar cells using organic materials as hole transfer layers. (**D**) Hybrid solar cells with tandem structures.

Type	Device Architecture	V_{oc} (V)	J_{sc} (mA/cm^2)	FF (%)	PCE (%)	Published Year	Ref.
A	ITO/P3HT: CdSe NC/Al	0.70	5.70	40.0	1.7	2002	[18]
	ITO/PEDOT: PSS/MEH-PPV:CdSe NC/Al	0.90	2.03	47.0	0.85	2006	[22]
	ITO/PEDOT:PSS/P3HT:CdS/BCP/Mg:Ag	1.10	10.90	35.0	4.1	2011	[34]
	ITO/ PEDOT:PSS/PCPDTBT:CdSe/PFN/Al	0.69	10.17	57.0	3.99	2014	[36]
	ITO/PEDOT:PSS/P3HT:CdSe NC/Al	0.73	7.40	54.0	2.9	2013	[37]
	ITO/PEDOT:PSS/PCPDTBT:CdSe NC/Al	0.74	12.80	50.0	4.7	2013	[37]
	ITO/PEDOT:PSS/PDTPBT:PbS/TiO$_2$/LiF/Al	0.57	13.06	51.0	3.78	2011	[47]
	ITO/PEDOT:PSS/Si-PCPDTBT:PbS/ZnO/Al	0.48	18.20	55.0	4.78	2016	[50]
B	ITO/TiO$_2$/CdTe/MEH-PPV:CdTe/MoO$_3$/Au	0.60	13.56	51.7	4.20	2016	[57]
	ITO/TiO$_2$/CdTe/P3HT:CdTe/MoO$_3$/Au	0.54	16.59	47.2	4.32	2015	[58]
	ITO/TiO$_2$/CdTe:TiO$_2$/CdTe/PPV:CdTe/MoO$_3$/Au	0.615	18.90	51.7	6.01	2018	[59]
	ITO/ZnO/CdTe:ZnO/CdTe/PPV:CdTe/MoO$_3$/Au	0.62	19.50	53.9	6.51	2019	[60]
	ITO/ZnO/PbS/PBDTTT-E-T:IEICO/MoO$_3$/Ag	0.66	29.60	67.0	13.1	2019	[63]
C	Glass/SnO$_2$:F/SnO$_2$/CdS/CdTe/PEDOT:PSS/Au	0.71	21.42	60.0	9.1	2016	[71]
	ITO/TiO$_2$/CdTe NC/spiro-OMeTAD/Au	0.71	18.78	49.2	6.56	2016	[72]
	ITO/ZnO/CdSe/CdTe/Si-TPA/Au	0.66	23.38	54.1	8.34	2018	[73]
	ITO/ZnO/CdS/CdSe/CdTe/P-TPA/Au	0.72	25.31	50.5	9.20	2019	[75]
D	ITO/ZnO/PFN-Br/PBDB-T:F-M/M-PEDOT/ZnO/PTB7-Th:O6T-4F:PC$_{71}$BM/MoO$_3$/Ag	1.64	14.35	73.7	17.36	2018	[76]
	ITO/PbS/WO$_3$/Al/P3HT:PCBM/Al	0.89	3.90	53.0	1.8	2014	[80]
	FTO/TiO$_2$/PbS NC/MoO$_x$/ZnO/PFN/Polymer-:fullerene/MoO$_x$/Ag	1.30	5.76	68.1	5.25	2015	[81]
	ITO/ZnO/PbS/MoO$_3$/Au/ZnO/PTB7-Th/PC$_{71}$BM/MoO$_3$/Ag	1.27	10.36	63.0	8.27	2017	[82]
	ITO/ZnO/PTB7:PCBM/MoO$_x$/Au/AZO/PbS/-MoO$_x$/Au/Ag	1.31	12.50	56.7	9.4	2018	[85]
	ITO/ZnO/PbS/EDTPbS/Au/ZnO/PTBTTh:IEICO-4F/MoO$_3$/Ag	1.36	13.63	69.0	12.82	2020	[86]

Author Contributions: D.Q., S.X. and W.X. searched and classified the literature; D.Q., S.X., L.H., W.X., X.L., Y.J., R.Y., M.F., W.L. and Y.P. wrote the paper. All authors have read and agreed to the published version of the manuscript.

Funding: This research received no external funding.

Acknowledgments: This work is financially supported by the National Natural Science Foundation of China (No. 21875075, 61774077, 61274062, 61775061), Guangdong Natural Science Fund (No. 2018A0303130041, No. 2018A0303130211), Guangzhou Science and Technology Plan Project (No. 201804010295), National Undergraduate Innovative and Entrepreneurial Training Program (No. 201910561011), the Key Projects of the Joint Fund of Basic and Applied Basic Research Fund of Guangdong Province (2019B1515120073) and the Fundamental Research Funds for the Central Universities for financial support.

Conflicts of Interest: The authors declare no conflict of interest.

References

1. Kiani, A.; Sutherland, B.R.; Kim, Y.; Ouellette, O.; Levina, L.; Walters, G.; Dinh, C.T.; Liu, M.; Voznyy, O.; Lan, X.; et al. Single-step colloidal quantum dot films for infrared solar harvesting. *Appl. Phys. Lett.* **2016**, *109*, 183105. [CrossRef]

2. Zeng, Q.; Chen, Z.; Zhao, Y.; Du, X.; Liu, F.; Jin, G.; Dong, F.; Zhang, H.; Yang, B. Aqueous-Processed inorganic thin-film solar cells based on $CdSe_xTe_{1-x}$ nanocrystals: The impact of composition on photovoltaic performance. *ACS Appl. Mater. Interfaces* **2015**, *7*, 2322–2323. [CrossRef] [PubMed]

3. Beek, W.J.; Wienk, M.M.; Janssen, R.A. Efficient hybrid solar cells from zinc oxide nanoparticles and a conjugated polymer. *Adv. Mater.* **2004**, *16*, 1009–1013. [CrossRef]

4. Arici, E.; Sariciftci, N.S.; Meissner, D. Hybrid solar cells based on nanoparticles of $CuInS_2$ in organic matrices. *Adv. Funct. Mater.* **2003**, *13*, 165–171. [CrossRef]

5. Arici, E.; Hoppe, H.; Schäffler, F.; Meissner, D.; Malik, M.A.; Sariciftci, N.S. Morphology effects in nanocrystalline CuInSe 2-conjugated polymer hybrid systems. *Appl. Phys. A* **2004**, *79*, 59–64. [CrossRef]

6. Chaure, S. MEH-PPV/CdS Hybrid Nanowire Polymer Solar Cell Array. *J. Electron. Mater.* **2019**, *48*, 1074–1078. [CrossRef]

7. Dayal, S.; Kopidakis, N.; Olson, D.C.; Ginley, D.S.; Rumbles, G. Photovoltaic devices with a low band gap polymer and CdSe nanostructures exceeding 3% efficiency. *Nano Lett.* **2010**, *10*, 239–242. [CrossRef]

8. Jin, G.; Wei, H.T.; Na, T.Y.; Sun, H.Z.; Zhang, H.; Yang, B. High-efficiency aqueous-processed hybrid solar cells with an enormous Herschel infrared contribution. *ACS Appl. Mater. Interfaces* **2014**, *6*, 8606–8612. [CrossRef]

9. Beek, W.J.; Wienk, M.M.; Kemerink, M.; Yang, X.; Janssen, R.A. Hybrid zinc oxide conjugated polymer bulk heterojunction solar cells. *J. Phys. Chem. B* **2005**, *109*, 9505–9516. [CrossRef] [PubMed]

10. Wu, M.C.; Lo, H.H.; Liao, H.C.; Chen, S.; Lin, Y.Y.; Yen, W.C.; Zeng, T.W.; Chen, Y.F.; Chen, C.W.; Su, W.F. Using scanning probe microscopy to study the effect of molecular weight of poly (3-hexylthiophene) on the performance of poly (3-hexylthiophene): TiO_2 nanorod photovoltaic devices. *Sol. Energy Mater. Sol. Cells* **2009**, *93*, 869–873. [CrossRef]

11. Wang, Z.; Qu, S.; Zeng, X.; Liu, J.; Zhang, C.; Shi, M.; Tan, F.; Wang, Z. The synthesis of MDMO-PPV capped PbS nanorods and their application in solar cells. *Curr. Appl. Phys.* **2009**, *9*, 1175–1179. [CrossRef]

12. Jung, J. Preparation of anisotropic CdSe-P3HT core-shell nanorods using directly synthesized Br-functionalized CdSe nanorods. *Surf. Coat. Technol.* **2019**, *362*, 84–89. [CrossRef]

13. Tulsiram, N.; Kerr, C.; Chen, J.I. Photoinduced charge transfer in poly (3-hexylthiophene)/TiO_2 hybrid inverse opals: Photonic vs. interfacial effects. *J. Phys. Chem. C* **2017**, *121*, 26987–26996. [CrossRef]

14. Greaney, M.J.; Brutchey, R.L. Ligand engineering in hybrid polymer: Nanocrystal solar cells. *Mater. Today* **2015**, *18*, 31–38. [CrossRef]

15. Tsang, S.W.; Fu, H.; Wang, R.; Lu, J.; Yu, K.; Tao, Y. Highly efficient cross-linked PbS nanocrystal/C 60 hybrid heterojunction photovoltaic cells. *Appl. Phys. Lett.* **2009**, *95*, 183505. [CrossRef]

16. Xie, Y.; Huang, W.; Liang, Q.; Zhu, J.; Cong, Z.; Lin, F.; Yi, S.; Luo, G.; Yang, T.; Liu, S.; et al. High-performance fullerene-free polymer solar cells featuring efficient photocurrent generation from dual pathways and low nonradiative recombination loss. *ACS Energy Lett.* **2018**, *4*, 8–16. [CrossRef]

17. Yang, T.; Cai, W.; Qin, D.; Wang, E.; Lan, L.; Gong, X.; Cao, Y. Solution-processed zinc oxide thin film as a buffer layer for polymer solar cells with an inverted device structure. *J. Phys. Chem. C* **2010**, *114*, 6849–6853. [CrossRef]

18. Huynh, W.U.; Dittmer, J.J.; Alivisatos, A.P. Hybrid nanorod-polymer solar cells. *Science* **2002**, *295*, 2425–2427. [CrossRef]

19. Couderc, E.; Greaney, M.J.; Brutchey, R.L.; Bradforth, S.E. Direct spectroscopic evidence of ultrafast electron transfer from a low band gap polymer to CdSe quantum dots in hybrid photovoltaic thin films. *J. Am. Chem. Soc.* **2013**, *135*, 18418–18426. [CrossRef]

20. Sun, B.; Marx, E.; Greenham, N.C. Photovoltaic devices using blends of branched CdSe nanoparticles and conjugated polymers. *Nano Lett.* **2003**, *3*, 961–963. [CrossRef]

21. Sun, B.; Snaith, H.J.; Dhoot, A.S.; Westenhoff, S.; Greenham, N.C. Vertically segregated hybrid blends for photovoltaic devices with improved efficiency. *J. Appl. Phys.* **2005**, *97*, 014914. [CrossRef]

22. Han, L.; Qin, D.; Jiang, X.; Liu, Y.; Wang, L.; Chen, J.; Cao, Y. Synthesis of high -quality zinc-blende CdSe nanocrystals and their application in hybrid solar cells. *Nanotechnology* **2006**, *17*, 4736. [CrossRef] [PubMed]

23. Wang, L.; Liu, Y.; Jiang, X.; Qin, D.; Cao, Y. Enhancement of photovoltaic characteristics using a suitable solvent in hybrid polymer/multiarmed CdS nanorods solar cells. *J. Phys. Chem. C* **2007**, *111*, 9538–9542. [CrossRef]

24. Peng, Z.A.; Peng, X. Formation of high quality CdTe, CdSe, and CdS nanocrystals using CdO as precursor. *J. Am. Chem. Soc.* **2001**, *123*, 183–184. [CrossRef]

25. Peng, Z.A.; Peng, X. Nearly monodisperse and shape-controlled CdSe nanocrystals via alternative routes: Nucleation and growth. *J. Am. Chem. Soc.* **2002**, *124*, 3343–3353. [CrossRef]

26. Peng, X.; Manna, L.; Yang, W.; Wickham, J.; Scher, E.; Kadavanich, A.; Alivisatos, A.P. Shape control of CdSe nanocrystals. *Nature* **2000**, *404*, 59–61. [CrossRef]

27. Huynh, W.U.; Peng, X.; Alivisatos, A.P. CdSe nanocrystal rods/poly (3-hexylthiophene) composite photovoltaic devices. *Adv. Mater.* **1999**, *11*, 923–927. [CrossRef]

28. Kramer, I.J.; Sargent, E.H. The architecture of colloidal quantum dot solar cells: Materials to devices. *Chem. Rev.* **2014**, *114*, 863–882. [CrossRef]

29. Liu, J.; Tanaka, T.; Sivula, K.; Alivisatos, A.P.; Fréchet, J.M. Employing end-functional polythiophene to control the morphology of nanocrystal– polymer composites in hybrid solar cells. *J. Am. Chem. Soc.* **2004**, *126*, 6550–6551. [CrossRef]

30. Lan, X.; Voznyy, O.; García de Arquer, F.P.; Liu, M.; Xu, J.; Proppe, A.H.; Walters, G.; Fan, F.; Tan, H.; Liu, M.; et al. 10.6% certified colloidal quantum dot solar cells via solvent-polarity-engineered halide passivation. *Nano Lett.* **2016**, *16*, 4630–4634. [CrossRef]

31. Chuang, C.H.M.; Brown, P.R.; Bulović, V.; Bawendi, M.G. Improved performance and stability in quantum dot solar cells through band alignment engineering. *Nat. Mater.* **2014**, *13*, 796–801. [CrossRef] [PubMed]

32. Tang, J.; Kemp, K.W.; Hoogland, S.; Jeong, K.S.; Liu, H.; Levina, L.; Furukawa, M.; Wang, X.; Debnath, R.; Chou, K.W.; et al. Colloidal-quantum-dot photovoltaics using atomic-ligand passivation. *Nat. Mater.* **2011**, *10*, 765–771. [CrossRef] [PubMed]

33. Ning, Z.; Voznyy, O.; Pan, J.; Hoogland, S.; Adinolfi, V.; Xu, J.; Li, M.; Kirmani, A.R.; Sun, J.P.; Kemp, K.W.; et al. Air-stable n-type colloidal quantum dot solids. *Nat. Mater.* **2014**, *13*, 822–828. [CrossRef] [PubMed]

34. Ren, S.; Chang, L.Y.; Lim, S.K.; Zhao, J.; Smith, M.; Zhao, N.; Bulovic, V.; Bawendi, M.; Gradecak, S. Inorganic–organic hybrid solar cell: Bridging quantum dots to conjugated polymer nanowires. *Nano Lett.* **2011**, *11*, 3998–4002. [CrossRef]

35. Cui, C.; He, Z.; Wu, Y.; Cheng, X.; Wu, H.; Li, Y.; Wong, W.Y.; Cao, Y. High-performance polymer solar cells based on a 2D-conjugated polymer with an alkylthio side-chain. *Energy Environ. Sci.* **2016**, *9*, 885–891. [CrossRef]

36. Fu, W.; Wang, L.; Ling, J.; Li, H.; Shi, M.; Xue, J.; Chen, H. Highly efficient hybrid solar cells with tunable dipole at the donor–acceptor interface. *Nanoscale* **2014**, *6*, 10545–10550. [CrossRef]

37. Zhou, R.; Stalder, R.; Xie, D.; Cao, W.; Zheng, Y.; Yang, Y.; Plaisant, M.; Holloway, P.H.; Schanze, K.S.; Xue, J.; et al. Enhancing the efficiency of solution-processed polymer: Colloidal nanocrystal hybrid photovoltaic cells using ethanedithiol treatment. *ACS Nano* **2013**, *7*, 4846–4854. [CrossRef]

38. Kerr, C.S.; Kryukovskiy, A.; Chen, J.I. Effects of Surface Passivation on Trap States, Band Bending, and Photoinduced Charge Transfer in P3HT/TiO$_2$ Hybrid Inverse Opals. *J. Phys. Chem. C* **2018**, *122*, 17301–17308. [CrossRef]

39. Liao, H.C.; Lee, C.H.; Ho, Y.C.; Jao, M.H.; Tsai, C.M.; Chuang, C.M.; Shyue, J.J.; Chen, Y.F.; Su, W.F. Diketopyrrolopyrrole-based oligomer modified TiO$_2$ nanorods for air-stable and all solution processed poly (3-hexylthiophene): TiO$_2$ bulk heterojunction inverted solar cell. *J. Mater. Chem.* **2012**, *22*, 10589–10596. [CrossRef]

40. Blachowicz, T.; Ehrmann, A. Recent Developments of Solar Cells from PbS Colloidal Quantum Dots. *Appl. Sci.* **2020**, *10*, 1743. [CrossRef]

41. Zhang, X.; Zhang, Y.; Wu, H.; Yan, L.; Wang, Z.; Zhao, J.; Yu, W.W.; Rogach, A.L. PbSe quantum dot films with enhanced electron mobility employed in hybrid polymer/nanocrystal solar cells. *RSC Adv.* **2016**, *6*, 17029–17035. [CrossRef]

42. Song, T.; Cheng, H.; Fu, C.; He, B.; Li, W.; Xu, J.; Tang, Y.; Yang, S.; Zou, B. Influence of the active layer nanomorphology on device performance for ternary PbS x Se1− x quantum dots based solution-processed infrared photodetector. *Nanotechnology* **2016**, *27*, 165202. [CrossRef] [PubMed]

43. Ma, W.; Luther, J.M.; Zheng, H.; Wu, Y.; Alivisatos, A.P. Photovoltaic devices employing ternary PbS$_x$Se$_{1-x}$ nanocrystals. *Nano Lett.* **2009**, *9*, 1699–1703. [CrossRef]

44. Xu, J.; Voznyy, O.; Liu, M.; Kirmani, A.R.; Walters, G.; Munir, R.; Abdelsamie, M.; Proppe, A.H.; Sarkar, A.; Wei, M.; et al. 2D matrix engineering for homogeneous quantum dot coupling in photovoltaic solids. *Nat. Nanotechnol.* **2018**, *13*, 456–462. [CrossRef] [PubMed]

45. Midgett, A.G.; Luther, J.M.; Stewart, J.T.; Smith, D.K.; Padilha, L.A.; Klimov, V.I.; Nozik, A.J.; Beard, M.C. Size and composition dependent multiple exciton generation efficiency in PbS, PbSe, and PbS$_x$Se$_{1-x}$ alloyed quantum dots. *Nano Lett.* **2013**, *13*, 3078–3085. [CrossRef] [PubMed]

46. Noone, K.M.; Strein, E.; Anderson, N.C.; Wu, P.T.; Jenekhe, S.A.; Ginger, D.S. Broadband absorbing bulk heterojunction photovoltaics using low-bandgap solution-processed quantum dots. *Nano Lett.* **2010**, *10*, 2635–2639. [CrossRef] [PubMed]

47. Seo, J.; Cho, M.J.; Lee, D.; Cartwright, A.N.; Prasad, P.N. Efficient heterojunction photovoltaic cell utilizing nanocomposites of lead sulfide nanocrystals and a low-bandgap polymer. *Adv. Mater.* **2011**, *23*, 3984–3988. [CrossRef]

48. Carey, G.H.; Abdelhady, A.L.; Ning, Z.; Thon, S.M.; Bakr, O.M.; Sargent, E.H. Colloidal quantum dot solar cells. *Chem. Rev.* **2015**, *115*, 12732–12763. [CrossRef]

49. Piliego, C.; Manca, M.; Kroon, R.; Yarema, M.; Szendrei, K.; Andersson, M.R.; Heiss, W.; Loi, M.A. Charge separation dynamics in a narrow band gap polymer–PbS nanocrystal blend for efficient hybrid solar cells. *J. Mater. Chem.* **2012**, *22*, 24411–24416. [CrossRef]

50. Lu, H.; Joy, J.; Gaspar, R.L.; Bradforth, S.E.; Brutchey, R.L. Iodide-passivated colloidal PbS nanocrystals leading to highly efficient polymer: Nanocrystal hybrid solar cells. *Chem. Mater.* **2016**, *28*, 1897–1906. [CrossRef]

51. Su, Y.W.; Lin, W.H.; Hsu, Y.J.; Wei, K.H. Conjugated polymer/nanocrystal nanocomposites for renewable energy applications in photovoltaics and photocatalysis. *Small* **2014**, *10*, 4427–4442. [CrossRef] [PubMed]

52. Gur, I.; Fromer, N.A.; Chen, C.P.; Kanaras, A.G.; Alivisatos, A.P. Hybrid solar cells with prescribed nanoscale morphologies based on hyperbranched semiconductor nanocrystals. *Nano Lett.* **2007**, *7*, 409–414. [CrossRef] [PubMed]

53. Van Bavel, S.S.; Bärenklau, M.; de With, G.; Hoppe, H.; Loos, J. P3HT/PCBM bulk heterojunction solar cells: Impact of blend composition and 3D morphology on device performance. *Adv. Funct. Mater.* **2010**, *20*, 1458–1463. [CrossRef]

54. Campoy-Quiles, M.; Ferenczi, T.; Agostinelli, T.; Etchegoin, P.G.; Kim, Y.; Anthopoulos, T.D.; Stavrinou, P.N.; Bradley, D.D.C.; Nelson, J. Morphology evolution via self-organization and lateral and vertical diffusion in polymer: Fullerene solar cell blends. *Nat. Mater.* **2008**, *7*, 158–164. [CrossRef]

55. Xue, J.; Rand, B.P.; Uchida, S.; Forrest, S.R. A hybrid planar–mixed molecular heterojunction photovoltaic cell. *Adv. Mater.* **2005**, *17*, 66–71. [CrossRef]

56. Wang, Z.; Yokoyama, D.; Wang, X.F.; Hong, Z.; Yang, Y.; Kido, J. Highly efficient organic p–i–n photovoltaic cells based on tetraphenyldibenzoperiflanthene and fullerene C 70. *Energy Environ. Sci.* **2013**, *6*, 249–255. [CrossRef]

57. Liu, F.; Chen, Z.; Du, X.; Zeng, Q.; Ji, T.; Cheng, Z.; Jin, G.; Yang, B. High efficiency aqueous-processed MEH-PPV/CdTe hybrid solar cells with a PCE of 4.20%. *J. Mater. Chem. A* **2016**, *4*, 1105–1111. [CrossRef]

58. Yao, S.; Chen, Z.; Li, F.; Xu, B.; Song, J.; Yan, L.; Jin, G.; Wen, S.; Wang, C.; Tian, W.; et al. High-efficiency aqueous-solution-processed hybrid solar cells based on P3HT Dots and CdTe nanocrystals. *ACS Appl. Mater. Interfaces* **2015**, *7*, 7146–7152. [CrossRef]

59. Jin, G.; Chen, N.; Zeng, Q.; Liu, F.; Yuan, W.; Xiang, S.; Feng, T.; Du, X.; Ji, T.; Wang, L.; et al. Aqueous-Processed Polymer/Nanocrystal Hybrid Solar Cells with Double-Side Bulk Heterojunction. *Adv. Energy Mater.* **2018**, *8*, 1701966. [CrossRef]

60. Chen, N.N.; Jin, G.; Wang, L.J.; Sun, H.N.; Zeng, Q.S.; Yang, B.; Sun, H.Z. Highly efficient aqueous-processed hybrid solar cells: Control depletion region and improve carrier extraction. *Adv. Energy Mater.* **2019**, *9*, 1803849. [CrossRef]

61. Liu, Z.; Sun, Y.; Yuan, J.; Wei, H.; Huang, X.; Han, L.; Wang, W.; Wang, H.; Ma, W. High-efficiency hybrid solar cells based on polymer/PbS$_x$Se$_{1-x}$ nanocrystals benefiting from vertical phase segregation. *Adv. Mater.* **2013**, *25*, 5772–5778. [CrossRef] [PubMed]

62. Yuan, J.; Gallagher, A.; Liu, Z.; Sun, Y.; Ma, W. High-efficiency polymer–PbS hybrid solar cells via molecular engineering. *J. Mater. Chem. A* **2015**, *3*, 2572–2579. [CrossRef]

63. Baek, S.W.; Jun, S.; Kim, B.; Proppe, A.H.; Ouellette, O.; Voznyy, O.; Kim, C.; Kim, J.; Walters, G.; Jeong, S.; et al. Efficient hybrid colloidal quantum dot/organic solar cells mediated by near-infrared sensitizing small molecules. *Nat. Energy* **2019**, *4*, 969–976. [CrossRef]

64. Kumar, S.G.; Rao, K.K. Physics and chemistry of CdTe/CdS thin film heterojunction photovoltaic devices: Fundamental and critical aspects. *Energy Environ. Sci.* **2014**, *7*, 45–102. [CrossRef]

65. Lin, H.; Xia, W.; Wu, H.N.; Tang, C.W. CdS/CdTe solar cells with MoO$_x$ as back contact buffers. *Appl. Phys. Lett.* **2010**, *97*, 123504. [CrossRef]

66. Shen, K.; Yang, R.; Wang, D.; Jeng, M.; Chaudhary, S.; Ho, K.; Wang, D. Stable CdTe solar cell with V$_2$O$_5$ as a back contact buffer layer. *Sol. Energy Mater. Sol. Cells* **2016**, *144*, 500–508. [CrossRef]

67. Xiao, D.; Li, X.; Wang, D.; Li, Q.; Shen, K.; Wang, D. CdTe thin film solar cell with NiO as a back contact buffer layer. *Sol. Energy Mater. Sol. Cells* **2017**, *169*, 61–67. [CrossRef]

68. Paudel, N.R.; Xiao, C.; Yan, Y. CdS/CdTe thin-film solar cells with Cu-free transition metal oxide/Au back contacts. *Prog. Photovolt. Res. Appl.* **2015**, *23*, 437–442. [CrossRef]

69. Paudel, N.R.; Yan, Y. Application of copper thiocyanate for high open-circuit voltages of CdTe solar cells. *Prog. Photovolt. Res. Appl.* **2016**, *24*, 94–101. [CrossRef]

70. Paudel, N.R.; Yan, Y. CdTe thin-film solar cells with cobalt-phthalocyanine back contacts. *Appl. Phys. Lett.* **2014**, *104*, 143507. [CrossRef]

71. Wang, W.; Paudel, N.R.; Yan, Y.; Duarte, F.; Mount, M. PEDOT: PSS as back contact for CdTe solar cells and the effect of PEDOT: PSS conductivity on device performance. *J. Mater. Sci. Mater. Electron.* **2016**, *27*, 1057–1061. [CrossRef]

72. Du, X.; Chen, Z.; Liu, F.; Zeng, Q.; Jin, G.; Li, F.; Yao, D.; Yang, B. Improvement in open-circuit voltage of thin film solar cells from aqueous nanocrystals by interface engineering. *ACS Appl. Mater. Interfaces* **2016**, *8*, 900–907. [CrossRef]

73. Guo, X.; Tan, Q.; Liu, S.; Qin, D.; Mo, Y.; Hou, L.; Liu, A.; Wu, H.; Ma, Y. High-efficiency solution-processed CdTe nanocrystal solar cells incorporating a novel crosslinkable conjugated polymer as the hole transport layer. *Nano Energy* **2018**, *46*, 150–157. [CrossRef]

74. Paudel, N.R.; Poplawsky, J.D.; Moore, K.L.; Yan, Y. Current enhancement of CdTe-based solar cells. *IEEE J. Photovolt.* **2015**, *5*, 1492–1496. [CrossRef]

75. Rong, Z.; Guo, X.; Lian, S.; Liu, S.; Qin, D.; Mo, Y.; Xu, W.; Wu, H.; Zhao, H.; Hou, L. Interface engineering for both cathode and anode enables low-cost highly efficient solution-processed CdTe nanocrystal solar cells. *Adv. Funct. Mater.* **2019**, *29*, 1904018. [CrossRef]

76. Meng, L.; Zhang, Y.; Wan, X.; Li, C.; Zhang, X.; Wang, Y.; Ke, X.; Xiao, Z.; Ding, L.; Yip, H.L.; et al. Organic and solution-processed tandem solar cells with 17.3% efficiency. *Science* **2018**, *361*, 1094–1098. [CrossRef] [PubMed]

77. Gilot, J.; Wienk, M.M.; Janssen, R.A. Optimizing polymer tandem solar cells. *Adv. Mater.* **2010**, *22*, E67–E71. [CrossRef]

78. Dou, L.; You, J.; Yang, J.; Chen, C.C.; He, Y.; Murase, S.; Moriarty, T.; Emery, K.; Li, G.; Yang, Y. Tandem polymer solar cells featuring a spectrally matched low-bandgap polymer. *Nat. Photonics* **2012**, *6*, 180–185. [CrossRef]

79. Zheng, Z.; Zhang, S.; Zhang, M.; Zhao, K.; Ye, L.; Chen, Y.; Yang, B.; Hou, J. Highly efficient tandem polymer solar cells with a photovoltaic response in the visible light range. *Adv. Mater.* **2015**, *27*, 1189–1194. [CrossRef]

80. Speirs, M.J.; Groeneveld, B.G.H.M.; Protesescu, L.; Piliego, C.; Kovalenko, M.V.; Loi, M.A. Hybrid inorganic–organic tandem solar cells for broad absorption of the solar spectrum. *Phys. Chem. Chem. Phys.* **2014**, *16*, 7672–7676. [CrossRef]

81. Kim, T.; Gao, Y.; Hu, H.; Yan, B.; Ning, Z.; Jagadamma, L.K.; Zhao, K.; Kirmani, A.R.; Eid, J.; Sargent, E.H.; et al. Hybrid tandem solar cells with depleted-heterojunction quantum dot and polymer bulk heterojunction subcells. *Nano Energy* **2015**, *17*, 196–205. [CrossRef]

82. Aqoma, H.; Azmi, R.; Oh, S.H.; Jang, S.Y. Solution-processed colloidal quantum dot/organic hybrid tandem photovoltaic devices with 8.3% efficiency. *Nano Energy* **2017**, *31*, 403–409. [CrossRef]

83. Tong, J.; Yang, X.; Xu, Y.; Li, W.; Tang, J.; Song, H.; Zhou, Y. Efficient top-illuminated organic-quantum dots hybrid tandem solar cells with complementary absorption. *ACS Photonics* **2017**, *4*, 1172–1177. [CrossRef]

84. Kim, T.; Palmiano, E.; Liang, R.Z.; Hu, H.; Murali, B.; Kirmani, A.R.; Firdaus, Y.; Gao, Y.; Sheikh, A.; Mohammed, O.F.; et al. Hybrid tandem quantum dot/organic photovoltaic cells with complementary near infrared absorption. *Appl. Phys. Lett.* **2017**, *110*, 223903. [CrossRef] [PubMed]

85. Kim, T.; Firdaus, Y.; Kirmani, A.R.; Liang, R.Z.; Hu, H.; Liu, M.; Labban, A.E.; Hoogland, S.; Beaujuge, P.M.; Amassian, A.; et al. Hybrid tandem quantum dot/organic solar cells with enhanced photocurrent and efficiency via ink and interlayer engineering. *ACS Energy Lett.* **2018**, *3*, 1307–1314. [CrossRef]

86. Aqoma, H.; Imran, I.F.; Mubarok, M.A.; Hadmojo, W.T.; Do, Y.R.; Jang, S.Y. Efficient hybrid tandem solar cells based on optical reinforcement of colloidal quantum dots with organic bulk heterojunctions. *Adv. Energy Mater.* **2020**, *10*, 1903294. [CrossRef]

87. Green, M.A.; Dunlop, E.D.; Hohl-Ebinger, J.; Yoshita, M.; Kopidakis, N.; Ho-Baillie, A.W. Solar cell efficiency tables (Version 55). *Prog. Photovolt. Res. Appl.* **2019**, *28*. [CrossRef]

MDPI

St. Alban-Anlage 66

4052 Basel

Switzerland

Tel. +41 61 683 77 34

Fax +41 61 302 89 18

www.mdpi.com

Applied Sciences Editorial Office

E-mail: applsci@mdpi.com

www.mdpi.com/journal/applsci